INTRODUCTORY GROUP THEORY

and Its Application to Molecular Structure

SECOND EDITION

INTRODUCTORY GROUP THEORY

and Its Application to Molecular Structure

SECOND EDITION

John R. Ferraro
Argonne National Laboratory

and

Joseph S. Ziomek
Argonne National Laboratory, Consultant

Plenum Press · New York and London

Library of Congress Cataloging in Publication Data

Ferraro, John R • 1918-
Introductory group theory and its application to molecular structure.

Bibliography: p.
Includes index.
1. Molecular theory. 2. Molecular spectra. 3. Groups, Theory of. I. Ziomek, Joseph S., 1913- joint author. II. Title.
QD461.F38 1975 539'.12'0151222 75-33752
ISBN 0-306-30786-5

A Division of Plenum Publishing Corporation
227 West 17th Street, New York, N.Y. 10011

United Kingdom edition published by Plenum Press, London
A Division of Plenum Publishing Company, Ltd.
Davis House (4th Floor), 8 Scrubs Lane, Harlesden, London, NW10 6SE, England

Printed in the United States of America

To our families

PREFACE TO THE SECOND EDITION

The success of the first edition of this book has encouraged us to revise and update it. In the second edition we have attempted to further clarify portions of the text in reference to point symmetry, keeping certain sections and removing others. The ever-expanding interest in solids necessitates some discussion on space symmetry. In this edition we have expanded the discussion on point symmetry to include space symmetry. The selection rules include space group selection rules (for $k = 0$). Numerous examples are provided to acquaint the reader with the procedure necessary to accomplish this. Recent examples from the literature are given to illustrate the use of group theory in the interpretation of molecular spectra and in the determination of molecular structure.

The text is intended for scientists and students with only a limited theoretical background in spectroscopy. For this reason we have presented detailed procedures for carrying out the selection rules and normal coordinate treatment of molecules. We have chosen to exclude discussion on symmetry aspects of molecular orbital theory and ligand field theory. It has been our approach to highlight vibrational data only, primarily to keep the size and cost of the book to a reasonable limit.

The authors wish to thank Miss Mary Ellen Matthews for her secretarial assistance on the book and several of our colleagues who have made innumerable suggestions, namely Dr. Priscilla LaBonville, Dr. Louis J. Basile, and Dr. Monica Choca of ANL; Dr. Gary Long of the University of Missouri, Rolla, Missouri; Mr. Barry Scheetz of Pennsylvania State University; and Dr. Bruce B. Murray of Wisconsin State University at River Falls.

JOHN R. FERRARO
JOSEPH S. ZIOMEK

November 1975

PREFACE

This volume is a consequence of a series of seminars presented by the authors at the Infrared Spectroscopy Institute, Canisius College, Buffalo, New York, over the last nine years. Many participants on an intermediate level lacked a sufficient background in mathematics and quantum mechanics, and it became evident that a nonmathematical or nearly nonmathematical approach would be necessary. The lectures were designed to fill this need and proved very successful. As a result of the interest that was developed in this approach, it was decided to write this book.

The text is intended for scientists and students with only limited theoretical background in spectroscopy, but who are sincerely interested in the interpretation of molecular spectra. The book develops the *detailed* selection rules for fundamentals, combinations, and overtones for molecules in several point groups. *Detailed* procedures used in carrying out the normal coordinate treatment for several molecules are also presented. Numerous examples from the literature illustrate the use of group theory in the interpretation of molecular spectra and in the determination of molecular structure.

The authors wish to thank Professor Herman A. Szymanski, Director of the Infrared Spectroscopy Institute, for affording them the opportunity to develop the approach presented in this book. One of them (J.R.F.) wishes to express his thanks and appreciation to Argonne National Laboratory for allowing him to participate in the Institutes; to Professor V. Caglioti, President of the Consiglio Nazionale delle Ricerche, Rome, Italy, for inviting him to spend his sabbatical at the University of Rome, where a good portion of this book was written; and to Professor G. Sartori, for his aid and hospitality in making available the facilities of the Instituto di Chimica Generale ed Inorganica at the University.

The other author (J.S.Z.) wishes to thank Professor Szymanski for extending the invitation to direct the theoretical section at the Institute.

JOHN R. FERRARO
JOSEPH S. ZIOMEK

November, 1968

CONTENTS

SYMBOLS AND ABBREVIATIONS

Symbol	Meaning	
IR	— infrared	
R	— Raman	
C	— coincidence	
p	— polarized	
d	— depolarized	
ϱ	— depolarization ratio in reference to a Raman band	
ν	— frequency or vibration	
ν_s	— symmetric stretching frequency or vibration	
ν_{as}	— asymmetric stretching frequency or vibration	
δ	— bending vibration	
ν_d	— degenerate vibration	
S	— strong	in reference to intensities of vibrations
M	— medium	
W	— weak	
V	— very	
B	— broad	
Sh	— shoulder	
p, q, r	— branch of vibration in a gas	
a	— active	in reference to infrared or Raman activity
ia	— inactive	
μ	— reciprocal of mass	in reference to g elements
ϱ	— reciprocal of bond distance	

NOTE: a right-hand XYZ coordinate system is used throughout the text.

Chapter 1

SYMMETRY

1-1. INTRODUCTION

The concept of symmetry is extremely old, and most individuals are well aware of the important role that symmetry plays in our physical environment. Symmetry has always been present in nature, and as man's culture developed, it made a dramatic and decisive contribution to his way of life. This chapter will attempt to demonstrate the concept of symmetry as applied to structural chemistry.

1-2. DEFINITION OF SYMMETRY

The word *symmetry* comes from the Greek word *symmetria*, and may be defined as harmony or balance in the proportion of parts to the whole. In the nonmathematical sense, symmetry is associated with beauty—with pleasing proportions or regularity in form, harmonious arrangement, or regular repetition of certain characteristics (periodicity). In a narrower mathematical or geometric sense, symmetry refers to the correspondence of elements on opposite sides of a point, line, or plane, which we call the center, axis, or plane of symmetry.

1-3. SYMMETRY IN SCIENCE

Symmetry concepts find numerous applications in the various scientific disciplines. In biology, for example, animals can be distinguished by a system involving symmetry. Four types are defined: (1) *the radially symmetric type* (wheel type), exemplified by jellyfish and sponges, which has an infinite number of planes and rotational axes of symmetry; (2) *the bilaterally symmetric type*, exemplified by mammals, birds, fish, etc., in which the external features of the body have a single plane of symmetry only; (3) *the serially symmetric type*, exemplified by earthworms, in which elements repeat at regular intervals; (4) *the asymmetric type*, exemplified by paramecium, where no symmetry exists at all.

Other examples of symmetry in science are found in Mendeleev's periodic table, in X-ray interference patterns, in crystal symmetry, in Raman spectra (Stokes–anti-Stokes bands), in crystal and ligand field theory, in Bose–Einstein statistics, which govern the behavior of nuclei of atoms with an even number of fundamental particles, in Fermi–Dirac statistics, which apply to systems with an odd number of fundamental particles, in the odd–even effect in rare-earth chemistry etc.

1-4. SYMMETRY IN STRUCTURAL CHEMISTRY

A knowledge of the symmetry concepts in chemistry affords one a better understanding of the ever-increasing battery of tools available for the solution of structural problems. Raman, infrared, and ultraviolet spectroscopy, and X-ray, electron, and neutron diffraction methods are some of the powerful tools that are based on symmetry considerations. Knowing the symmetry of a molecule, one can predict the infrared or Raman spectrum, and knowing the spectrum, one can arrive at the symmetry or structure of the molecule.

Previously, we indicated that symmetry can be defined both in a nonmathematical and in a mathematical (or geometric) sense. At this point, we will develop the geometric definition of symmetry as related to the isolated molecule, and which may be called point symmetry.

A. Point Symmetry Elements

The spatial arrangement of the atoms in a molecule is called its equilibrium configuration or structure. This configuration is invariant under a set of geometric operations called a group. The molecule is oriented in a coordinate system (a right-hand *xyz* coordinate system is used throughout the discussion in this text). If by carrying out certain geometric operations on the original configuration, the molecule can be transformed into another configuration that is superimposable on the original (i.e., indistinguishable from it), although its orientation may be changed, the molecule is said to contain a symmetry element. The following symmetry elements can be cited.

Identity

The symmetry element that transforms the original equilibrium configuration into another one superimposable on the original without change in orientation, in such a manner that each atom goes into itself, is called the identity, and is denoted by I or E (E from the German *Einheit* meaning "unit" or, loosely, "identical").

Rotation Axes

When a molecule can be rotated about an axis to a new configuration, and the new configuration is indistinguishable from the original one, the molecule is said to possess a rotational axis of symmetry. The rotation can be clockwise or counterclockwise, depending on the molecule. For example, the same configuration is obtained for water whether one rotates the molecule clockwise or counterclockwise. However, for the ammonia molecule, different configurations are obtained, depending on which way the rotation is performed. The angle of rotation may be $2\pi/n$, or $360°/n$, where n can be 1–6 or ∞ for various molecules existing in nature. The order of the rotational axis is called n (sometimes p), and the notation C_n is used, where C denotes rotation, etc. In cases where several axes of rotation exist, the highest order of rotation is chosen as the principal axis. Linear molecules have an infinite-fold axis of symmetry.

The selection of the axes in a coordinate system can be confusing. To avoid this, the following rules can be used for the selection of the z axis of a molecule:

(1) In molecules with only one rotational axis, this axis is taken as the z axis.

(2) In molecules where several rotational axes exist, the highest-order axis is selected as the z axis.

(3) If a molecule possesses several axes of the highest order, the axis passing through the greatest number of atoms is taken as the z axis.

For the selection of the x axis the following rules can be cited:

(1) For a planar molecule where the z axis lies in this plane, the x axis can be selected to be normal to this plane.

(2) In a planar molecule where the z axis is chosen to be perpendicular to the plane, the x axis must lie in the plane, and is chosen to pass through the largest number of atoms in the molecule.

(3) In nonplanar molecules the plane going through the largest number of atoms is located as if it were in the plane of the molecule and rule (1) or (2) is used. For complex molecules where a selection is difficult, one chooses the x and y axes arbitrarily.

Planes of Symmetry

If a plane divides the equilibrium configuration of a molecule into two parts which are mirror images of each other, then the plane is called a symmetry plane. If a molecule has two such planes which intersect in a line, this line is an axis of rotation (see above), the molecule is said to have a vertical rotation axis C, and the two planes are referred to as vertical planes of symmetry, denoted by σ_v. Another case involving two planes of symmetry and their intersection arises when a molecule has more than one axis of symmetry. For example, planes intersecting in an n-fold axis perpendicular to n two-fold axes, with each of the planes bisecting the angle between two successive two-fold axes, are called diagonal and are denoted by the symbol σ_d. Figure 1-1 illustrates the diagonal planes of symmetry for the AB_4 molecule (e.g., $PtCl_4^{2-}$ ion). If a plane of symmetry is perpendicular to the rotational axis, it is called horizontal and is denoted by σ_h.

Center of Symmetry

If a straight line drawn from each atom of a molecule through a certain point meets an equivalent atom equidistant from the point, we call the point the center of symmetry of the molecule. The center of symmetry may or may not coincide with the position of an atom. The designation for the center of symmetry, or center of inversion, is i. If the center of symmetry is situated on an atom, the total number of atoms in the molecule is odd. If the center of symmetry is not on an atom, the number of atoms in the molecule is even.

Rotation–Reflection Axes

If a molecule is rotated 360°/n about an axis and then reflected in a plane perpendicular to this axis, and if the operation produces a configuration

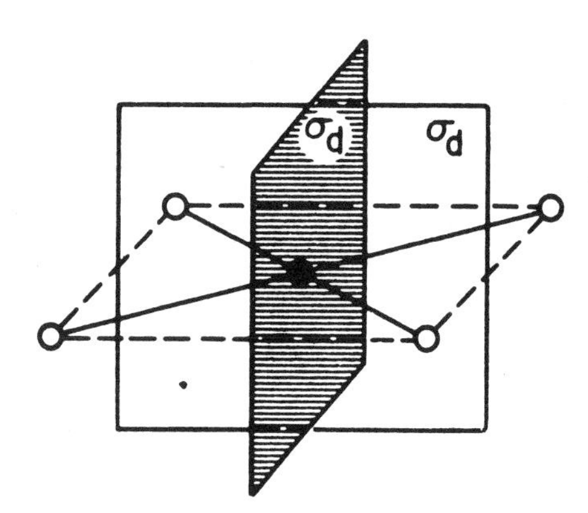

Fig. 1-1. Diagonal planes of symmetry for the AB_4 molecule (e.g., $PtCl_4^{2-}$ ion).

Table 1-1. Point Symmetry Elements and Symmetry Operations

Symmetry element	Symmetry operation
1. Identity (E or I)	Molecule unchanged
2. Axis of rotation (C_n)	Rotation about axis by $2\pi/n$
3. Center of symmetry or center of inversion (i)	Inversion of all atoms through center
4. Plane (σ)	Reflection in the plane
5. Rotation–reflection axis (S_n)	Rotation about axis by $2\pi/n$, followed by reflection in a plane perpendicular to the axis

indistinguishable from the original one, the molecule has the symmetry element of rotation–reflection, which is designated by S_n.

Table 1-1 lists the point symmetry elements and the corresponding symmetry operations. The notation used by spectroscopists and chemists, and used here, is the so-called Schoenflies system, which deals only with point groups. Crystallographers generally use the Hermann–Mauguin system, which applies to both point and space groups.

Several examples will serve to clarify the concept of the symmetry operations and symmetry elements. Consider first a linear, homopolar, diatomic molecule (e.g., H_2 or Cl_2). The molecule possesses a center of symmetry at i (Fig. 1-2). It also has an infinite-fold rotational axis C_∞ along the internuclear axis (Fig. 1-3), and an S_∞ element of symmetry. The molecule further possesses an infinite number of two-fold axes of symmetry (Fig. 1-4) perpendicular to the infinite-fold axis of rotation and an infinite number of symmetry planes σ_v parallel to the internuclear axis (Fig. 1-5). Moreover, there is a plane of symmetry σ_h perpendicular to the internuclear axis (Fig. 1-6).

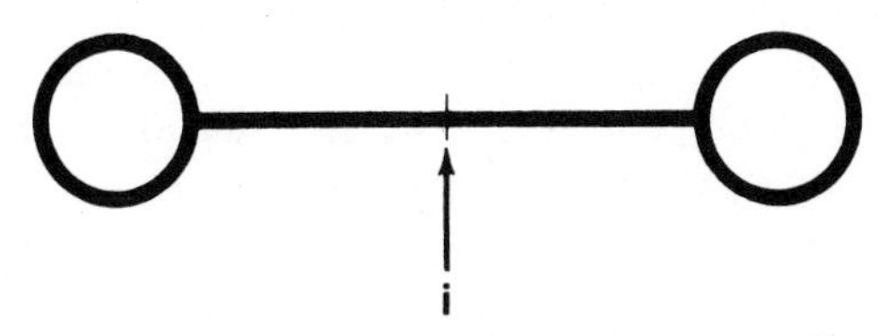

Fig. 1-2. Linear, homopolar, diatomic molecule, illustrating center of symmetry.

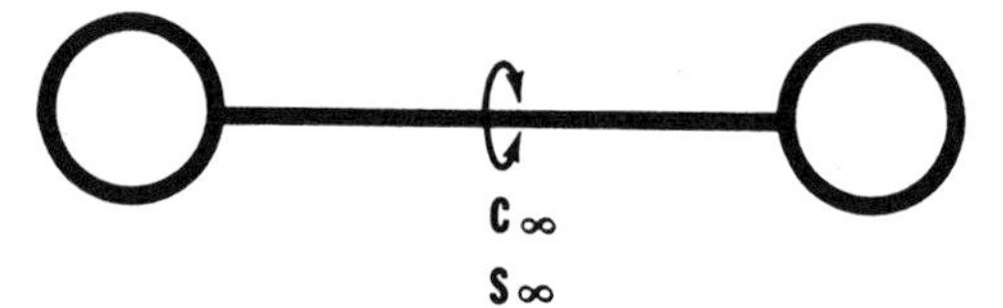

Fig. 1-3. Linear, homopolar, diatomic molecule, illustrating C_∞ and S_∞ elements of symmetry.

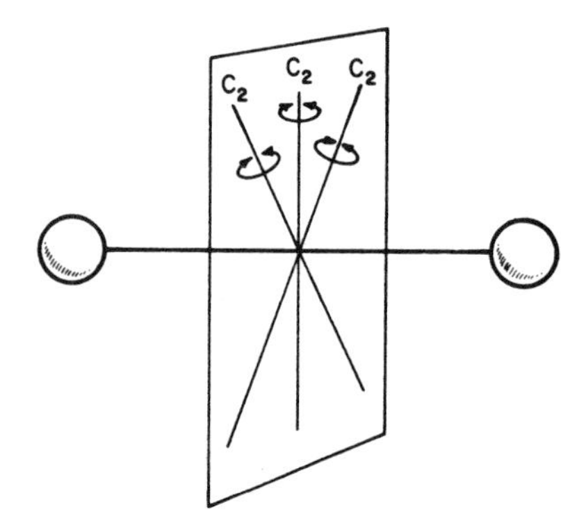

Fig. 1-4. Linear, homopolar, diatomic molecule, illustrating infinite number of C_2 axes of symmetry.

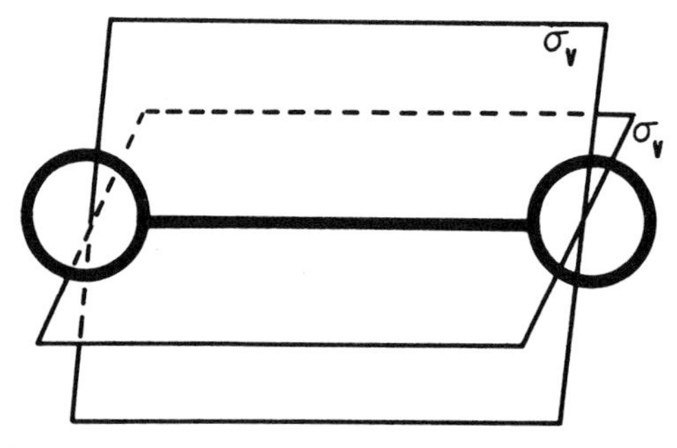

Fig. 1-5. Linear, homopolar, diatomic molecule, illustrating infinite number of σ_v symmetry planes.

Now consider a linear, heteropolar, diatomic molecule (e.g., HCl). This is a less symmetric molecule, and can be shown to possess fewer elements of symmetry than the homopolar, diatomic molecule. The molecule possesses only the infinite-fold axis of symmetry (Fig. 1-7) and an infinite number of planes of symmetry (Fig. 1-8).

Next, consider a planar molecule of type AB_3, where the B atoms are at the corners of an imaginary equilaterial triangle and A is at the center

(Fig. 1-9). BF_3 is an example of such a molecule. The molecule possesses three C_2 axes of rotation, as illustrated in Fig. 1-10. It also has a C_3 axis, about which the rotation can be performed clockwise or counterclockwise (Fig. 1-11), and an S_3 element of symmetry. In addition, the molecule has three σ_v planes of symmetry (Fig. 1-12) and one σ_h plane of symmetry (Fig. 1-13).

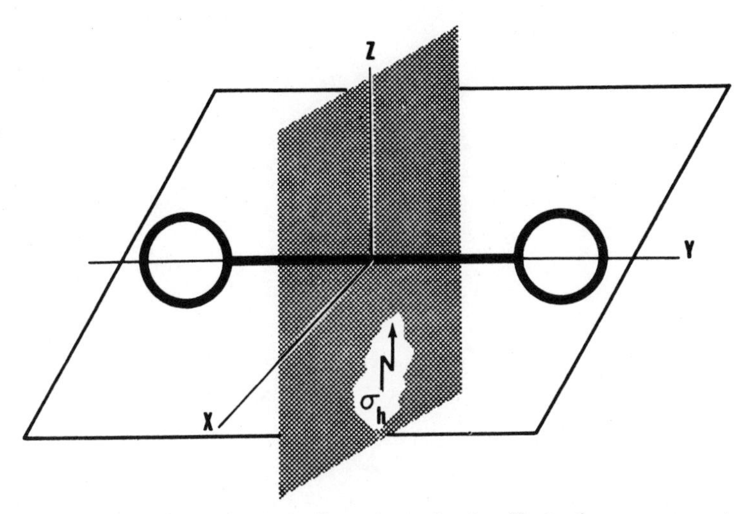

Fig. 1-6. Linear, homopolar, diatomic molecule, illustrating σ_h symmetry.

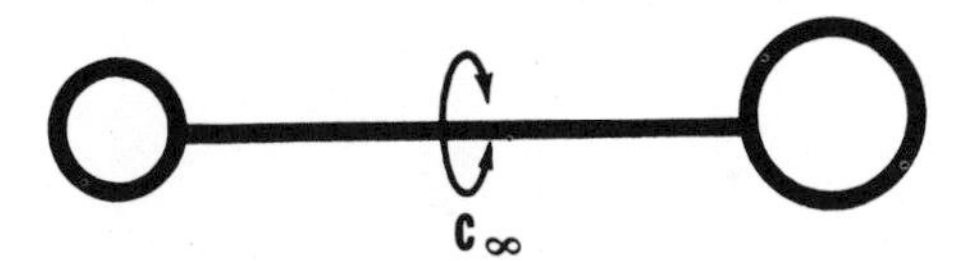

Fig. 1-7. Linear, heteropolar, diatomic molecule, illustrating the C_∞ symmetry element.

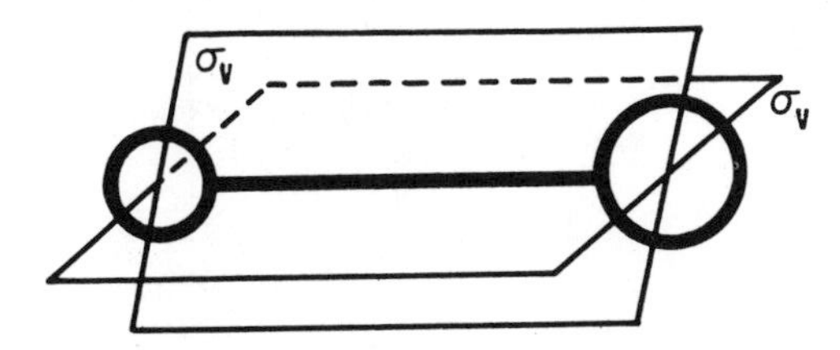

Fig. 1-8. Linear, heteropolar, diatomic molecule, illustrating σ_v symmetry elements.

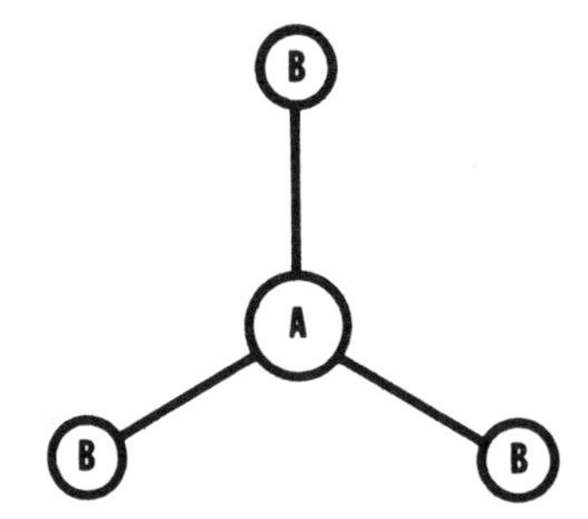

Fig. 1-9. Planar AB_3 type molecule.

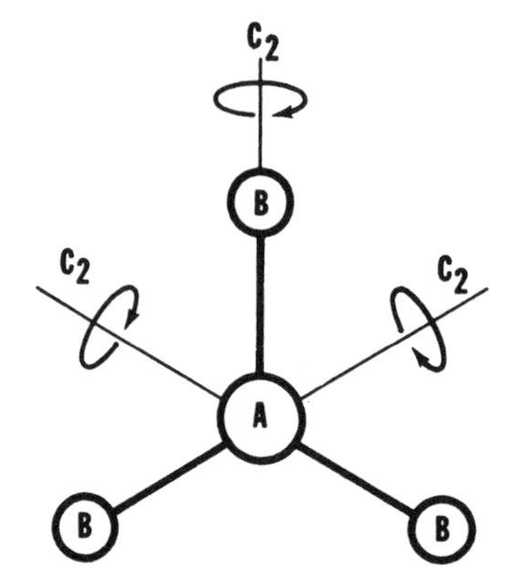

Fig. 1-10. Illustration of the C_2 elements of symmetry in AB_3 molecule.

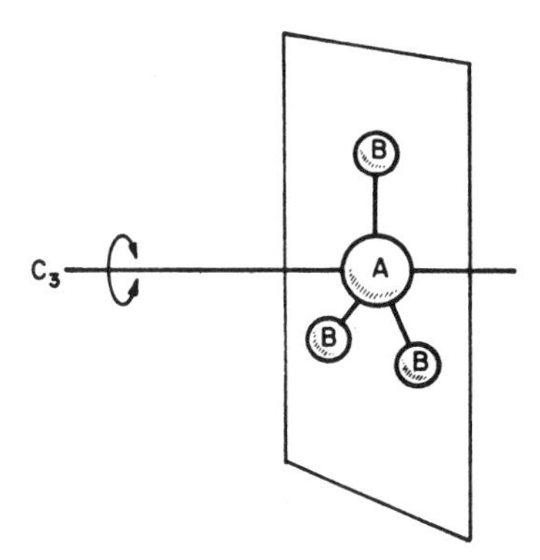

Fig. 1-11. The C_3 symmetry element in AB_3 molecule.

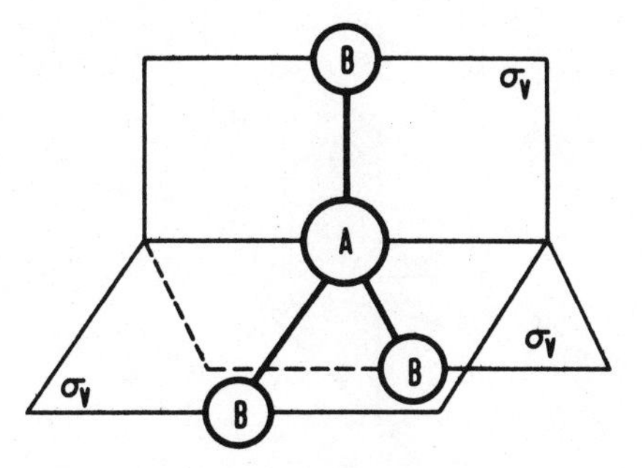

Fig. 1-12. The σ_v symmetry element in AB_3 molecule.

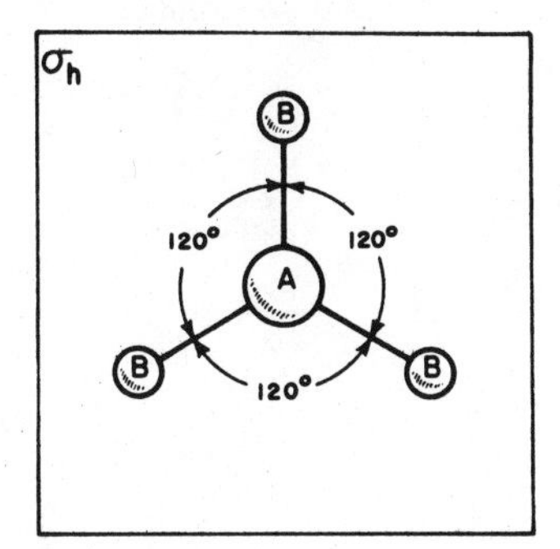

Fig. 1-13. The σ_h symmetry element in AB_3 molecule.

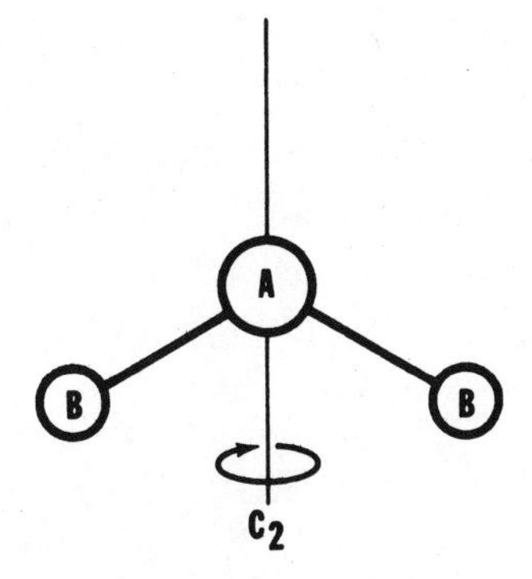

Fig. 1-14. The C_2 symmetry element in a bent AB_2 molecule.

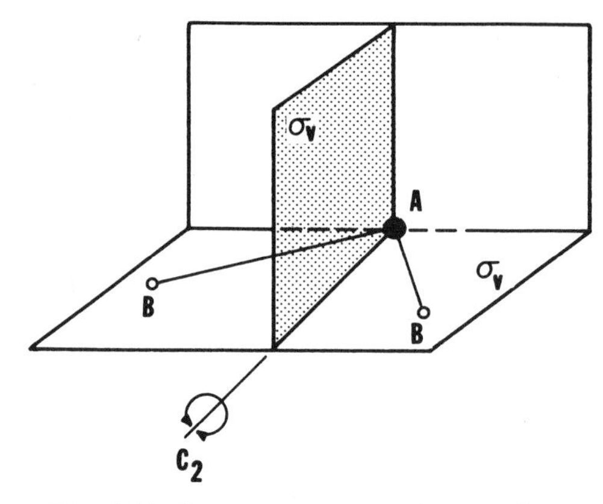

Fig. 1-15. The σ_v symmetry elements in a bent AB_2 molecule.

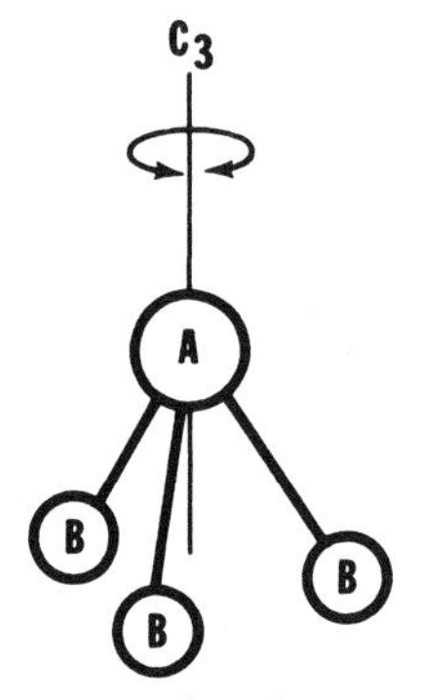

Fig. 1-16. The C_3 symmetry element in pyramidal AB_3 molecule.

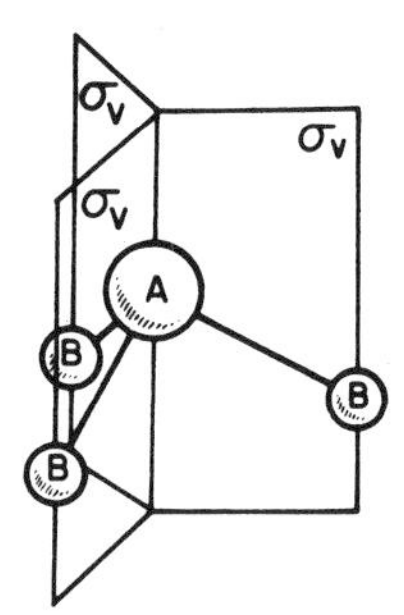

Fig. 1-17. The σ_v symmetry elements in pyramidal AB_3 molecule.

In a bent AB_2 molecule, such as H_2O, we find a two-fold axis of rotation (Fig. 1-14) and two vertical planes of symmetry (Fig. 1-15).

A pyramidal AB_3 molecule, such as NH_3, has a three-fold axis of rotation (Fig. 1-16), which can be clockwise or counterclockwise, and three vertical planes of symmetry (Fig. 1-17).

A planar, hexagonal A_6B_6 molecule (e.g., benzene) possesses a greater variety of symmetry elements. Figure 1-18 shows the six two-fold axes of rotation in the plane of the molecule, the center of symmetry, and the six-fold axis of rotation perpendicular to the plane of the molecule. Also shown are the C_2, C_3, S_6, and S_3 elements coincident with the C_6 axis. The planes

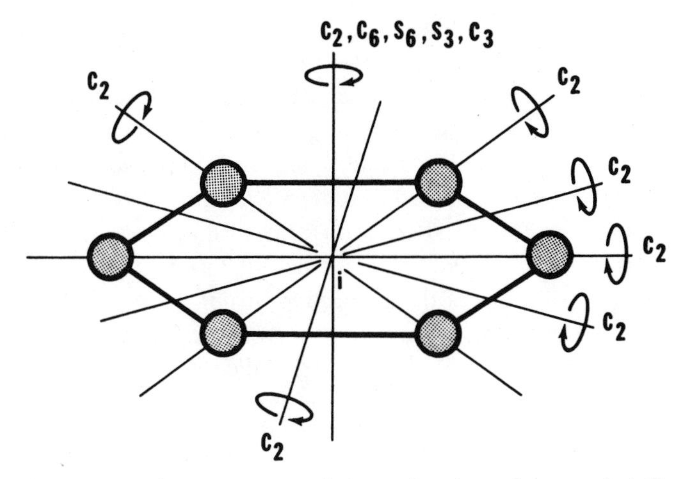

Fig. 1-18. Various symmetry elements in planar hexagonal A_6B_6 molecule.

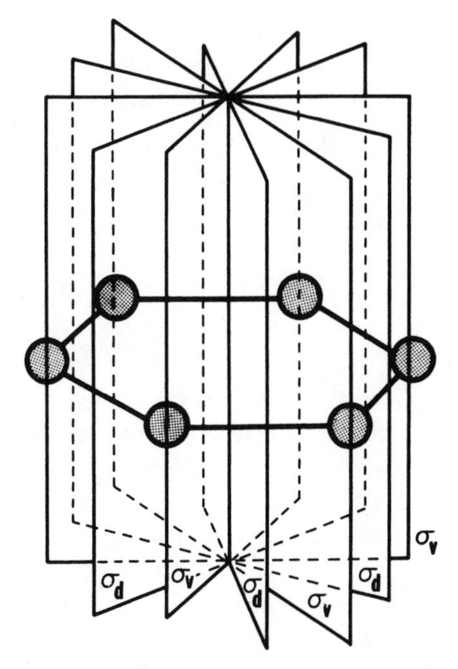

Fig. 1-19. Various planes of symmetry in planar hexagonal A_6B_6 molecule.

of symmetry are shown in Fig. 1-19. There are three σ_v and three σ_d planes. A σ_h plane coincides with the plane of the molecule.

In general, the above-named symmetry elements transform an equilibrium configuration into one which is indistinguishable from the original. This can be a set of points on a line, a tetrahedron, or some other type of polyhedron. For such a set of symmetry elements that obey the group postulates and transform a given configuration into itself, it can be shown that there is at least one point of the configuration that transforms into itself under all operations of the group. Groups of this type are called point groups. Groups that contain translations, translation–rotations, or translation–reflections as symmetry elements are called space groups and will be discussed later.

B. Point Groups

The set of all symmetry elements of a molecule can be made to form a mathematical system called a group once a binary operation is defined for it, and the set of these elements conforms to the following definition in terms of axioms. If the operation is taken to be the performing of one symmetry operation after another in succession, and the result of these operations is equivalent to a single symmetry operation in the set, then the set will be a group. The postulates for the set of elements A, B, C, ... are as follows:

(1) For every pair of elements A and B there exists a binary operation that yields the product AB belonging to the set.
(2) This binary operation is associative, which implies that $A(BC) = (AB)C$.
(3) There exists an identity element E such that for every A, $AE = EA = A$.
(4) There is an inverse A^{-1} for each element A such that $AA^{-1} = A^{-1}A = E$.

For molecules the number of elements is finite and the equilibrium configuration has at least one point (not necessarily on an atom) that is invariant with respect to the group. This latter aspect is emphasized by indicating that the group is a point group. These pertinent point groups are given in Table 1-2.

Satisfaction of these requirements can readily be tested with the water molecule, which has C_{2v} symmetry:

(1) For water, the first requirement is fulfilled; e.g., if a two-fold rotation C_2 is performed, followed by a reflection in the xz plane,

Table 1-2. The 32 Crystallographic Point Groups

Symbol	Plane (C_S)	Axes of symmetry 6(C_6)	4(C_4)	3(C_3)	2(C_2)	Center $i(C_i)$	Example
C_1	—	—	—	—	—	—	CH_3CHO
C_2	—	—	—	—	1	—	H_2O_2
C_3	—	—	—	1	—	—	—
C_4	—	—	1	—	—	—	$H_2S(s)$
C_6	—	1	—	—	—	—	—
C_h	1	—	—	—	—	—	—
C_{2h}	1	—	—	—	1	1	*trans*-CHCl=CHCl
C_{3h}	1	—	—	1	—	—	—
C_{4h}	1	—	1	—	—	1	—
C_{6h}	1	1	—	—	—	1	—
D_2	—	—	—	—	3	—	—
D_3	—	—	—	1	3	—	—
D_4	—	—	1	—	4	—	—
D_6	—	1	—	—	6	—	—
D_{2h}	3	—	—	—	3	1	C_2H_4
D_{3h}	4	—	—	1	3	—	BCl_3
D_{4h}	5	—	1	—	4	1	$PtCl_4^{2-}$
D_{6h}	7	1	—	—	6	1	C_6H_6
S_2	—	—	—	—	—	1	—
S_4	—	—	—	—	1	—	—
S_6	—	—	—	1	—	1	—
D_{2d}	2	—	—	—	3	—	$CH_2{=}C{=}CH_2$
D_{3d}	3	—	—	1	3	1	Cyclohexane
C_{2v}	2	—	—	—	1	—	$H_2O(g)$
C_{3v}	3	—	—	1	—	—	$NH_3(g)$
C_{4v}	4	—	1	—	—	—	IF_5
C_{6v}	6	1	—	—	—	—	—
T	—	—	—	4	3	—	$NH_3(s)$
O	—	—	3	4	6	—	—
T_h	3	—	—	4	3	1	$CO_2(s)$
O_h	9	—	3	4	6	1	SF_6
T_d	6	—	—	4	3	—	CH_4

the resulting configuration for water is the same as if a reflection in the yz plane had taken place:

$$C_2(z)\sigma_v(xz) = \sigma_v(yz)$$

(2) The second requirement is also fulfilled, since a two-fold rotation C_2 followed by reflection in the xz plane and then in the yz plane is equal to reflection in the xz plane followed by a two-fold rotation and then by a reflection in the yz plane:

$$C_2(z)[\sigma_v(xz)\sigma_v(yz)] = [C_2(z)\sigma_v(xz)]\sigma_v(yz)$$

(3) The third requirement is fulfilled since

$$EC_2(z) = C_2(z)E = C_2(z)$$

(4) The fourth requirement is also met since each of the elements E, $C_2(z)$, $\sigma_v(xz)$, and $\sigma_v(yz)$ is its own inverse; for example,

$$C_2(z)C_2(z)^{-1} = E$$

C. Rules for Classification of Molecules into Point Groups

The method for the classification of molecules into different point groups suggested by Zeldin[1] is outlined in Table 1-3. The method can be described as follows:

(1) Determine whether the molecule belongs to a special group such as $D_{\infty h}$, $C_{\infty v}$, T_d, O_h, or I_h. If the molecule is linear, it will be either $D_{\infty h}$ or $C_{\infty v}$. If the molecule has an infinite number of two-fold rotation axes perpendicular to the C_∞ axis, it will fall into point group $D_{\infty h}$. If not, it is $C_{\infty v}$.

(2) If the molecule is not linear, it may belong to a point group of extremely high symmetry such as T_d, O_h, or I_h.

(3) If (1) or (2) is not found to be the case, look for a proper axis of rotation of the highest order in the molecule. If none is found, the molecule is of low symmetry, falling into point group C_3, C_i, or C_1. The presence in the molecule of a plane of symmetry or an inversion center will distinguish among these point groups.

(4) If C_n axes exist, select the one of highest order. If the molecule also has an S_{2n} axis, with or without an inversion center, the point group is S_n.

Table 1-3. Method of Classifying Molecules into Point Groups[1]

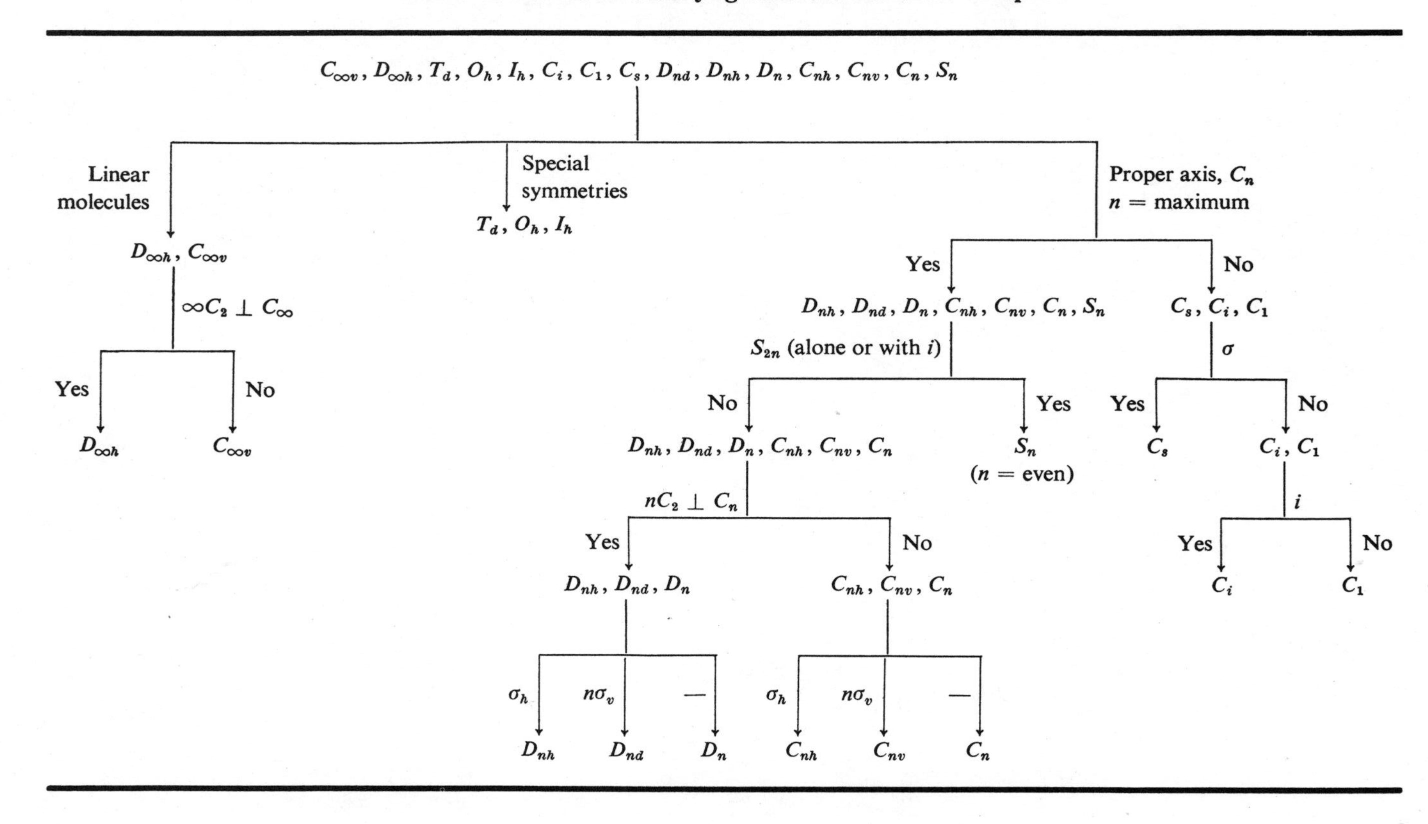

(5) If no S_n exists, look for a set of n two-fold axes lying perpendicular to the major C_n axis. If no such set is found, the molecule belongs to C_{nh}, C_{nv}, or C_n. If a σ_h plane exists, the molecule is of C_{nh} symmetry even if other planes of symmetry are present. If no σ_h plane exists and a σ_v plane is found, the molecule is of C_{nv} symmetry. If no planes exist, it is of C_n symmetry.

(6) If in (5) $nC_2 \perp C_n$ are found, the molecule belongs to the D_{nh}, D_{nd}, or D_n point group. These can be differentiated by the presence (or absence) of symmetry planes (σ_h, σ_v, or no σ, respectively).

Several examples will be considered to illustrate the classification of molecules into point groups. Consider, for instance, the bent triatomic molecule of type AB_2 (H_2O) shown in Fig. 1-20. Following the rules and Table 1-3, it can be determined that the molecule is not of a special symmetry. It does have a C_2 axis of rotation but no S_4 axis. There are no $nC_2 \perp C_n$, and therefore the molecule is either C_{2h}, C_{2v}, or C_2 (see Figs. 1-14 and 1-15). The molecule possesses two vertical planes of symmetry but no σ_h plane, and therefore belongs to the C_{2v} point group.

Now consider the pyramidal molecule of type AB_3 (NH_3) shown in Fig. 1-21. This molecule also is not of a special symmetry. It has a C_3 axis of rotation but no S_6 axis. There are no nC_n axes perpendicular to the C_3 axis, and therefore the molecule belongs to the C classification. Since three vertical planes of symmetry are found but no σ_h plane, the molecule can be classified into C_{3v} (see Figs. 1-16 and 1-17).

Next, consider the square planar AB_4 molecule ($PtCl_4^{2-}$) shown in Fig. 1-22. This molecule is not of a special symmetry. It has a C_4 axis of rotation perpendicular to the plane of molecule but no S_8 axis. Since four C_2 axes are found perpendicular to the C_4 axis, the molecule belongs to one of the D groups. The molecule possesses a σ_h plane perpendicular to the C_4 axis, and therefore it belongs to the point group D_{4h}. Although this molecule possesses two σ_v and two σ_d planes, it is still classified D_{4h}, for the σ_h plane of symmetry predominates in our definition.

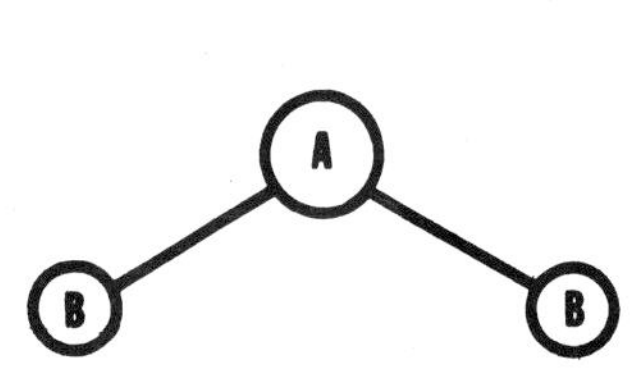

Fig. 1-20. Bent triatomic molecule AB_2.

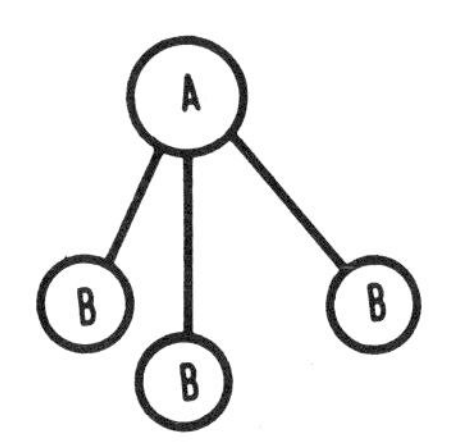

Fig. 1-21. Pyramidal molecule AB_3.

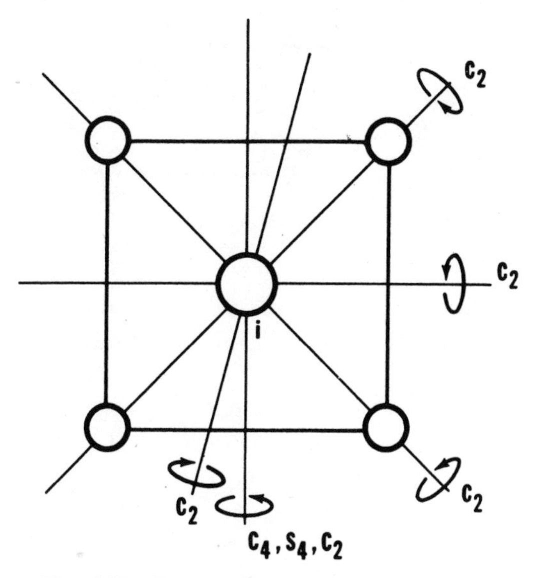

Fig. 1-22. Square planar molecule AB_4.

Next, consider the planar AB_3 molecule (BF_3) shown in Fig. 1-23. This molecule has no special symmetry. It has a C_3 axis of rotation without a collinear S_6 axis. It has three C_2 axes perpendicular to the C_3 axis, and therefore falls into the D classification. It has a σ_h plane of symmetry perpendicular to the C_3 axis and three σ_v planes of symmetry. However, the σ_h plane predominates and the molecule is of D_{3h} symmetry (see Figs. 1-10–1-13).

Our next example is the hexagonal planar molecule of type A_6 or A_6B_6 (benzene) shown in Fig. 1-24. The molecule is not of a special symmetry. It has a center of symmetry and a C_6 axis of symmetry. No S_2 axis exists. Since six C_2 axes perpendicular to the C_6 axis are found, this molecule also falls

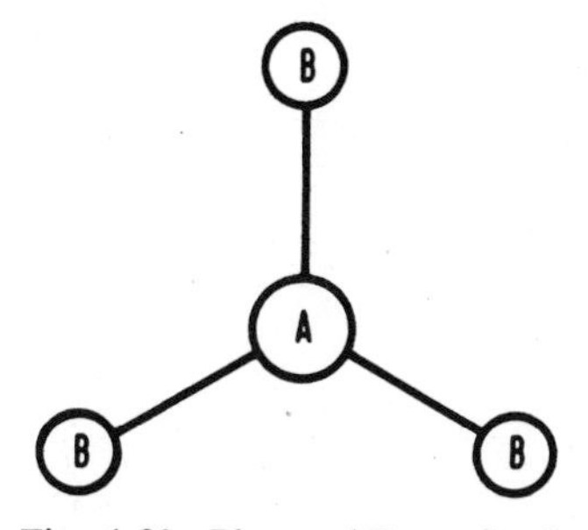

Fig. 1-23. Planar AB_3 molecule.

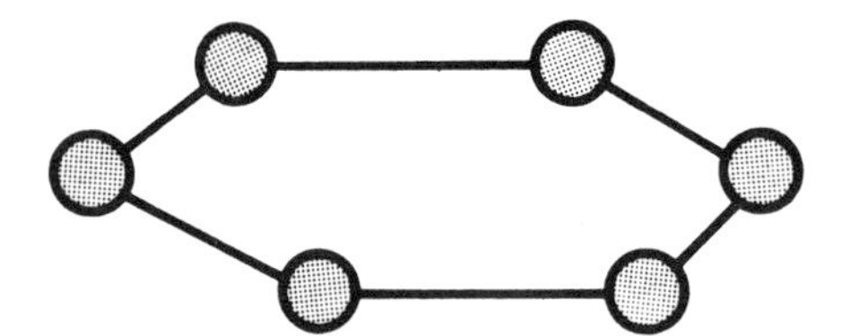

Fig. 1-24. Planar hexagonal A_6B_6 molecule.

into the D classification. Since it has a horizontal plane of symmetry perpendicular to the C_6 axis, the molecule belongs to the D_{6h} point group (see Figs. 1-18 and 1-19).

As our last example, we take the AB_5 trigonal bipyramid (e.g., gaseous PCl_5) shown in Fig. 1-25. This molecule does not belong to a special symmetry. The axis of highest order is C_3. There is no S_6 collinear with C_3. There are three C_2 axes perpendicular to the C_3 axis, and therefore the molecule belongs to one of the D groups. Since it possesses a σ_h plane perpendicular to the C_3 axis, the proper classification is D_{3h}.

D. The Character Table

The derivation of the character table for the C_{2v} point group will be made later in this section. It will be demonstrated in Chapter 2 that character tables can be used for the determination of the selection rules for point and space symmetry. Table 1-4 shows a typical character table for point group C_{3v} and Fig. 1-26 diagrammatically illustrates the significance of the various parts of the character table for C_{3v}. The character table is used to classify the displacements of the atoms of molecules from their equilibrium posi-

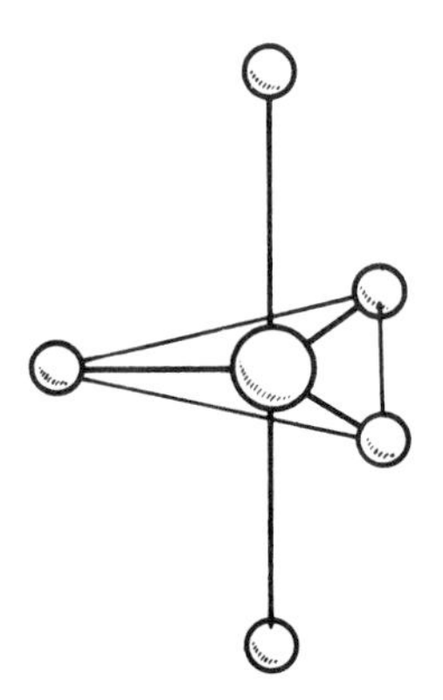

Fig. 1-25. Trigonal bipyramid AB_5 molecule.

Table 1-4. Character Table for the C_{3v} Point Group

C_{3v}	E	$2C_3$	$3\sigma_v$
A_1	1	1	1
A_2	1	1	-1
E	2	-1	0

tions according to the irreducible representation of the symmetry group. The first column of the character table lists the types of representations, or species of vibrations, possible for the given point group. The most symmetric species are placed near the top of the table, and the least symmetric species near the bottom. The symmetry classes pertinent to the point group form the column headings.

Types of Representation

1. *Nonlinear Molecules.* A species is designated by the letter A if the transformation of the molecule is symmetric with respect to the ro-

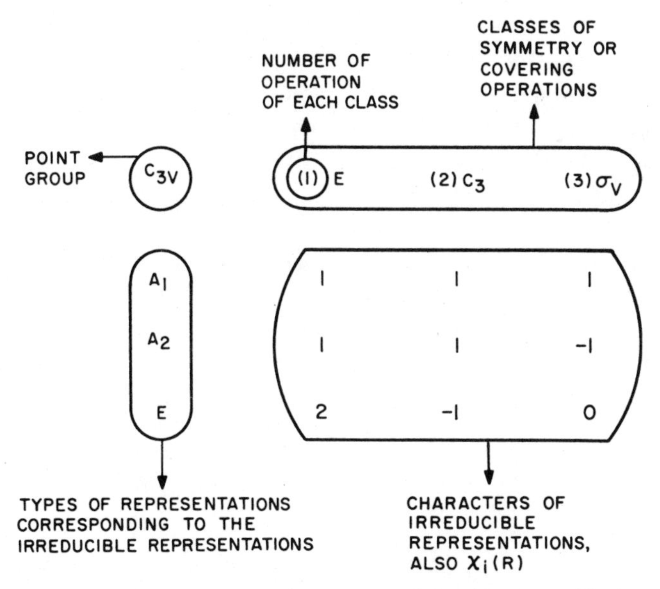

Fig. 1-26. Diagrammatic interpretation of the character table for the C_{3v} point group.

Table 1-5. Character Table for the $C_{\infty v}$ Point Group

$C_{\infty v}$	E	$2C_{\infty}^{\varphi}$	$2C_{\infty}^{2\varphi}$	...	$\infty\sigma_v$
Σ^+	+1	+1	+1	...	+1
Σ^-	+1	+1	+1	...	−1
π	+2	$2\cos\varphi$	$2\cos 2\varphi$	...	$2\cos 3\varphi$
Δ	+2	$2\cos 2\varphi$	$2\cos 2\times 2\varphi$	...	$2\cos 3\times 2\varphi$
φ	+2	$2\cos 3\varphi$	$2\cos 2\times 3\varphi$	...	$2\cos 3\times 3\varphi$

tation about the principal axis of symmetry. In Fig. 1-26 this axis is C_3, and, as can be seen, A_1 is totally symmetric, being labeled with positive 1's for all symmetry classes. A species which is symmetric with respect to the rotation, but is antisymmetric with respect to a rotation about the C_2 axis perpendicular to the principal axis or the vertical plane of reflection, is designated by A_2.

If a species of vibration belongs to the antisymmetric (−1) representation, it is designated by the letter B. If it is symmetric with respect to a rotation about the C_2 axis perpendicular to the principal axis of symmetry or to the vertical plane of reflection, it is a B_1 vibration, and if it is antisymmetric, it is a B_2 vibration. The letter E designates a two-fold degenerate* vibration and the letter F denotes† a triply degenerate vibration. The character under the class of identity gives the degeneracy of the vibration, 1 for singly degenerate, 2 for doubly degenerate, and 3 for triply degenerate. For point groups containing a σ_h operation, primes (e.g., A') and double primes (e.g., A'') are used. The single prime indicates symmetry and the double prime antisymmetry with respect to σ_h. In molecules with a center of symmetry i, the symbols g and u are used, g standing for the German word *gerade* (which means even) and u for *ungerade* (or uneven). The symbol g goes with the species that transforms symmetrically with respect to i, and the symbol u goes with the species that transforms antisymmetrically with respect to i.

2. *Linear Molecules.* Different symbols are used for linear molecules belonging to the point groups $C_{\infty v}$ and $D_{\infty h}$, namely Greek letters

* The bending vibration of CO_2 is an example of a degenerate vibration. The frequency and character of the vibrations are the same, but they occur perpendicular to one another.

† Some texts use the symbol T for the triply degenerate vibration.

identical with the designations used for the electronic states of homonuclear diatomic molecules. The symbols σ or Σ are used for species symmetry with respect to the principal axis. A superscript plus sign (σ^+ or Σ^+) is used for species that are symmetric, and a superscript minus sign (σ^- or Σ^-) for species that are antisymmetric with respect to a plane of symmetry through the molecular axis. The symbols π, Δ, and φ are used for degenerate vibrations, with the degree of degeneracy increasing in this order. This is illustrated in Table 1-5.

Table 1-6 summarizes the symbolism used for various species of vibrations.

Table 1-6. Summary of Symbolisms Used for Various Species of Vibrations

	Symbol	Remarks
Nonlinear molecules	A	One-dimensional representations which are symmetric with respect to the rotation about the principal axis of rotation. $(C_n) = 1$.
	B	One-dimensional representations which are antisymmetric with respect to the rotation about the principal axis of rotation. $(C_n) = 1$.
	E	Two-dimensional representations. Occur in molecules having an axis higher than C_2.
	F	Three-dimensional representations. Occur in molecules having more than two C_3 axes.
	Subscripts 1 and 2 to A and B	Symmetric or antisymmetric with respect to a C_2 axis (or a vertical plane of symmetry) perpendicular to the the principal axis.
	Subscripts g and u to A and B	Symmetric or antisymmetric with respect to a center of symmetry (i).
	Primes and double primes with A and B	Symmetric or antisymmetric with respect to σ_h.
Linear molecules	σ^+ or Σ^+	Symmetric with respect to a plane of symmetry through the molecular axis.
	σ^- or Σ^-	Antisymmetric with respect to a plane of symmetry through the molecular axis.
	π, Δ, φ	Degenerate vibrations, with the degree of degeneracy increasing in this order.

Table 1-7. Classes of Symmetry Operations

Class	Definition
E or I	Identity—rotation through 0°, in effect leaving the molecule untouched
C_n	An n-fold rotation through an angle $2\pi/n$
S_n	An n-fold rotation–reflection axis of symmetry
i	Center of symmetry
σ_v	Vertical plane of symmetry
σ_h	Horizontal plane of symmetry
σ_d	Diagonal plane of symmetry

Classes of Symmetry Operations

The symmetry operations have been previously discussed. Table 1-7 summarizes the various symmetry operations encountered with molecules.

Characters of the Irreducible Representations

The method of determining the characters of the irreducible representations can be illustrated by using water as an example. The effect of the C_2, $\sigma_v(xz)$, and $\sigma_v(yz)$ symmetry operations on the normal coordinates for water is illustrated in Table 1-8. Section II of this table can be simplified as follows:

	E	C_2	$\sigma_v(xz)$	$\sigma_v(yz)$
A_1	+1	+1	+1	+1
A_2	+1	+1	−1	−1
B_1	+1	−1	+1	−1
B_2	+1	−1	−1	+1

This, of course is the character table for the C_{2v} point group. Similar procedures can be used for character tables involving degenerate vibrations. Mutually degenerate vibrations always behave in the same way with respect to an inversion. For an E-type vibration, the character can be either +2, when both components are symmetric, or −2, when they are antisymmetric. For an F-type vibration it can be +3 or −3. Likewise, for a reflection at a σ_h plane perpendicular to C_n, the characters can be +2 or −2 for E and +3 or −3 for F. It is also possible for one component to be antisymmetric and the other symmetric with respect to a symmetry operation, which allows characters of zero for E-type vibrations (e.g., for the C_2,

S_4, and σ_d operations in O_h symmetry). Similarly, one can obtain characters of $+1$, -1, or 0 for *F*-type vibrations.

In Table 1-8 it can be observed that the normal coordinates for vibrations Q_1 and Q_2 correspond to species A_1, in which all symmetry properties are preserved during the vibration. These then are totally symmetric vibrations. By contrast, vibration Q_3 corresponds to species B_2, which is antisymmetric with respect to the C_2 and $\sigma_v(xz)$ operations, and is thus classified as an antisymmetric vibration.

Other symbolisms appear in the literature. For instance, for water the following notation is in use:

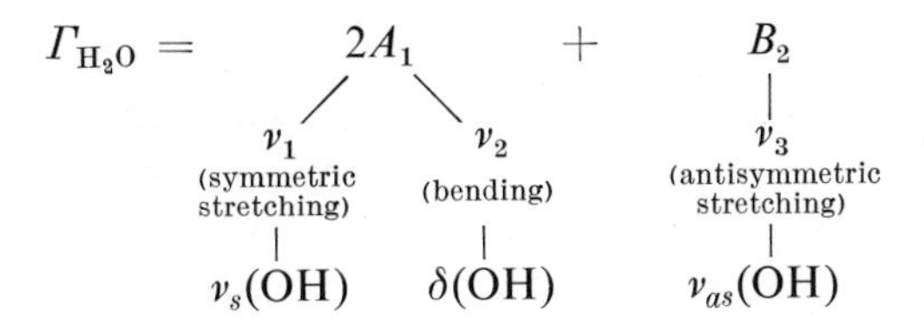

Table 1-8. Effect of the Symmetry Operations on the Normal Coordinates of Water (Q = Vibration, R = Rotation, T = Translation)

Normal coordinate	I				II			
	C_2	$\sigma_v(xz)$	$\sigma_v(yz)$	Species	E	C_2	$\sigma_v(xz)$	$\sigma_v(yz)$
Q_1	Q_1	Q_1	Q_1	A_1	+1	+1	+1	+1
Q_2	Q_2	Q_2	Q_2	A_1	+1	+1	+1	+1
Q_3	$-Q_3$	$-Q_3$	Q_3	B_2	+1	−1	−1	+1
R_x	$-R_x$	$-R_x$	R_x	B_2	+1	−1	−1	+1
R_y	$-R_y$	R_y	$-R_y$	B_1	+1	−1	+1	−1
R_z	R_z	$-R_z$	$-R_z$	A_2	+1	+1	−1	−1
T_x	$-T_x$	T_x	$-T_x$	B_1	+1	−1	+1	−1
T_y	$-T_y$	$-T_y$	T_y	B_2	+1	−1	−1	+1
T_z	T_z	T_z	T_z	A_1	+1	+1	+1	+1

Here ν stands for a stretching mode and δ for a bending mode. For degenerate vibrations, the symbol d is sometimes used. Furthermore, the symbols ϱ_w (wagging), ϱ_r (rocking), ϱ_t (twisting), and π (out-of-plane bending) may be used. Similar notations are employed for other molecules.

The character tables for the common point groups are tabulated in Appendix 1.

E. Space Symmetry

Symmetry Elements for a Molecule in the Condensed State

If one takes into account symmetry elements combined with translations, one obtains operations or elements that can be used to define the symmetry of space. Here translation is defined as the superposition of atoms or molecules from one neighborhood onto the same atoms or molecules in another neighborhood without the use of a rotation. The symmetry element called the *screw axis* involves an operation combining a translation with a rotation. The symmetry element called the *glide plane* involves an operation combining a reflection with a translation. These operations can be used to describe the symmetry of homogeneous spatial materials like crystals and polymers, as well as space itself if one takes the identity element to be the set of all translations. This implies that these translations are those that describe the lattice in contrast to those that are fractional translations associated with some rotation (screw motions and glides).

The screw axis and the glide plane are further defined as follows:

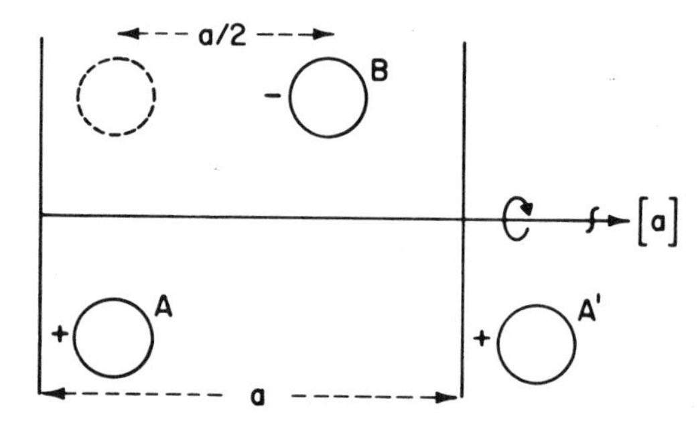

Fig. 1-27. The operation of a two-fold screw axis. [From P. J. Wheatley, *The Determination of Molecular Structure*, 2nd Ed. (1968). Courtesy of the Clarendon Press, Oxford.]

Table 1-9. Symmetry Elements and Symmetry Operations Point and Space

Symmetry element	Symmetry operation
1. Identity (E or I)	1. Molecule or unit cell unchanged
2. Axis of rotation (C_n)	2. Rotation about axis by $2\pi/n$
3. Center of symmetry or inversion center (i)	3. Inversion of all atoms through center
4. Plane (σ)	4. Reflection on the plane
5. Rotation–reflection axis (S_n)	5. Rotation about axis by $2\pi/n$ followed by reflection in a plane perpendicular to the axis
6. Screw axis (n_p)	6. Rotation followed by a translation
7. Glide plane (c)	7. Reflection followed by a translation

1. **Screw Axis—n_p:**

   Rotation followed by a translation;

   n = order of axis;

   p/n = fraction of the unit cell* over which translation occurs;

   $n = 2, 3, 4$ or 6; $p = 1, 2, 3 \ldots, n - 1$;

   2_1 = two-fold screw axis, translation one-half the distance of the unit cell;

   3_1 = three-fold screw axis, translation one-third the distance of the unit cell.

   Figure 1-27 shows the operation of a two-fold screw axis. Figure 1-28 shows a comparison of rotation and screw axes.

2. **Glide Plane:** Reflection followed by a translation. Figure 1-29 shows the operation of a glide plane.

The new symmetry elements that are added to the point symmetry elements are thus the screw axis and glide plane. Table 1-9 shows the point

* A crystal may be considered to be made up of a large number of blocks of the same size and shape. One such block is defined as a unit cell. The unit cell must be capable of repetition in space without leaving any gap. The unit cell may be primitive or a centered or nonprimitive unit cell. The distinction between these will be made later in this section.

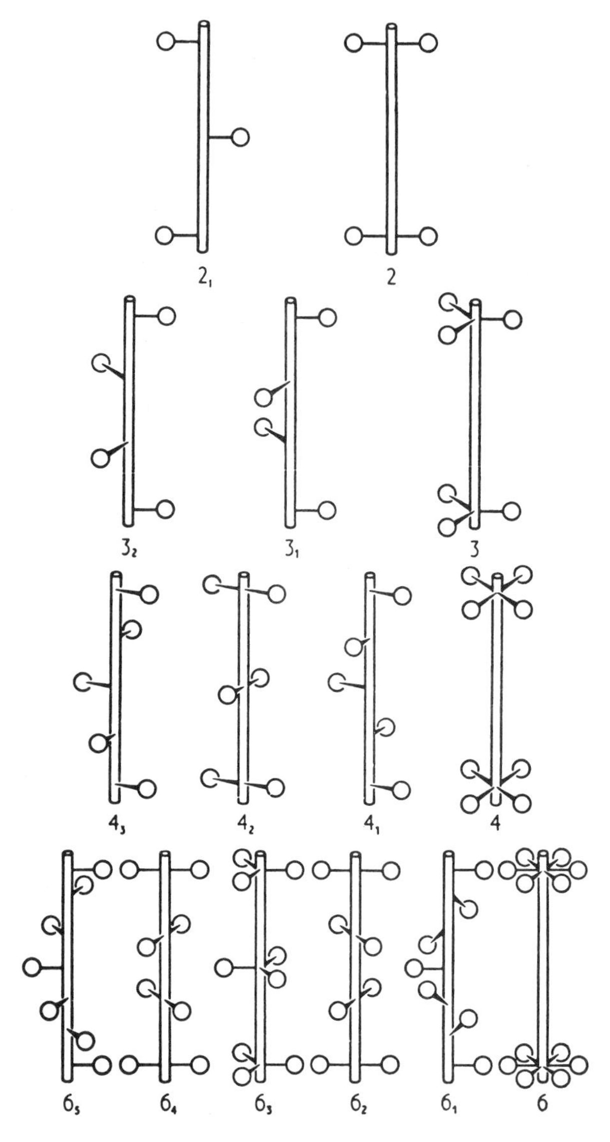

Fig. 1-28. Comparison of pure rotation axes and screw axes. [From P. J. Wheatley, *The Determination of Molecular Structure*, 2nd Ed. (1968). Courtesy of the Clarendon Press, Oxford.]

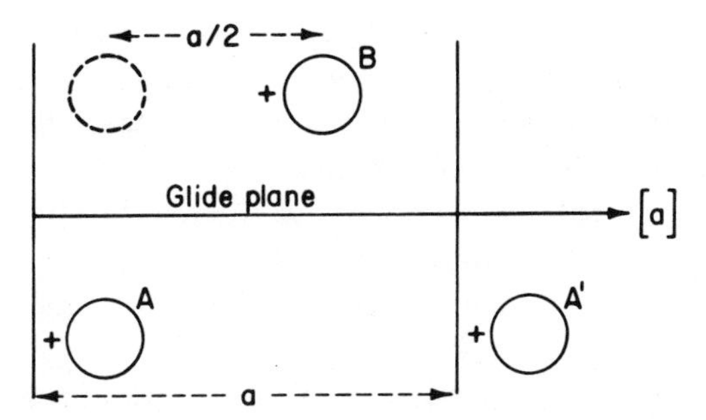

Fig. 1-29. The operation of a glide plane. [From P. J. Wheatley, *The Determination of Molecular Structure*, 2nd Ed. (1968). Courtesy of the Clarendon Press, Oxford.]

and space symmetry elements. As a result, 230 different combinations of symmetry elements become possible in what are called space groups. Table 1-10 shows the distribution of the space groups among the seven crystal systems. Some of the space groups are never found in actual crystals and about one-half of the crystals belong to the 13 space groups of the monoclinic system.

Space Group

Just as the point group collects all of the point symmetry elements, the space group is seen as collecting all of the space symmetry elements in mol-

Table 1-10. Distribution of Space Groups Among the Seven Crystal Systems

Crystal system*	Number of space groups
Triclinic	2
Monoclinic	13
Orthorhombic	59
Trigonal	25
Hexagonal	27
Tetragonal	68
Cubic	36

* As defined by A. F. Wells, *Structural Inorganic Chemistry*, Clarendon Press (1950).

ecules involving translation. For the space group selection rules it is necessary to work on molecules for which space groups are known from X-ray studies, or where sufficient information is available to make a choice of structure. Alternatively, the structure may be assumed and the space group selection rules can serve as a test of this assumption. In deriving space group selection rules, one must deal with a primitive or Bravais unit cell. A primitive unit cell is the smallest unit in a crystal which, by a series of translations, would build up the whole crystal.

The Hermann–Mauguin notation is generally used to describe the space group. Tables exist to convert this notation to the Schoenflies notation. The first symbol is a capital letter and indicates whether the lattice is primitive. The next symbol refers to the principal axis, whether it is rotation, inversion, or screw, e.g.,

$P2_1$ = primitive lattice with a two-fold axis of rotation, translation one-half the distance of the unit cell;

$C2$ = nonprimitive centered lattice with a two-fold axis of rotation.

A mirror plane is symbolized as m, and a glide plane by c, e.g.,

$P2_1/m$, m = mirror plane perpendicular to principal axis;

$C2/c$, c = glide plane perpendicular to principal axis.

Table 1-11 shows the space group symbolism used and Appendix 2 contains a description of the space group D_{2h}^{15}.

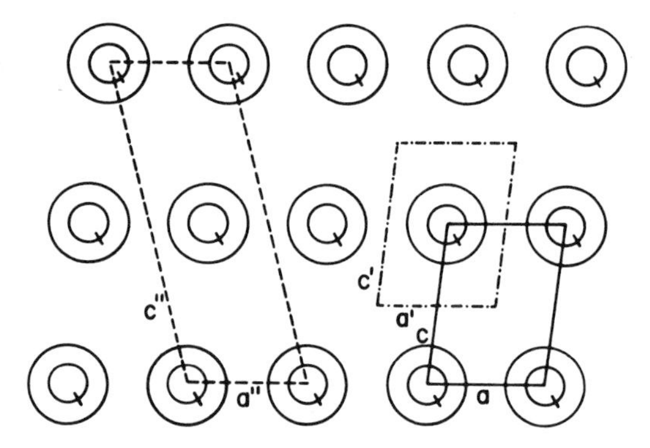

Fig. 1-30. Differentiation between primitive and centered unit cells. [From P. J. Wheatley, *The Determination of Molecular Structure*, 2nd Ed. (1968). Courtesy of the Clarendon Press, Oxford.]

Table 1-11. Space Group Symbolism

First symbol refers to the Bravais lattice
- P = primitive lattice
- C = centered lattice
- F = face-centered lattice
- I = body-centered lattice
- R = rhombohedral (unit cell can be primitive or nonprimitive; see notes to Table 1-12)

Principal axis of rotation given number n = order
- e.g., 2 = two-fold axis of rotation

For screw axis p/n = fraction of primitive lattice over which translation parallel to screw axis occurs
- e.g., $P2_1$ = primitive lattice with a two-fold axis of rotation, translation one-half unit cell

Mirror plane = m

Glide planes = symbols a, b, c along $[a]$, $[b]$, $[c]$ axes
- symbol $n = (b + c)/2$ or $(a + b)/2$
- symbol $d = (a + b)/4$ or $(b + c)/4$ or $(a + c)/4$
- e.g., $P2_1/m$, m = mirror plane perpendicular to principal axis
- $C2/c$, c = glide plane axis perpendicular to principal axis

As previously mentioned, the primitive unit cell is the smallest unit of a crystal that reproduces itself by translations. Figure 1-30 illustrates the difference between a primitive and a centered or nonprimitive cell. The primitive cell can be defined by the lines a and c. Alternatively, we could have defined it by the lines a' and c'. Choosing the cell defined by the lines a'' and c'' gives us a nonprimitive cell or centered cell, which has twice the volume and two repeat units. Table 1-12 illustrates the symbolism used for the various types of lattices, and records the number of repeat units in the cell for a primitive and a nonprimitive lattice. The spectroscopist is concerned with the primitive or Bravais unit cell in dealing with lattice vibrations. For factor group selection rules it is necessary to convert the number of molecules per crystallographic unit cell Z to Z', which is the number of molecules per Bravais or primitive cell. For example,

Table 1-12. Primitive and Centered Lattices

Type	Symbol	Number of repeat units in cell
Primitive	P	1
Rhombohedral*	R	3 or 1†
Body centered	I	2
Side centered	A, B, or C	2
Face centered	F	4

* Also called trigonal.
† There are cases in which the number of repeat units in the crystallographic cell may be three or one. For the cases where it is three, Z will be divisable by three. For example, for TiS, D_{3d}^5–$R\bar{3}1m$ (No. 166), $Z = 9$, and therefore $Z' = 9/3 = 3$. However, for Cr_2O_3, D_{3d}^6–$R\bar{3}c$ (No. 167), $Z = 2$, and $Z' = 2/1 = 2$. Thus in the latter crystal the cell can be considered to be primitive.

$$Z' = \text{number of molecules in the Bravais or primitive cell}$$
$$= \frac{Z\text{ (number of molecules in crystallographic cell)}}{\text{repeat units in cell}}$$

If $Z = 4$ for an F-type lattice, then

$$Z' = 4/4 = 1$$

In the site symmetry compilation for the 230 space groups given in Appendix 3, the data are for a primitive cell and can be used directly.

Factor Group

It is necessary to define a factor group and describe how it relates to a space group. In a crystal one primitive cell or unit cell can be carried into another primitive cell or unit cell by a translation. The number of translations of unit cells then would seem to be infinite since a crystal is composed of many such units. If, however, one considers only one translation and consequently only two unit cells, and defines the translation that takes a point in one unit cell to an equivalent point in the other unit cell as the identity, one can define a finite group, which is called a *factor group* of the space group.[2]

The factor groups are isomorphic (one-to-one correspondence) with the 32 point groups and, consequently, the character table of the factor group can be obtained from the corresponding isomorphic point group.

Site Group

It also becomes necessary to define a site group. A unit cell of a crystal is composed of points (molecules or ions) located at particular positions in the cell. It turns out, however, that the points can only be located at certain positions in the lattice which are called sites, that is, they can only be located on one of the symmetry elements of the factor group and thus remain invariant under that operation independent of translation. The point has fewer symmetry elements than the parent factor group and belongs to what is called a "site group," which is a subgroup of the factor group. In general, factor groups can have a variety of different sites possible, that is, many subgroups can be formed from the factor group. Also, a number of distinct sites of the same site group are possible.

PROBLEMS

1. Determine the point symmetry of the following molecules:
   a) DCH_3 (methane-like structure)
   b) D_2CH_2 (methane-like structure)
   c) B_2H_6 (bridged hydrogen structure)
   d) PF_5 (trigonal bipyramid)
   e) ClF_3 (T-shaped planar)
   f) Pyrazine (planar six-membered ring with nitrogens at 1,4 positions)
   g) IF_5 (tetragonal pyramid)
   h) IF_7 (pentagonal bipyramid)
   i) N_2O (linear asymmetric)
   j) $Ni(CN)_4^{2-}$ (planar with cyanide groups at corners of a square around a central atom).

2. Show the effect of the symmetry operations on the normal coordinates of ammonia and methane.

Answers

1. a) C_{3v}; b) C_{2v}; c) D_{2h}; d) D_{3h}; e) C_{2v}; f) D_{2h}; g) C_{4v}; h) D_{5h}; i) $C_{\infty v}$; j) D_{4h}.

REFERENCES

1. M. Zeldin, *J. Chem. Ed.*, **43**:17 (1956).
2. W. H. Zachariasen, *Theory of X-Ray Diffraction in Crystals*, J. Wiley & Sons, New York (1945).

BIBLIOGRAPHY

G. Herzberg, *Molecular Spectra and Molecular Structure*, II, *Infrared and Raman Spectra of Polyatomic Molecules*, D. Van Nostrand Co., New York (1945).

K. Nakamoto, *Infrared Spectra of Inorganic and Coordination Compounds*, J. Wiley & Sons, New York (1970).

N. B. Colthup, L. H. Daly, and S. E. Wiberley, *Introduction to Infrared and Raman Spectroscopy*, Academic Press, New York (1975).

F. Albert Cotton, *Chemical Application of Group Theory*, J. Wiley & Sons, New York (1971).

H. H. Jaffè and M. Orchin, *Symmetry in Chemistry*, J. Wiley & Sons, New York (1965).

G. M. Barrow, *Introduction to Molecular Spectroscopy*, McGraw-Hill Book Co., New York (1962).

N. L. Alpert, W. E. Keiser, and H. A. Szymanski, *IR—Theory and Practice of Infrared Spectroscopy*, Plenum Press, New York (1970).

A. J. Sonnessa, *Introduction to Molecular Spectroscopy*, Reinhold Publishing Corp., New York (1966).

D. N. Kendall (ed.), *Applied Infrared Spectroscopy*, Reinhold Publishing Corp., New York (1966).

S. K. Freeman (ed.), *Interpretative Spectroscopy*, Reinhold Publishing Corp., New York (1965).

J. R. Ferraro and J. S. Ziomek, *Introductory Group Theory and Its Application to Molecular Structure*, Plenum Press, New York (1969).

L. H. Jones, *Inorganic Vibrational Spectroscopy*, M. Dekker, New York (1971).

J. M. Hollas, *Symmetry in Molecules*, Chapman and Hall Ltd., London (1972).

J. E. White, *J. Chem. Ed.*, **44**:128 (1967).

H. Weyl, *Symmetry*, Princeton University Press, Princeton, N. J. (1952).

R. M. Hochstrasser, *Molecular Aspects of Symmetry*, W. A. Benjamin, New York (1966).

Chapter 2

DERIVATION OF SELECTION RULES*

2-1. SELECTION RULES FOR ISOLATED MOLECULES

The complete selection rules for nonlinear molecules of the point group T_d of the type AB_4 are presented in this chapter. The selection rules govern which bands will appear in the infrared and Raman spectra and will apply to any molecule with the above-mentioned symmetry. The selection rules for linear molecules of the point group $C_{\infty v}$ are also presented. For the derivation of selection rules we have employed the notation of Meister, Cleveland, and Murray.[1]

A. The T_d Point Group

The symmetry elements of the T_d point group are illustrated in Fig. 2-1. The characters for the irreducible representations for T_d are given in Table 2-1. Two main types of symmetry operations are possible, and these are called proper and improper rotations. A proper rotation is defined as a rotation through an angle $\pm\varphi$ about some axis of symmetry. An improper rotation is a rotation followed by a reflection in a plane perpendicular to the axis of rotation. Thus, a reflection may be considered an improper rotation through an angle of 0°. The center of symmetry (i) operation can be considered an S_2 operation and, thus, may also be considered an improper rotation. The difference between a proper and an improper rotation is shown in Fig. 2-2.

Number of Fundamentals of Each Type

The quantity $\Xi(R)$ from Table 2-1 for CCl_4 is used to determine the number of fundamentals of each type,[1,2] where

* The authors would like to extend their thanks and appreciation to Dr. Alan Walker, Scarborough College, University of Toronto, Scarborough, Canada, for checking some of the selection rule computations presented in this chapter.

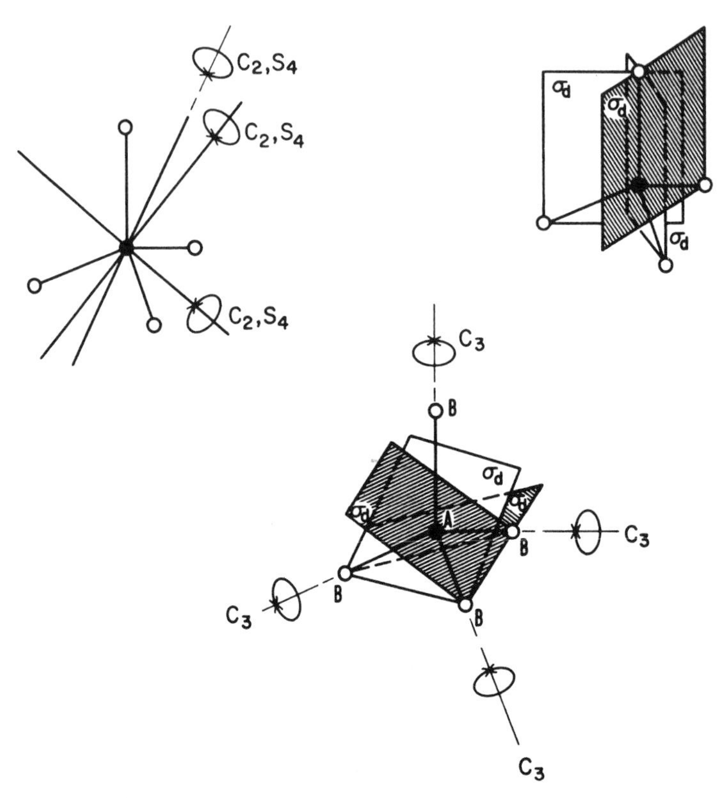

Fig. 2-1. Symmetry elements for the T_d point group ($4C_3$, $3C_2 \approx S_4$, $6\sigma_d$).

$$\begin{aligned}\Xi(R) &= (\mu_R - 2)(1 + 2\cos\varphi) \text{ for proper rotations} \\ &= (\mu_R)(-1 + 2\cos\varphi) \text{ for improper rotations}\end{aligned}$$

These relationships are applicable for molecules of different symmetries. Here φ is the angle associated with the proper or improper movement and μ_R is obtained from Table 2-1 and is the number of atoms left unchanged by the operation R. The number of vibrations of each type depends on the number of atoms contained in the molecule[1,2] and the symmetry involved.

The reduction formula used to obtain the number of frequencies of each species is

$$N_i = \frac{1}{N_G} \sum n_e \Xi(R) \chi_i(R) \tag{2-1}$$

where N_G is the number of elements in the group (e.g., 1 + 8 + 3 + 6 + 6

$= 24$), n_e is the number of elements in each class (e.g., for E, $n_e = 1$), $\chi_i(R)$ is the character of the vibration species.

For T_d symmetry, and specifically for the CCl_4 molecule, the following results are obtained when the proper values are substituted into Eq. (2-1):

$$N_{A_1} = \tfrac{1}{24}[(1\times9\times1) + (8\times0\times1) + (3\times1\times1) + (6\times3\times1) + (6\times-1\times1)] = 1$$

$$N_{A_2} = \tfrac{1}{24}[(1\times9\times1) + (8\times0\times1) + (3\times1\times1) + (6\times3\times-1) + (6\times-1\times-1)] = 0$$

$$N_{E} = \tfrac{1}{24}[(1\times9\times2) + (8\times0\times-1) + (3\times1\times2) + (6\times3\times0) + (6\times-1\times0)] = 1$$

$$N_{F_2} = \tfrac{1}{24}[(1\times9\times3) + (8\times0\times0) + (3\times1\times-1) + (6\times3\times1) + (6\times-1\times-1)] = 2$$

$$N_{F_1} = \tfrac{1}{24}[(1\times9\times3) + (8\times0\times0) + (3\times1\times-1) + (6\times3\times-1) + (6\times-1\times1)] = 0$$

(Note that in the expansions here and below to save space and avoid confusion negative values have not been enclosed in parentheses; thus,

Table 2-1. Character Table for T_d Symmetry

T_d	E	$8C_3$	$3C_2$	$6\sigma_d$	$6S_4$	
A_1	1	1	1	1	1	$\chi_i(R)$
A_2	1	1	1	−1	−1	
E	2	−1	2	0	0	
F_2	3	0	−1	1	−1	
F_1	3	0	−1	−1	1	
φ	0°	120°	180°	0°	90°	
$2\cos\varphi$	2	−1	−2	2	0	
$\pm1 + 2\cos\varphi$	3	0	−1	1	−1	$\chi_M(R)$
2φ	0°	240°	360°	0°	180°	
$2\cos 2\varphi$	2	−1	2	2	−2	
$2 \pm 2\cos\varphi + 2\cos 2\varphi$	6	0	2	2	0	$\chi_\alpha(R)$
μ_R	5	2	1	3	1	
$\Xi(R)$	9	0	1	3	−1	e.g., CCl_4
	Proper rotations			Improper rotations		

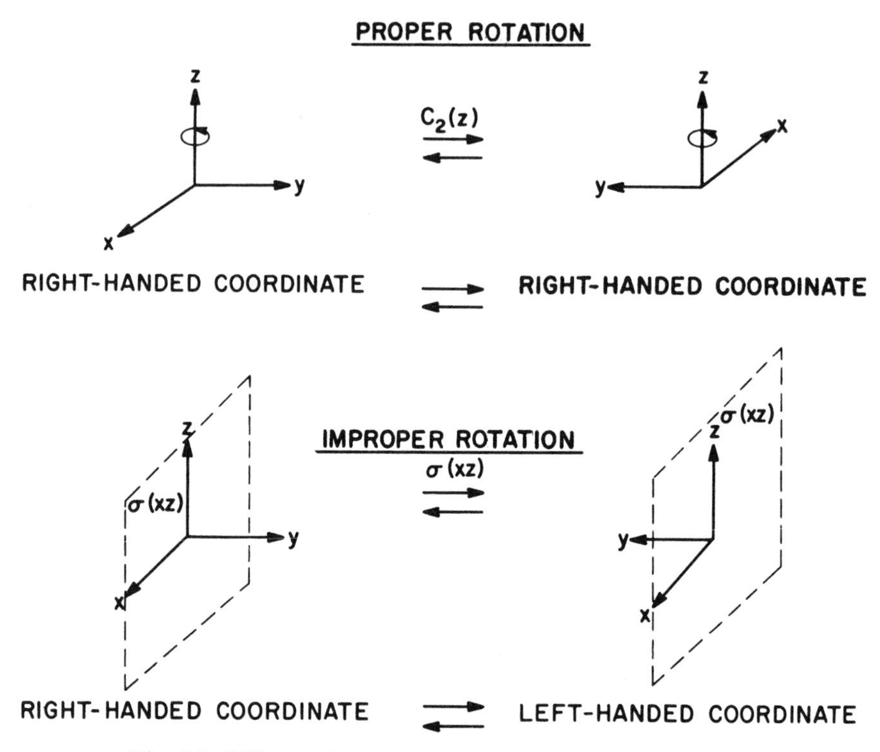

Fig. 2-2. Difference between a proper and an improper rotation.

$(6\times-1\times-1)$ has been written instead of $[6\times(-1)\times(-1)]$). Positive integers are obtained for A_1, E, and F_2, and the numbers indicate how many fundamentals are found for each type. This can be written as

$$\Gamma = A_1 + E + 2F_2$$

Thus in the case of CCl_4 there are 3 degenerate fundamentals and one nondegenerate fundamental. The total number of fundamentals is 9, which checks with the well-known formula $3n - 6$, where $n = 5$ is the number of atoms in the molecule.

Number of Fundamental Vibrations Allowed in the Infrared

To determine which fundamentals will be active in the infrared, the character of the dipole moment $\chi_M(R)$ is used,[1,2] where

$$\chi_M(R) = \pm 1 + 2\cos\varphi$$

The angle φ is associated with the proper or improper rotation R. The positive sign is used with the proper rotation, in the case of E, C_3, or C_2. The negative sign is used with the improper rotation, in the case of σ_d or S_4.

The reduction formula

$$N_i = \frac{1}{N_G} \sum n_e \chi_M(R) \chi_i(R) \qquad (2\text{-}2)$$

is used to determine the vibrations allowed in the infrared. For the case of T_d the following sums are obtained when the proper values are substituted into Eq. (2-2):

$$N_{A_1} = \tfrac{1}{24}[(1\times3\times1) + (8\times0\times1) + (3\times-1\times1) + (6\times1\times1) + (6\times-1\times1)] = 0$$

$$N_{A_2} = \tfrac{1}{24}[(1\times3\times1) + (8\times0\times1) + (3\times-1\times1) + (6\times1\times-1) + (6\times-1\times-1)] = 0$$

$$N_E = \tfrac{1}{24}[(1\times3\times2) + (8\times0\times-1) + (3\times-1\times2) + (6\times1\times0) + (6\times-1\times0)] = 0$$

$$N_{F_2} = \tfrac{1}{24}[(1\times3\times3) + (8\times0\times0) + (3\times-1\times-1) + (6\times1\times1) + (6\times-1\times-1)] = 1$$

$$N_{F_1} = \tfrac{1}{24}[(1\times3\times3) + (8\times0\times0) + (3\times-1\times-1) + (6\times1\times-1) + (6\times-1\times1)] = 0$$

In this case, the magnitude of the result obtained for each species has no special significance, but infrared activity is indicated whenever a positive integer is obtained from Eq. (2-2). This is the case only for $\chi_{F_2}(R)$, and therefore here $\chi_M(R) = \chi_{F_2}(R)$. Thus, for a molecule possessing T_d symmetry, the vibrations of types A_1, A_2, E, and F_1 are inactive in the infrared, and only the F_2 species is infrared active.

Number of Fundamental Vibrations Allowed in the Raman Spectrum

To determine the Raman-active fundamentals the character of the polarizability α, $\chi_\alpha(R)$, is used,[1,2] where

$$\chi_\alpha(R) = 2 \pm 2\cos\varphi + 2\cos 2\varphi$$

The values of $\chi_\alpha(R)$ appear in Table 2-1. Again, the positive sign is used with proper rotations and the negative sign with improper rotations. The reduction formula

$$N_i = \frac{1}{N_G} \Sigma\, n_e \chi_\alpha(R) \chi_i(R) \qquad (2\text{-}3)$$

is used. This is similar to Eq. (2-2), except that $\chi_\alpha(R)$ replaces $\chi_M(R)$. For the case of T_d the following results are obtained:

$$\begin{aligned}
N_{A_1} &= \tfrac{1}{24}[(1\times 6\times 1) + (8\times 0\times 1) + (3\times 2\times 1) + (6\times 2\times 1) \\
&\quad + (2\times 0\times 1)] = 1 \\
N_{A_2} &= \tfrac{1}{24}[(1\times 6\times 1) + (8\times 0\times 1) + (3\times 2\times 1) + (6\times 2\times -1) \\
&\quad + (6\times 0\times -1)] = 0 \\
N_E &\quad \tfrac{1}{24}[(1\times 6\times 2) + (8\times 0\times -1) + (3\times 2\times 2) + (6\times 2\times 0) \\
&\quad + (6\times 0\times 0)] = 1 \\
N_{F_2} &= \tfrac{1}{24}[(1\times 6\times 3) + (8\times 0\times 0) + (3\times 2\times -1) + (6\times 2\times 1) \\
&\quad + (6\times 0\times -1)] = 1 \\
N_{F_1} &= \tfrac{1}{24}[(1\times 6\times 3) + (8\times 0\times 0) + (3\times 2\times -1) + (6\times 2\times -1) \\
&\quad + (6\times 0\times 1)] = 0
\end{aligned}$$

As in the case of the infrared allowed vibrations, the species is Raman active if a positive integer is obtained from Eq. (2-3). Therefore, A_1, E, and F_2 appear in $\chi_\alpha(R)$, where $\chi_\alpha(R) = \chi_{A_1}(R) + \chi_E(R) + \chi_{F_2}(R)$, and only the types A_1, E, and F_2 are Raman active. The selection rules for the fundamentals in the T_d symmetry are summarized in Table 2-2.

Combinations Allowed

To obtain the selection rules for combination or difference frequencies ($\nu_1 \pm \nu_2$) the characters of the vibration types are multiplied. For example, for the combination* frequency ($A_1 \pm A_2$),

	E	C_3	C_2	σ_d	S_4	
$\chi_{A_1}(R)$	1	1	1	1	1	
$\chi_{A_2}(R)$	1	1	1	-1	-1	multiply
$\chi_{A_1}(R)\chi_{A_2}(R)$	1	1	1	-1	-1	

* Henceforth, whenever the term "combination" appears in the text, "difference" is also implied.

Table 2-2. Selection Rules for the Fundamental Vibrations for the Case of T_d

Type	Fundamentals*	Activity	
		Raman	Infrared
A_1	1	a	ia
A_2	0	ia	ia
E	1	a	ia
F_2	2	a	a
F_1	0	ia	ia

* For the case of CCl_4.
a = active.
ia = inactive.

The character of irreducible representation for $(A_1 \pm A_2)$ is thus identical with that for A_2, and the combination is active where A_2 is active. Since A_2 is inactive in T_d, this combination or difference tone is inactive also. The same procedure can be used for other combinations. The reduction formula can be used in cases where it is not easily determined from Table 2-1 which species are contained in the combination.

For example, for the combination $(E \times F_2)$,

	E	C_3	C_2	σ_d	S_4	
$\chi_E(R)$	2	−1	2	0	0	
$\chi_{F_2}(R)$	3	0	−1	1	−1	multiply
$\chi_E(R)\chi_{F_2}(R)$	6	0	−2	0	0	

The reduction formula

$$N_i = \frac{1}{N_G} \Sigma\, n_e[\chi_E(R)\chi_{F_2}(R)]\chi_i(R) \tag{2-4}$$

is used:

$$N_{A_1} = \tfrac{1}{24}[(1\times6\times1) + (8\times0\times1) + (3\times-2\times1) + (6\times0\times1) + (6\times0\times1)] = 0$$

$$N_{A_2} = \tfrac{1}{24}[(1\times6\times1) + (8\times0\times1) + (3\times-2\times1) + (6\times0\times-1) + (6\times0\times-1)] = 0$$

$$N_E = \frac{1}{24}[(1\times6\times2) + (8\times0\times-1) + (3\times-2\times2) + (6\times0\times0) + (6\times0\times0)] = 0$$

$$N_{F_2} = \frac{1}{24}[(1\times6\times3) + (8\times0\times0) + (3\times-2\times-1) + (6\times0\times1) + (6\times0\times-1)] = 1$$

$$N_{F_1} = \frac{1}{24}[(1\times6\times3) + (8\times0\times0) + (3\times-2\times-1) + (6\times0\times-1) + (6\times0\times1)] = 1$$

This means that when E combines with F_2 the combination is active whenever any one of the components F_2 and F_1 is active. Table 2-3 summarizes the selection rules for combinations for the case of T_d.

Overtones Allowed

To obtain the selection rules for the overtones of nondegenerate vibrations, the same procedure used for combination frequencies is followed. Thus, for the overtones of A_1, the following characters are obtained:

$\chi_{A_1}(R)$	1	1	1	1	1	
$\chi_{A_1}(R)$	1	1	1	1	1	
						multiply
$\chi^2_{A_1}(R)$	1	1	1	1	1	same as A_1
$\chi_{A_1}(R)$	1	1	1	1	1	
						multiply
$\chi^3_{A_1}(R)$	1	1	1	1	1	same as A_1

Therefore, the overtones of A_1 are active wherever A_1 is active. For the overtones of A_2 the characters obtained are:

$\chi_{A_2}(R)$	1	1	1	-1	-1	
$\chi_{A_2}(R)$	1	1	1	-1	-1	
						multiply
$\chi^2_{A_2}(R)$	1	1	1	1	1	same as A_1
$\chi_{A_2}(R)$	1	1	1	-1	-1	
						multiply
$\chi^3_{A_2}(R)$	1	1	1	-1	-1	same as A_2

Thus, for A_1 overtones, $A_1^n = A_1$. For A_2 overtones, A_2^n equals A_1 for n even, and is active where A_1 is active, and equals A_2 for n odd and is inactive since A_2 is inactive.

For degenerate vibrations[1] obtaining the selection rules is more difficult. For E type vibrations, the equation used is

$$\chi_{E^n}(R) = \tfrac{1}{2}[\chi_{E^{n-1}}(R)\chi_E(R) + \chi_E(R^n)] \quad (2\text{-}5)$$

Table 2-3. Selection Rules for Combination Frequencies of a Molecule in T_d Symmetry

Combination	A_1	A_2	E	F_2	F_1	Activity	
						Raman	Infrared
$A_1 \times A_1$	1	0	0	0	0	a	ia
$A_1 \times A_2$	0	1	0	0	0	ia	ia
$A_1 \times E$	0	0	1	0	0	a	ia
$A_1 \times F_2$	0	0	0	1	0	a	a
$A_1 \times F_1$	0	0	0	0	1	ia	ia
$A_2 \times E$	0	0	1	0	0	a	ia
$A_2 \times F_2$	0	0	0	0	1	ia	ia
$A_2 \times F_1$	0	0	0	1	0	a	a
$E \times E$	1	1	1	0	0	a	ia
$A_2 \times A_2$	1	0	0	0	0	a	ia
$F_2 \times F_2$	1	0	1	1	1	a	a
$F_2 \times F_1$	0	1	1	1	1	a	a
$F_1 \times F_1$	1	0	1	1	1	a	a
$A_1 \times A_2 \times E$	0	0	1	0	0	a	ia
$A_1 \times A_2 \times F_2$	0	0	0	0	1	ia	ia
$A_1 \times A_2 \times F_1$	0	0	0	1	0	a	a
$A_1 \times E \times F_2$	0	0	0	1	1	a	a
$A_1 \times E \times F_1$	0	0	0	1	1	a	a
$A_2 \times E \times F_2$	0	0	0	1	1	a	a
$A_2 \times E \times F_1$	0	0	0	1	1	a	a
$E \times F_2$	0	0	0	1	1	a	a
$E \times F_1$	0	0	0	1	1	a	a
$A_1 \times F_2 \times F_2$	1	0	1	1	1	a	a
$A_1 \times F_1 \times F_2$	0	1	1	1	1	a	a
$A_2 \times F_1 \times F_2$	1	0	1	1	1	a	a
$A_2 \times F_2 \times F_2$	0	1	1	1	1	a	a

For the first overtone $n = 2$, and Eq. (2-5) becomes

$$\chi_{E^2}(R) = \tfrac{1}{2}[\chi_E(R)\chi_E(R) + \chi_E(R^2)] \tag{2-6}$$

$\chi_E(R^2)$ is the character corresponding to the operation R performed twice in succession. Table 2-4 contains $\chi_E(R^n)$ for the different operations. This tabulation is possible because the following operations can be shown to be true:

$$\chi_E(2C_3) = \chi_E(-C_3) = -1$$
$$\chi_E(3C_3) = \chi_E(E) \quad = \quad 2$$

Table 2-4. Character Table of $\chi_E(R^n)$

	E	C_3	C_2	σ_d	S_4
$\chi_E(R^2)$	2	-1	2	2	2
$\chi_E(R^3)$	2	2	2	0	0
$\chi_E(R^4)$	2	-1	2	2	2

$$\begin{aligned}
\chi_E(4C_3) &= \chi_E(C_3) &&= -1\\
\chi_E(2C_2) &= \chi_E(E) &&= 2\\
\chi_E(3C_2) &= \chi_E(C_2) &&= 2\\
\chi_E(4C_2) &= \chi_E(E) &&= 2\\
\chi_E(2\sigma_d) &= \chi_E(E) &&= 2\\
\chi_E(3\sigma_d) &= \chi_E(\sigma_d) &&= 0\\
\chi_E(4\sigma_d) &= \chi_E(E) &&= 2\\
\chi_E(2S_4) &= \chi_E(C_2) &&= 2\\
\chi_E(3S_4) &= \chi_E(S_4) &&= 0\\
\chi_E(4S_4) &= \chi_E(E) &&= 2
\end{aligned}$$

For $\chi_{E^2}(R)$, from Eq. (2-6),

$\chi_E(R)$	2	-1	2	0	0	
$\chi_E(R)$	2	-1	2	0	0	
						multiply
$[\chi_E(R)]^2$	4	1	4	0	0	
$\chi_E(R^2)$	2	-1	2	2	2	
						add
	6	0	6	2	2	
						divide by 2
$\chi_{E^2}(R)$	3	0	3	1	1	

Hence, $E^2 = A_1 + E$; therefore, it is Raman active only. This result is obtained by inspection from the character table for T_d (Table 2-1), or by use of a reduction formula similar to Eq. (2-1), substituting $\chi_{E^2}(R)$ for

$\Xi(R)$:

$$N_i = \frac{1}{N_G} \sum n_e \chi_{E^2}(R) \chi_i(R) \tag{2-7}$$

For the second overtone, Eq. (2-5) becomes

$$\chi_{E^3}(R) = \tfrac{1}{2}[\chi_{E^2}(R)\chi_E(R) + \chi_E(R^3)] \tag{2-8}$$

$\chi_{E^2}(R)$	3	0	3	1	1	
$\chi_E(R)$	2	−1	2	0	0	
						multiply
$\chi_{E^2}(R)\chi_E(R)$	6	0	6	0	0	
$\chi_E(R^3)$	2	2	2	0	0	(from Table 2-4)
						add
	8	2	8	0	0	
						divide by 2
$\chi_{E^3}(R)$	4	1	4	0	0	

$E^3 = A_1 + A_2 + E$ and, therefore, is Raman active only.

For the third overtone, Eq. (2-5) becomes

$$\chi_{E^4}(R) = \tfrac{1}{2}[\chi_{E^3}(R)\chi_E(R) + \chi_E(R^4)] \tag{2-9}$$

$\chi_{E^3}(R)$	4	1	4	0	0	
$\chi_E(R)$	2	−1	2	0	0	
						multiply
$\chi_{E^3}(R)\chi_E(R)$	8	−1	8	0	0	
$\chi_E(R^4)$	2	−1	2	2	2	(from Table 2-4)
						add
	10	−2	10	2	2	
						divide by 2
$\chi_{E^4}(R)$	5	−1	5	1	1	

$E^4 = A_1 + 2E$ and, therefore, is Raman active only.

For triply degenerate vibrations, the equation used[(1)] is

$$\chi_{F^n}(R) = \tfrac{1}{3}\{2\chi_F(R)\chi_{F^{n-1}}(R) - \tfrac{1}{2}\chi_{F^{n-2}}(R)[\chi_F(R)]^2 + \tfrac{1}{2}\chi_F(R^2)\chi_{F^{n-2}}(R) + \chi_F(R^n)\} \tag{2-10}$$

Table 2-5. Character Table of $\chi_{F_1}(R^n)$ and $\chi_{F_2}(R^n)$

	E	C_3	C_2	σ_d	S_4
$\chi_{F_1}(R^2)$	3	0	3	3	-1
$\chi_{F_1}(R^3)$	3	3	-1	-1	1
$\chi_{F_1}(R^4)$	3	0	3	3	3
$\chi_{F_2}(R^2)$	3	0	3	3	-1
$\chi_{F_2}(R^3)$	3	3	-1	1	-1
$\chi_{F_2}(R^4)$	3	0	3	3	3

For the F_2 and F_1 type vibration, the values of $\chi_{F_1}(R^n)$ and $\chi_{F_2}(R^n)$ are given in Table 2-5. For $\chi_{F_1^2}(R)$,

	$\chi_{F_1}(R)$	3	0	-1	-1	1	
	$\chi_{F_1}(R)$	3	0	-1	-1	1	multiply
	$\chi_{F_1}(R)\chi_{F_1}(R)$	9	0	1	1	1	multiply by 2
1st term*	$2\chi_{F_1}(R)\chi_{F_1}(R)$	18	0	2	2	2	
2nd term	$-\frac{1}{2}[\chi_{F_1}(R)]^2$	$-\frac{9}{2}$	0	$-\frac{1}{2}$	$-\frac{1}{2}$	$-\frac{1}{2}$	
3rd term	$\frac{1}{2}\chi_{F_1}(R^2)$	$\frac{3}{2}$	0	$\frac{3}{2}$	$\frac{3}{2}$	$-\frac{1}{2}$	
4th term	$\chi_{F_1}(R^2)$	3	0	3	3	-1	add
3rd term + 4th term		$\frac{9}{2}$	0	$\frac{9}{2}$	$\frac{9}{2}$	$-\frac{3}{2}$	
	1st term	18	0	2	2	2	
	2nd term	$-\frac{9}{2}$	0	$-\frac{1}{2}$	$-\frac{1}{2}$	$-\frac{1}{2}$	
	3rd + 4th terms	$\frac{9}{2}$	0	$\frac{9}{2}$	$\frac{9}{2}$	$-\frac{3}{2}$	
	Sum of 4 terms	18	0	6	6	0	divide by 3
$\chi_{F_1^2}(R)$		6	0	2	2	0	

* Of Eq. (2-10).

$F_1^2 = A_1 + E + F_2$ and, therefore, is both Raman and infrared active.

For $\chi_{F_1^3}(R)$, from Eq. (2-10),

$$\chi_{F_1^3}(R) = \tfrac{1}{3}\{2\chi_{F_1}(R)\chi_{F_1^2}(R) - \tfrac{1}{2}\chi_{F_1}(R)[\chi_{F_1}(R)]^2 + \tfrac{1}{2}\chi_{F_1}(R^2)\chi_{F_1}(R) + \chi_{F_1}(R^3)\} \tag{2-11}$$

	$\chi_{F_1}(R)$	3	0	-1	-1	1	
	$\chi_{F_1^2}(R)$	6	0	2	2	0	multiply
	$\chi_{F_1}(R)\chi_{F_1^2}(R)$	18	0	-2	-2	0	multiply by 2
1st term	$2\chi_{F_1}(R)\chi_{F_1^2}(R)$	36	0	-4	-4	0	
	$\chi_{F_1}(R)$	3	0	-1	-1	1	
	$[\chi_{F_1}(R)]^2$	9	0	1	1	1	multiply
		27	0	-1	-1	1	multiply by $-\frac{1}{2}$
2nd term	$-\frac{1}{2}\chi_{F_1}(R)[\chi_{F_1}(R)]^2$	$-\frac{27}{2}$	0	$\frac{1}{2}$	$\frac{1}{2}$	$-\frac{1}{2}$	
	$\chi_{F_1}(R^2)$	3	0	3	3	-1	
	$\chi_{F_1}(R)$	3	0	-1	-1	1	multiply
		9	0	-3	-3	-1	multiply by $\frac{1}{2}$
3rd term	$\frac{1}{2}\chi_{F_1}(R^2)\chi_{F_1}(R)$	$\frac{9}{2}$	0	$-\frac{3}{2}$	$-\frac{3}{2}$	$-\frac{1}{2}$	
4th term	$\chi_{F_1}(R^3)$	3	3	-1	-1	1	
Sum of 4 terms		30	3	-6	-6	0	divide by 3
$\chi_{F_1^3}(R)$		10	1	-2	-2	0	

$F_1^3 = 2F_1 + F_2 + A_2$ and, therefore, is both Raman and infrared active.

For $\chi_{F_1^4}(R)$ Eq. (2-10) becomes

$$\chi_{F_1^4}(R) = \tfrac{1}{3}\{2\chi_{F_1}(R)\chi_{F_1^3}(R) - \tfrac{1}{2}\chi_{F_1^2}(R)[\chi_{F_1}(R)]^2 + \tfrac{1}{2}\chi_{F_1}(R^2)\chi_{F_1^2}(R) + \chi_{F_1}(R^4)\} \tag{2-12}$$

$\chi_{F_1}(R)$	3	0	-1	-1	1	
$\chi_{F_1^3}(R)$	10	1	-2	-2	0	multiply

	$\chi_{F_1}(R)\chi_{F_1^3}(R)$	30	0	2	2	0	
							multiply by 2
1st term	$2\chi_{F_1}(R)\chi_{F_1^3}(R)$	60	0	4	4	0	
	$\chi_{F_1^2}(R)$	6	0	2	2	0	
	$[\chi_{F_1}(R)]^2$	9	0	1	1	1	
							multiply
		54	0	2	2	0	
							multiply by $-\frac{1}{2}$
2nd term	$-\frac{1}{2}\chi_{F_1^2}(R)[\chi_{F_1}(R)]^2$	-27	0	-1	-1	0	
	$\chi_{F_1}(R^2)$	3	0	3	3	-1	
	$\chi_{F_1^2}(R)$	6	0	2	2	0	
							multiply
		18	0	6	6	0	
							multiply by $\frac{1}{2}$
3rd term	$\frac{1}{2}\chi_{F_1}(R^2)\chi_{F_1^2}(R)$	9	0	3	3	0	
4th term	$\chi_{F_1}(R^4)$	3	0	3	3	3	
Sum of 4 terms		45	0	9	9	3	
							divide by 3
$\chi_{F_1^4}(R)$		15	0	3	3	1	

To find the species involved in $\chi_{F_1^4}(R)$ the reduction formula

$$N_i = \frac{1}{N_G} \Sigma\, n_e[(\chi_{F_1^4}(R)\chi_i(R)] \tag{2-13}$$

can be used:

$$N_{A_1} = \tfrac{1}{24}[(1\times 15\times 1) + (8\times 0\times 1) + (3\times 3\times 1) + (6\times 3\times 1) + (6\times 1\times 1)] = 2$$

$$N_{A_2} = \tfrac{1}{24}[(1\times 15\times 1) + (8\times 0\times 1) + (3\times 3\times 1) + (6\times 3\times -1) + (6\times 1\times -1)] = 0$$

$$N_E = \tfrac{1}{24}[(1\times 15\times 2) + (8\times 0\times -1) + (3\times 3\times 2) + (6\times 3\times 0) + (6\times 1\times 0)] = 2$$

$$N_{F_2} = \tfrac{1}{24}[(1\times 15\times 3) + (8\times 0\times 0) + (3\times 3\times -1) + (6\times 3\times 1) + (6\times 1\times -1)] = 2$$

$$N_{F_1} = \tfrac{1}{24}[(1\times 15\times 3) + (8\times 0\times 0) + (3\times 3\times -1) + (6\times 3\times -1) + (6\times 1\times 1)] = 1$$

Consequently $F_1^4 = F_1 + 2F_2 + 2E + 2A_1$ and is therefore both Raman and infrared active.

Continuing for $\chi_{F_2^2}(R)$,

	$\chi_{F_2}(R)$	3	0	−1	1	−1	
	$\chi_{F_2}(R)$	3	0	−1	1	−1	multiply
	$\chi_{F_2}(R)\chi_{F_2}(R)$	9	0	1	1	1	multiply by 2
1st term	$2\chi_{F_2}(R)\chi_{F_2}(R)$	18	0	2	2	2	
2nd term	$-\frac{1}{2}[\chi_{F_2}(R)]^2$	$-\frac{9}{2}$	0	$-\frac{1}{2}$	$-\frac{1}{2}$	$-\frac{1}{2}$	
3rd term	$\frac{1}{2}\chi_{F_2}(R^2)$	$\frac{3}{2}$	0	$\frac{3}{2}$	$\frac{3}{2}$	$-\frac{1}{2}$	
4th term	$\chi_{F_2}(R^2)$	3	0	3	3	−1	
Sum of 4 terms		18	0	6	6	0	divide by 3
$\chi_{F_2^2}(R)$		6	0	2	2	0	

$F_2^2 = A_1 + E + F_2$; therefore, it is both Raman and infrared active.

For $\chi_{F_2^3}(R)$,

	$\chi_{F_2}(R)$	3	0	−1	1	−1	
	$\chi_{F_2^2}(R)$	6	0	2	2	0	multiply
	$\chi_{F_2}(R)\chi_{F_2^2}(R)$	18	0	−2	2	0	multiply by 2
1st term	$2\chi_{F_2}(R)\chi_{F_2^2}(R)$	36	0	−4	4	0	
	$\chi_{F_2}(R)$	3	0	−1	1	−1	
	$[\chi_{F_2}(R)]^2$	9	0	1	1	1	multiply
		27	0	−1	1	−1	multiply by $-\frac{1}{2}$
2nd term	$-\frac{1}{2}\chi_{F_2}(R)[\chi_{F_2}(R)]^2$	$-\frac{27}{2}$	0	$\frac{1}{2}$	$-\frac{1}{2}$	$\frac{1}{2}$	

	$\chi_{F_2}(R^2)$	3	0	3	3	−1	
	$\chi_{F_2}(R)$	3	0	−1	1	−1	
							multiply
		9	0	−3	3	1	
							multiply by 2
3rd term	$\frac{1}{2}\chi_{F_2}(R^2)\chi_{F_2}(R)$	$\frac{9}{2}$	0	$-\frac{3}{2}$	$\frac{3}{2}$	$\frac{1}{2}$	
4th term	$\chi_{F_2}(R^3)$	3	3	−1	1	−1	
Sum of 4 terms		30	3	−6	6	0	
							divide by 3
$\chi_{F_2^3}(R)$		10	1	−2	2	0	

$F_2^3 = 2F_2 + F_1 + A_1$; therefore, it is both infrared and Raman active. For $\chi_{F_2^4}(R)$

	$\chi_{F_2}(R)$	3	0	−1	1	−1	
	$\chi_{F_2^3}(R)$	10	1	−2	2	0	
							multiply
	$\chi_{F_2}(R)\chi_{F_2^3}(R)$	30	0	2	2	0	
							multiply by 2
1st term	$2\chi_{F_2}(R)\chi_{F_2^3}(R)$	60	0	4	4	0	
	$\chi_{F_2^2}(R)$	6	0	2	2	0	
	$[\chi_{F_2}(R)]^2$	9	0	1	1	1	
							multiply
		54	0	2	2	1	
							multiply by $-\frac{1}{2}$
2nd term	$-\frac{1}{2}\chi_{F_2^2}(R)[\chi_{F_2}(R)]^2$	−27	0	−1	−1	0	
	$\chi_{F_2}(R^2)$	3	0	3	3	−1	
	$\chi_{F_2^2}(R)$	6	0	2	2	0	
							multiply
		18	0	6	6	0	
							multiply by $\frac{1}{2}$
3rd term	$\frac{1}{2}\chi_{F_2}(R^2)\chi_{F_2^2}(R)$	9	0	3	3	0	

Table 2-6. Selection Rules for the Overtones of a Molecule with T_d Symmetry

Overtone	A_1	A_2	E	F_2	F_1	Activity	
						Raman	Infrared
A_1^n	1	0	0	0	0	a	ia
A_2^n (n even)	1	0	0	0	0	a	ia
A_2^n (n odd)	0	1	0	0	0	ia	ia
E^2	1	0	1	0	0	a	ia
E^3	1	1	1	0	0	a	ia
E^4	1	0	2	0	0	a	ia
F_2^2	1	0	1	1	0	a	a
F_2^3	1	0	0	2	1	a	a
F_2^4	2	0	2	2	1	a	a
F_1^2	1	0	1	1	0	a	a
F_1^3	0	1	0	1	2	a	a
F_1^4	2	0	2	2	1	a	a

4th term	$\chi_{F_2}(R^4)$	3	0	3	3	3
Sum of 4 terms		45	0	9	9	3
divide by 3						
$\chi_{F_2^4}(R)$		15	0	3	3	1

$F_2^4 = 2A_1 + F_1 + 2F_2 + 2E_1$ and, therefore, is both infrared and Raman active. If inspection of the character table (Table 2-1) does not readily indicate which species are contained in the overtone, then a reduction formula similar to Eq. (2-13) can be used.

The selection rules for the overtones of a T_d molecule are tabulated in Table 2-6.

A molecule in the T_d symmetry has fundamentals of the A_1, E, and F_2 type, where A_1 and E are Raman active and F_2 is both Raman and infrared active. The A_2 and F_1 species do not exist as fundamentals. However, combinations and overtones of these species become active as indicated below:

	Fundamental	*Combination*	*Activity*
A_2	Not allowed	$A_2 \times A_2$	Raman
		$A_2 \times E$	Raman
		$A_2 \times F_1$	Raman and infrared
F_1	Not allowed	$F_1 \times F_2$	Raman and infrared
		$F_1 \times F_1$	Raman and infrared
		$F_1 \times E$	Raman and infrared
		Overtone	
A_2	Not allowed	A_2^n, n even	Raman
F_1	Not allowed	F_1^n, n odd or even	Raman and infrared

These examples illustrate the usefulness of the selection rules where combinations and overtones of forbidden fundamentals become active.

B. Linear Molecules

The method used for linear molecules has to be modified due to the infinite number of symmetry operations and the continuity of the group. The consequence of these two features is a change in the presentation of the elements of the group, and as a result the summation expressions are replaced by integrals. For linear molecules there are two continuous groups, $C_{\infty v}$ and $D_{\infty h}$. These are necessary for the description of the symmetry properties of such molecules.

For $C_{\infty v}$ (H—C≡N or C—O) the symmetry operations are divided into two sets of symmetry elements. The first, $C(\varphi)$, is a rotation about the axis of the molecule through an angle φ. Further considerations lead one to the result that there is a rotation about that axis for every value of φ in the interval $0 \leq \varphi \leq 2\pi$. Consequently, φ can be used as a parameter to list all the rotations. For example, $C(0)$ is the identity operation. Furthermore, a rotation about the nuclear axis through an angle φ followed by a reflection in a plane containing this axis is the second set of symmetry operations and is denoted by $\sigma_v \cdot C(\varphi)$. In this case $\sigma_v \cdot C(0)$ corresponds to the σ_v operation. Therefore, all elements of the $C_{\infty v}$ group are presented in terms of one parameter φ, such that φ takes on all values $0 \leq \varphi \leq 2\pi$. Such a group is called a one-parameter continuous group.

In a similar manner the elements of the group $D_{\infty h}$ can be presented. Here one has $C(\varphi)$ and $\sigma_v \cdot C(\varphi)$, and the group is completed with $C_2 \cdot C(\varphi)$

and $S(\varphi)$. Here $C_2 \cdot C(\varphi)$ is the rotation of the molecule about the nuclear axis through angle φ followed by a rotation about an axis perpendicular to the nuclear axis through an angle of 180° with $C_2 = C_2 \cdot C(0)$. Also, $S(\varphi)$ is a rotation about a nuclear axis through an angle φ followed by a reflection in a plane perpendicular to the nuclear axis. Here for $\varphi = 0$, $S(0)$ represents a reflection in a plane perpendicular to the nuclear axis and for $\varphi = \pi$, $S(\pi)$ represents a rotation about the nuclear axis through 180° followed by a reflection in a plane perpendicular to the nuclear axis or simply the inversion. Here again, $D_{\infty h}$ is a one-parameter (φ) continuous group. Tables 2-7A and 2-7B present the characters for the $C_{\infty v}$ and $D_{\infty h}$ groups, respectively.

For the determination of the selection rules the reduction formula becomes

$$N_j = \frac{\int \chi(R)\bar{\chi}^j(R)\, dR}{\int dR}$$

where N_j is the number of times the jth irreducible representation appears in the representation under consideration specified by $\chi(R)$ and $\bar{\chi}^j(R)$ is the complex conjugate of $\chi(R)$; $j = A_1, A_2, E_2, \ldots, E_k$ for C_{2v} and $j = A_{1g}$, A_{1u}, E_{1g}, E_{1u}, ... for $D_{\infty h}$. The integration is taken over the range of the continuously changing parameter R (in this case φ) to cover all the operations of the group.

Thus with the above modifications inserted, the method for linear molecules is quite similar then to that illustrated previously for the T_d mole-

Table 2-7A. Characters for the $C_{\infty v}$ Group

$C_{\infty v}$	$C(\varphi)$	$\sigma_v \cdot C(\varphi)$
A_1	1	1
A_2	1	-1
E_k	$2 \cos k\varphi$	0
$\chi_M(R)$	$1 + 2 \cos \varphi$	1
$\chi_\alpha(R)$	$2 + 2 \cos \varphi + 2 \cos 2\varphi$	2
μ_R	N	N
$\Xi(R)$	$2(N-2) \cos \varphi + N - 1$	$N - 1$

Table 2-7B. Characters for the $D_{\infty h}$ Group

$D_{\infty h}$	$C(\varphi)$	$C_2 \cdot C(\varphi)$	$S(\varphi)$	$\sigma_v \cdot C(\varphi)$
A_{1g}	1	1	1	1
A_{1u}	1	-1	-1	1
A_{2g}	1	-1	1	-1
A_{2u}	1	1	-1	-1
E_{kg}	$2 \cos k\varphi$	0	$2(-1)^k \cos k\varphi$	0
E_{ku}	$2 \cos k\varphi$	0	$-2(-1)^k \cos k\varphi$	0
$\chi_M(R)$	$1 + 2 \cos \varphi$	-1	$-1 + 2 \cos \varphi$	1
$\chi_\alpha(R)$	$2 + 2 \cos \varphi + 2 \cos 2\varphi$	2	$2 - 2 \cos \varphi + 2 \cos 2\varphi$	2
μ_R	N	m_0	m_0	N
$\Xi(R)$	$2(N-2) \cos \varphi + N - 1$	$-m_0 + 1$	$-m_0 + 1 + 2\, m_0 \cos \varphi$	$N - 1$

cule. The characters[3] for $\Xi(R)$, $\chi_M(R)$, and $\chi_\alpha(R)$ are then given in terms of the following expressions*:

$$\Xi(R) = 2(\mu_R - 2) \cos \varphi + (\mu_R - 1) \quad \text{for } C(\varphi) \tag{2-14}$$

$$\Xi(R) = (\mu_R - 1) \quad \text{for } \sigma_v \cdot C(\varphi) \tag{2-15}$$

$$\Xi(R) = (-\, m_0 + 1) \quad \text{for } C_2 \cdot C(\varphi) \tag{2-16}$$

$$\Xi(R) = -\, m_0 + 1 + 2m_0 \cos \varphi \quad \text{for } S(\varphi) \tag{2-17}$$

$$\chi_M(R) = 1 \quad \text{for } \sigma_v \cdot C(\varphi) \tag{2-18}$$

$$\chi_M(R) = 1 + 2 \cos \varphi \quad \text{for } C_2(\varphi) \tag{2-19}$$

$$\chi_M(R) = -\, 1 \quad \text{for } C_2 \cdot C(\varphi) \tag{2-20}$$

$$\chi_M(R) = -\, 1 + 2 \cos \varphi \quad \text{for } S(\varphi) \tag{2-21}$$

$$\chi_\alpha(R) = 2 + 2 \cos \varphi + 2 \cos 2\varphi \quad \text{for } C(\varphi) \tag{2-22}$$

$$\chi_\alpha(R) = 2 \quad \text{for } \sigma_v \cdot C(\varphi) \tag{2-23}$$

* In Eqs. (2-16) and (2-17), $m_0 = 1$ for molecules with a central atom and $m_0 = 0$ when there is no central atom.

$$\chi_\alpha(R) = 2 \qquad \text{for} \quad C_2 \cdot C(\varphi) \qquad (2\text{-}24)$$

$$\chi_\alpha(R) = 2 - 2\cos\varphi + 2\cos 2\varphi \qquad \text{for} \quad S(\varphi) \qquad (2\text{-}25)$$

The previous definitions hold for the symbols in Eqs. (2-14)–(2-25). The reduction formula makes use of a summation of an integral rather than strictly a summation and is expressed as follows:

$$N_i = \frac{\sum \int_0^{2\pi} n_e \chi_i(\varphi) \bar{\chi}^j(\varphi)\, d\varphi}{N_G \int_0^{2\pi} d\varphi} \qquad (2\text{-}26)$$

where $\chi_i(\varphi)$ is the character of the irreducible representation and $\bar{\chi}^j(\varphi)$ is either $\Xi(R)$, $\chi_M(R)$, or $\chi_\alpha(R)$, and N_G, n_e, and N_i have been defined earlier in this chapter. Table 2-7C illustrates the method for HC≡N belonging to the $C_{\infty v}$ point group.

For the calculation of the combination and overtones of linear molecules similar procedures[3] are used as were outlined for the nonlinear molecules earlier in this chapter. The activities are calculated as follows and appear in Table 2-7C.

Number of Fundamentals of Each Type

$$A_1 = \frac{\int_0^{2\pi} (1)(2 + 2\cos\varphi)\, d\varphi + \int_0^{2\pi} (2)(1)\, d\varphi}{2\int_0^{2\pi} d\varphi} = 2 \qquad (2\text{-}27)$$

Table 2-7C. Character Table for the HCN Molecule

$C_{\infty v}$	$C(\varphi)$	$\sigma_v \cdot C(\varphi)$	n_i	IR	R
$A_1 = \Sigma^+$	1	1	2	a	a
$A_2 = \Sigma^-$	1	-1	0	ia	ia
$E_1 = \pi$	$2\cos\varphi$	0	1	a	a
$E_2 = \Delta$	$2\cos 2\varphi$	0	0	ia	a
$\chi_M(R)$	$1 + 2\cos\varphi$	1			
$\chi_\alpha(R)$	$2 + 2\cos\varphi + 2\cos 2\varphi$	2			
μ_R	3	3			
$\Xi(R)$	$2(1 + \cos\varphi)$	2			

$$A_2 = \frac{\int_0^{2\pi}(1)(2 + 2\cos\varphi)\,d\varphi + \int_0^{2\pi}(-1)(2)\,d\varphi}{2\int_0^{2\pi}d\varphi} = 0 \tag{2-28}$$

$$E_1 = \frac{\int_0^{2\pi}(2\cos\varphi)(2 + 2\cos\varphi)\,d\varphi + 0}{2\int_0^{2\pi}d\varphi} = 1 \tag{2-29}$$

$$E_2 = \frac{\int_0^{2\pi}(2\cos 2\varphi)(2 + 2\cos\varphi)\,d\varphi}{2\int_0^{2\pi}d\varphi} = 0 \tag{2-30}$$

Number of Vibrations Allowed in the Infrared

$$A_1 = \frac{\int_0^{2\pi}(1)(1 + 2\cos\varphi)\,d\varphi + (1)\int_0^{2\pi}d\varphi}{2\int_0^{2\pi}d\varphi} = 1 \tag{2-31A}$$

$$A_2 = \frac{\int_0^{2\pi}(1)(1 + 2\cos\varphi)\,d\varphi + (-1)\int_0^{2\pi}d\varphi}{2\int_0^{2\pi}d\varphi} = 0 \tag{2-31B}$$

$$E_1 = \frac{\int_0^{2\pi}(2\cos\varphi)(1 + 2\cos\varphi)\,d\varphi + 0}{2\int_0^{2\pi}d\varphi} = 1 \tag{2-32A}$$

$$E_2 = \frac{\int_0^{2\pi}(2\cos 2\varphi)(1 + 2\cos\varphi)\,d\varphi}{2\int_0^{2\pi}d\varphi} = 0 \tag{2-32B}$$

Number of Raman-Active Fundamentals

$$A_1 = \frac{\int_0^{2\pi}(1)(2 + 2\cos\varphi + 2\cos 2\varphi)\,d\varphi + \int_0^{2\pi}(1)(2)\,d\varphi}{2\int_0^{2\pi}d\varphi} = 2 \tag{2-33}$$

$$A_2 = \frac{\int_0^{2\pi}(1)(2 + 2\cos\varphi + 2\cos 2\varphi)\,d\varphi + \int_0^{2\pi}-(1)(2)\,d\varphi}{2\int_0^{2\pi}d\varphi} = 0 \tag{2-34}$$

$$E_1 = \frac{\int_0^{2\pi}(2\cos\varphi)(2 + 2\cos\varphi + 2\cos 2\varphi)\,d\varphi + 0}{2\int_0^{2\pi}d\varphi} = 1 \tag{2.35A}$$

$$E_2 = \frac{\int_0^{2\pi}(2\cos 2\varphi)(2 + 2\cos\varphi + 2\cos 2\varphi)\,d\varphi}{2\int_0^{2\pi}d\varphi} = 1 \tag{2-35B}$$

Combinations in the $C_{\infty v}$ Point Group

To obtain the selection rules for combinations or differences $\nu_i \pm \nu_j$, the direct product of the characters of the vibrations ν_i and ν_j are formed Thus, for the combinations $A_1 \times E$

	$C(\varphi)$	$\sigma_v C(\varphi)$	
$\chi_{A_1}(R)$	1	1	
$\chi_{E_k}(R)$	$2\cos k\varphi$	0	multiply
$\chi_{A_1}(R)\chi_{E_k}(R)$	$2\cos k\varphi$	0	

It can be seen that $\chi_{A_1}(R)\chi_E(R) = \chi_E(R)$, which means that the combination is active whenever the E- type vibration is allowed; in this case it is allowed in both the Raman and infrared spectra. $A_1 \times E_2$ is active in the Raman. For other combinations of A_1 or A_2 with E_k, activity is unallowed in the Raman and infrared where $k > 2$.

Overtones in the $C_{\infty v}$ Point Group

Similar procedures as were used for the combinations can also be used for the overtone selection rules, provided one is dealing with nondegenerate vibrations. Thus, for $A_2{}^2$,

	$C(\varphi)$	$\sigma_v C(\varphi)$	
$\chi_{A_2}(R)$	1	-1	
$\chi_{A_2}(R)$	1	-1	multiply
$\chi_{A_2^2}(R)$	1	1	

which means that the first overtone of A_2 is active wherever A_1 is active.

For degenerate vibrations, the results indicate that for n even

$$\chi_{E_p^n}(R) = \chi_{A_1}(R) + \chi_{E_{2p}}(R) + \chi_{E_{4p}}(R) + \cdots + \chi_{E_{np}}(R) \qquad (2\text{-}36)$$

and these vibrations will have the activity of A_1 (for $E_1{}^2$, activity is observed in both the Raman and infrared spectra). For n odd

$$\chi_{E^n}(R) = \chi_{E_p}(R) + \chi_{E_{3p}}(R) + \chi_{E_{5p}}(R) + \cdots + \chi_{E_{np}}(R) \qquad (2\text{-}37)$$

and $E_1{}^3$ will have activity where E_1 has activity (both Raman and infrared active). The results are tabulated in Table 2-8.

Similar procedures can be used for the $D_{\infty h}$ point group.[3] Considering the diacetylene molecule (H—C≡C—C≡CH), with six atoms and $m_0 = 0$,

Table 2-8. Combination and Overtone Selection Rules $C_{\infty v}$ Symmetry

Combination or overtone	Activity	
	Infrared	Raman
$A_1 \times A_1$	a	a
$A_1 \times A_2$	ia	ia
$A_2 \times A_2$	a	a
$A_1 \times E_1$	a	a
$A_1 \times E_2$	ia	a
$A_1 \times E_k,\ k > 2$	ia	ia
$E_1 \times E_2$	a	a
$E_1 \times E_1$	a	a
$A_2 \times E_1$	a	a
$A_2 \times E_2$	ia	a
$A_2 \times E_k,\ k > 2$	ia	ia
A_1^n	a	a
A_2^n, n odd	ia	ia
n even	a	a
E_1^2	a	a
E_1^3	a	a

it can be demonstrated that the species $3A_{1g}$, $2A_{1u}$, $2E_{1g}$, and $2E_{1u}$ are the fundamentals existing for diacetylene. Combinations and overtones can be determined as was previously illustrated for the HCN molecule.

For further details the reader is referred to the paper by Ferigle and Meister.[(3)]

It should be noted that tables for the derivation of the selection rules for the fundamental modes in isolated molecules have been published by Herzberg,* although the derivations of the selection rules for combinations and overtones are not included. The method presented in this chapter, which uses the techniques of Meister, Cleveland, and Murray[(1)] and Ferigle and Meister,[(3)] allows the beginner to follow procedures for the determination of selection rules, which ordinarily are assumed.

For the deviations of $\chi_M(R)$, $\chi_\alpha(R)$, and $\Xi(R)$ necessary for the determination of the selection rules for the molecules of a point group, see Appendix 10.

* See reference in Bibliography of Chapter 1.

Table 2-8A. Selection Rules for A_3 to A_8 Type Molecules

Type of molecule	Point group	Structure	Fundamentals*			Number of Raman bands polarized	Examples
			IR	R	C		
A_3	$D_{\infty h}$	linear	2	1	0	1	N_3^-
AB_2	$D_{\infty h}$	linear, sym.	2	1	0	1	CO_2
	$C_{\infty v}$	linear, unsym.	3	3	3	1	NNO
	C_{2v}	bent, sym.	3	3	3	2	SO_2, H_2O
ABC	$C_{\infty v}$	linear	3	3	3	2	BrCN
	C_s	bent	3	3	3	3	ONCl
A_4	T_d	tetrahedral	1	3	1	1 ($\varrho = 0$)	P_4
	D_{4h}	planar square	2	3	0	1 ($\varrho \neq 0$)	
AB_3	D_{3h}	planar	3	3	2	1	NO_3^-
	C_{3v}	pyramid	4	4	4	2	NH_3
	C_{2v}	planar "T"	6	6	6	3	ClF_3
ABC_2	C_{2v}	planar "Y"	6	6	6	3	$COCl_2$
	C_s	pyramid	6	6	6	4	$SOCl_2$
AB_4	T_d	tetrahedral	2	4	2	1	CCl_4
	D_{4h}	planar square	3	3	0	1	$AuCl_4^-$
	C_{4v}	square pyramid	4	7	4	2	
	C_{3v}	quasi-T_d	6	6	6	3	$POCl_3$
	C_{2v}		8	9	8	4	SF_4
AB_5	D_{3h}	trigonal bipyramid	5	6	3	2	PCl_5†
	C_{4v}	tetragonal pyramid	6	9	6	3	BrF_5
AB_6	O_h	octahedral	2	3	0	1 ($\varrho = 0$)	‡
	D_{6h}	hexagonal	2	3	0	1 ($\varrho \neq 0$)	C_6H_6
AB_7	D_{5h}	pentagonal bipyramid	5	5	0	2	IF_7
A_8	D_{4h}	puckered octagon	3	7	0	2	S_8

* IR = infrared; R = Raman; C = coincidences.
† Gas or liquid.
‡ Group VI hexafluorides.

C. Selection Rules for A_3 to A_8 Molecules

The selection rules obtained for simple molecules of the type A_3 to those for complex molecules of the type AB_7 and A_8 are presented in Table 2-8A.

D. Forbidden Vibrations for Several Point Groups

The vibrations which are forbidden by the selection rules are tabulated for several point groups in Table 2-8B. This tabulation is useful in working problems like those which are included at the end of some of the chapters in this book.

2-2. SELECTION RULES FOR SYSTEMS INVOLVING TRANSLATIONS

By simply extending the methods used in Section 2-1 for the determination of the selection rules for an isolated molecule, one can obtain selection rules for molecules involving translations, translation–rotations, or translation–reflections, for $k = 0$ at the center of the Brillouin zone. A choice of two methods to accomplish this is available. The older method was developed by Bhagavantam and Venkatarayudu (BV), who considered the atoms in the unit cell as a large molecule.[4] Mitra[5] considers that this method is more suitable for classifying lattice modes. The other method, by Halford and Hornig (HH), can be used, and considers the local symmetry of a molecule group in the unit cell.[6–8]

A. The Method of Bhagavantam and Venkatarayudu[4]

The BV method will be considered first. Table 2-9 shows the group characters necessary to determine the total number of lattice vibrations in a solid (acoustical, rotational, and translational) containing linear and nonlinear groups. It is apparent that by applying extensions of the techniques for the development of the selection rules for the isolated molecule, one can derive the selection rules for the solid state.

The BV method will be illustrated by considering as an example $NaNO_3$,[9–11] which belongs to the $D_{3d}^6(R\bar{3}c)$ (No. 167)* space group, having a rhombohedral or pseudohexagonal crystal structure, and containing two

* The number refers to the space group number of the 230 crystallographic space groups.

Table 2-8B. Selection Rules (Forbidden Vibrations)

Point group	Infrared			Raman		
	Fund.	Combinations	Overtones	Fund.	Combinations	Overtones
C_1 C_2 C_{1v} C_{1h} C_s	none	none	none	none	none	none
C_{2v}	A_2	$A_1 \times A_2$, $B_1 \times B_2$	A_2^n $(n = \text{odd})$	none	none	none
C_{3v}	A_2	$A_1 \times A_2$	A_2^n $(n = \text{odd})$	A_2	$A_1 \times A_2$	A_2^n $(n = \text{odd})$
C_{4v}	A_2 B_1 B_2	$A_1 \times A_2$, $A_1 \times B_1$ $A_1 \times B_2$, $A_2 \times B_1$ $A_2 \times B_2$, $B_1 \times B_2$	A_2^n $(n = \text{odd})$ B_1^n $(n = \text{odd})$ B_2^n $(n = \text{odd})$	A_2	$A_1 \times A_2$, $B_1 \times B_2$	A_2^n $(n = \text{odd})$
C_{2h}	A_g B_g	$A_g \times A_g$, $A_g \times B_g$ $B_g \times B_g$, $A_u \times A_u$ $A_u \times B_u$, $B_u \times B_u$	A_g^n B_g^n A_u^n $(n = \text{even})$ B_u^n $(n = \text{even})$	A_u B_u	$A_g \times A_u$, $A_g \times B_u$ $B_g \times A_u$, $B_g \times B_u$	A_u^n $(n = \text{odd})$ B_u^n $(n = \text{odd})$
C_{3h}	A' E''	$A' \times A'$, $A' \times E''$ $A'' \times A''$, $A'' \times E'$	A'^n A''^n $(n = \text{even})$	A''	$A' \times A''$	A''^n $(n = \text{odd})$

Table 2-8B (*continued*)

Point group	Infrared			Raman		
	Fund.	Combinations	Overtones	Fund.	Combinations	Overtones
D_{2d}	A_1 A_2 B_1	$A_1 \times A_1$, $A_1 \times A_2$ $A_1 \times B_1$, $A_2 \times A_2$ $A_2 \times B_2$, $B_1 \times B_1$ $B_1 \times B_2$, $B_2 \times B_2$	A_1^n A_2^n B_1^n B_2^n (n = even)	A_2	$A_1 \times A_2$, $B_1 \times B_2$	A_2^n (n = odd)
D_{3d}	A_{1g} A_{1u} A_{2g} E_g	$A_{1g} \times A_{1g}$, $A_{1g} \times A_{1u}$ $A_{1g} \times A_{2g}$, $A_{1g} \times E_g$ $A_{1u} \times A_{1u}$, $A_{1u} \times A_{2u}$ $A_{1u} \times E_u$, $A_{2g} \times A_{2g}$ $A_{2g} \times A_{2u}$, $A_{2g} \times E_g$ $A_{2u} \times A_{2u}$, $A_{2u} \times E_u$ $E_g \times E_g$, $E_u \times E_u$	A_{1g}^n A_{1u}^n A_{2g}^n A_{2u}^n (n = even) E_g^n E_u^2 E_u^4	A_{1u} A_{2g} A_{2u} E_u	$A_{1g} \times A_{1u}$, $A_{1g} \times A_{2g}$ $A_{1g} \times A_{2u}$, $A_{1g} \times E_u$ $A_{1u} \times A_{2g}$, $A_{1u} \times A_{2u}$ $A_{1u} \times E_g$, $A_{2g} \times A_{2u}$ $A_{2g} \times E_u$, $A_{2u} \times E_g$ $E_g \times E_u$	A_{1u}^n (n = odd) A_{2g}^n (n = odd) A_{2u}^n (n = odd) E_u^3 E_u^5
D_{2h}	A_g A_u B_{1g} B_{2g} B_{3g}	$A_g \times A_u$, $A_g \times A_g$ $A_g \times B_{1g}$, $A_g \times B_{2g}$ $A_g \times B_{3g}$, $A_u \times A_u$ $A_u \times B_{1u}$, $A_u \times B_{2u}$ $A_u \times B_{3u}$, $B_{1g} \times B_{1g}$ $B_{1g} \times B_{1u}$, $B_{1g} \times B_{2g}$ $B_{1g} \times B_{2u}$, $B_{1g} \times B_{3g}$	A_g^n A_u^n B_{1g}^n B_{1u}^n (n = even) B_{2g}^n B_{2u}^n (n = even) B_{3g}^n	A_u B_{1u} B_{2u} B_{3u}	$A_g \times A_u$, $A_g \times B_{1u}$ $A_g \times B_{2u}$, $A_g \times B_{3u}$ $A_u \times B_{1g}$, $A_u \times B_{2g}$ $A_u \times B_{3g}$, $B_{1g} \times B_{1u}$ $B_{1g} \times B_{2u}$, $B_{1g} \times B_{3u}$ $B_{1u} \times B_{2g}$, $B_{1u} \times B_{3g}$ $B_{2g} \times B_{2u}$, $B_{2g} \times B_{3u}$	A_u^n (n = odd) B_{1u}^n (n = odd) B_{2u}^n (n = odd) B_{3u}^n (n = odd)

Table 2-8B (*continued*)

Point group	Infrared			Raman		
	Fund.	Combinations	Overtones	Fund.	Combinations	Overtones
		$B_{1u}\times B_{1u}$, $B_{1u}\times B_{2u}$ $B_{1u}\times B_{3u}$, $B_{2g}\times B_{2g}$ $B_{2g}\times B_{3g}$, $B_{2u}\times B_{2u}$ $B_{2u}\times B_{3u}$, $B_{3g}\times B_{3g}$ $B_{3u}\times B_{3u}$	B_{3u}^n (n = even)		$B_{2u}\times B_{3g}$, $B_{3g}\times B_{3u}$	
D_{3h}	A_1' A_1'' A_2' E''	$A_1'\times A_1'$, $A_1'\times A_1''$ $A_1'\times A_2'$, $A_1'\times E''$ $A_1''\times A_1''$, $A_1''\times A_2''$ $A_1''\times E'$, $A_2'\times A_2'$ $A_2'\times A_2''$, $A_2'\times E''$ $A_2''\times A_2''$, $A_2''\times E'$	$A_1'^n$ $A_1''^n$ $A_2'^n$ $A_2''^n$ (n = even)	A_1'' A_2' A_2''	$A_1'\times A_1''$, $A_1'\times A_2'$ $A_1'\times A_2''$, $A_1''\times A_2'$ $A_1''\times A_2''$, $A_2'\times A_2''$	$A_1''^n$ (n = odd) $A_2'^n$ (n = odd) A_2'' (n = odd)
D_{4h}	A_{1g} A_{1u} A_{2g} B_{1g} B_{1u} B_{2g} B_{2u} E_g	$A_{1g}\times A_{1g}$, $A_{1g}\times A_{1u}$ $A_{1g}\times A_{2g}$, $A_{1g}\times B_{1g}$ $A_{1g}\times B_{1u}$, $A_{1g}\times B_{2g}$ $A_{1g}\times B_{2u}$, $A_{1g}\times E_g$ $A_{1u}\times A_{1u}$, $A_{1u}\times A_{2u}$ $A_{1u}\times B_{1g}$, $A_{1u}\times B_{1u}$ $A_{1u}\times B_{2g}$, $A_{1u}\times B_{2u}$ $A_{1u}\times E_u$, $A_{2g}\times A_{2g}$ $A_{2g}\times A_{2u}$, $A_{2g}\times B_{1g}$ $A_{2g}\times B_{1u}$, $A_{2g}\times B_{2g}$	A_{1g}^n A_{1u}^n A_{2g}^n A_{2u}^n (n = even) B_{1g}^n B_{1u}^n B_{2g}^n B_{2u}^n E_g^n	A_{1u} A_{2g} A_{2u} B_{1u} B_{2u} E_u	$A_{1g}\times A_{1u}$, $A_{1g}\times A_{2g}$ $A_{1g}\times A_{2u}$, $A_{1g}\times E_u$ $A_{1g}\times B_{1u}$, $A_{1g}\times B_{2u}$ $A_{1u}\times A_{2u}$, $A_{1u}\times A_{2g}$ $A_{1u}\times B_{1g}$, $A_{1u}\times B_{2g}$ $A_{2g}\times A_{2u}$, $A_{2g}\times B_{1u}$ $A_{2g}\times B_{2u}$, $A_{2g}\times E_u$ $B_{2g}\times B_{2u}$, $B_{2g}\times E_u$ $B_{2u}\times E_g$, $E_g\times E_u$ $A_{1u}\times E_g$, $A_{1u}\times B_{1g}$	A_{1u}^n (n = odd) A_{2g}^n (n = odd) B_{1u}^n (n = odd) B_{2u}^n (n = odd) E_u^n (n = odd)

Table 2-8B (*continued*)

Point group	Infrared			Raman		
	Fund.	Combinations	Overtones	Fund.	Combinations	Overtones
		$A_{2g}\times B_{2u}$, $A_{2g}\times E_g$ $A_{2u}\times A_{2u}$, $A_{2u}\times B_{1g}$ $A_{2u}\times B_{1u}$, $A_{2u}\times B_{2g}$ $A_{2u}\times B_{2u}$, $A_{2u}\times E_u$ $B_{1g}\times B_{1g}$, $B_{1g}\times B_{1u}$ $B_{1g}\times B_{2g}$, $B_{1g}\times E_g$ $B_{1u}\times B_{1u}$, $B_{1u}\times B_{2u}$ $B_{1u}\times E_u$, $B_{2g}\times B_{2g}$ $B_{2g}\times B_{2u}$, $B_{2g}\times E_g$ $B_{2u}\times B_{2u}$, $B_{2u}\times E_u$ $E_g\times E_g$, $E_u\times E_u$			$A_{2u}\times B_{2g}$, $A_{2u}\times E_g$ $B_{1g}\times B_{1u}$, $B_{1g}\times B_{2g}$ $B_{1g}\times B_{2u}$, $B_{1g}\times E_u$ $B_{1u}\times B_{2g}$, $B_{1u}\times B_{2u}$ $B_{1u}\times E_g$	
T_d	A_1 A_2 E F_1	$A_1\times A_1$, $A_1\times A_2$ $A_1\times E$, $A_1\times F_1$ $A_2\times E$, $A_2\times F_2$ $E\times E$, $A_2\times A_2$	A_1^n A_2^n E^n	A_2 F_1	$A_1\times A_2$, $A_1\times F_1$ $A_2\times F_2$	A_2^n (n = odd)
$C_{\infty v}$	A_2 E_2 E_k $k > 2$	$A_1\times A_2$, $A_1\times E_2$ $A_1\times E_k$, $k > 2$ $A_2\times E_2$ $A_2\times E_k$, $k > 2$	A_2^n (n = odd)	A_2 E_k $k > 2$	$A_1\times A_2$ $A_1\times E_k$, $k > 2$ $A_2\times E_k$, $k > 2$	A_2^n (n = odd)

Table 2-8B (*continued*)

Point group	Infrared			Raman		
	Fund.	Combinations	Overtones	Fund.	Combinations	Overtones
$D_{\infty h}$	Σ_g^+	$\Sigma_g^+\times\Sigma_g^+$, $\Sigma_g^+\times\Sigma_g^-$,	Σ^{+n}, Σ^{-n} (n=even)	Σ_u^+	$\Sigma_g^+\times\Sigma_g^-$, $\Sigma_g^+\times\Sigma_u^+$,	Σ_u^{+n}, Σ_g^{-n}, Σ_u^{-n}
	Σ_g^-	$\Sigma_g^+\times\Sigma_u^-$, $\Sigma_g^+\times\pi_g$,	Σ^{-n}, Σ^{-n}, π_g^n, Δ_g^n,	Σ_g^-	$\Sigma_g^+\times\Sigma_u^-$, $\Sigma_g^+\times\Sigma_u^+$,	(n = odd)
	Σ_u^-	$\Sigma_g^+\times\Delta_g$, $\Sigma_g^+\times\Delta_u$,	Δ_u^n, ϕ_g^n, ϕ_u^n	Σ_u^-	$\Sigma_g^+\times\Sigma_u^-$, $\Sigma_g^-\times\Sigma_u^+$,	π_u^n, Δ_u^n, Δ_g^n
	π_g	$\Sigma_g^+\times\phi_g$, $\Sigma_g^+\times\phi_u$,	(n = even or odd)	π_u	$\Sigma_g^-\times\Sigma_u^-$, $\Sigma_g^-\times\pi_u$,	ϕ_u^n (n = odd)
	Δ_g	$\Sigma_g^-\times\Sigma_g^-$, $\Sigma_g^-\times\Sigma_u^+$,	π_u^n (n = even)	Δ_u	$\Sigma_g^-\times\Delta_u$, $\Sigma_g^-\times\phi_g$,	
	Δ_u	$\Sigma_g^-\times\pi_g$, $\Sigma_g^-\times\Delta_g$,		ϕ_g	$\Sigma_g^-\times\phi_u$, $\Sigma_u^+\times\Sigma_u^-$	
	ϕ_g	$\Sigma_g^-\times\Delta_u$, $\Sigma_g^-\times\phi_g$,		ϕ_u	$\Sigma_u^+\times\pi_g$, $\Sigma_u^+\times\Delta_g$	
	ϕ_u	$\Sigma_g^-\times\phi_u$, $\Sigma_u^+\times\Sigma_u^+$,			$\Sigma_u^+\times\phi_g$, $\Sigma_u^+\times\phi_u$	
		$\Sigma_u^+\times\Sigma_u^-$, $\Sigma_u^+\times\pi_u$,			$\Sigma_u^-\times\pi_g$, $\Sigma_u^-\times\Delta_g$,	
		$\Sigma_u^+\times\Delta_g$, $\Sigma_u^+\times\Delta_u$,			$\Sigma_u^-\times\phi_g$, $\Sigma_u^-\times\phi_u$	
		$\Sigma_u^+\times\phi_g$, $\Sigma_u^+\times\phi_u$,				
		$\pi_g\times\pi_g$, $\pi_g\times\Delta_g$			$\pi_g\times\pi_u$, $\pi_g\times\Delta_u$	
		$\pi_g\times\phi_g$, $\pi_u\times\pi_u$			$\pi_g\times\phi_u$, $\pi_u\times\Delta_g$	
		$\pi_u\times\Delta_u$, $\pi_u\times\phi_g$			$\pi_u\times\phi_g$, $\Delta_g\times\Delta_u$	
		$\pi_u\times\phi_u$, $\Delta_g\times\Delta_g$			$\Delta_g\times\phi_u$, $\Delta_u\times\phi_g$	
		$\Delta_g\times\phi_g$, $\Delta_u\times\Delta_u$				
		$\Delta_u\times\phi_u$, $\phi_g\times\phi_g$				
		$\phi_g\times\phi_u$, $\phi_u\times\phi_u$				

Table 2-8B (*continued*)

Point group	Infrared			Raman		
	Fund.	Combinations	Overtones	Fund.	Combinations	Overtones
O_h	A_{1g}	$A_{1g}\times A_{1g}$, $A_{1g}\times A_{1u}$	A_{1g}^n	A_{1u}	$A_{1g}\times A_{1u}$, $A_{1g}\times A_{2g}$	A_{1u}^n $(n = \text{odd})$
	A_{1u}	$A_{1g}\times A_{2g}$, $A_{1g}\times A_{2u}$	A_{1u}^n	A_{2g}	$A_{1g}\times A_{2u}$, $A_{1g}\times E_u$	A_{2g}^n $(n = \text{odd})$
	A_{2g}	$A_{1g}\times E_g$, $A_{1g}\times E_u$	A_{2g}^n	A_{2u}	$A_{1g}\times F_{1g}$, $A_{1g}\times F_{1u}$	A_{2u}^n $(n = \text{odd})$
	A_{2u}	$A_{1g}\times F_{1g}$, $A_{1g}\times F_{2g}$	A_{2u}^n	E_u	$A_{1g}\times F_{2u}$, $A_{1u}\times A_{2g}$	E_u^n $(n = \text{odd})$
	E_g	$A_{1g}\times F_{2u}$, $A_{1u}\times A_{1u}$	E_g^n	F_{1g}	$A_{1u}\times A_{2u}$, $A_{1u}\times E_g$	F_{1u}^n $(n = \text{odd})$
	E_u	$A_{1u}\times A_{2g}$, $A_{1u}\times A_{2u}$	E_u^n	F_{1u}	$A_{1u}\times F_{1g}$, $A_{1u}\times F_{1u}$	F_{2u}^n $(n = \text{odd})$
	F_{1g}	$A_{1u}\times E_g$, $A_{1u}\times E_u$	F_{1g}^n	F_{2u}	$A_{1u}\times F_{2g}$, $A_{2u}\times F_{2g}$	
	F_{2g}	$A_{1u}\times F_{1u}$, $A_{1u}\times F_{2g}$	F_{1u}^n $(n = \text{even})$		$A_{2g}\times A_{2u}$, $A_{2g}\times E_u$	
	F_{2u}	$A_{1u}\times F_{2u}$, $A_{2g}\times A_{2g}$	F_{2g}^n		$A_{2g}\times F_{1u}$, $A_{2g}\times F_{2g}$	
		$A_{2g}\times A_{2u}$, $A_{2g}\times E_g$	F_{2u}^n $(n = \text{even})$		$A_{2g}\times F_{2u}$, $A_{2u}\times E_g$	
		$A_{2g}\times E_u$, $A_{2g}\times F_{1g}$			$A_{2u}\times F_{1g}$, $A_{2u}\times F_{2g}$	
		$A_{2g}\times F_{1u}$, $A_{2g}\times F_{2g}$			$A_{2u}\times F_{2u}$, $E_g\times E_u$	
		$A_{2u}\times A_{2u}$, $A_{2u}\times E_g$			$E_g\times F_{1u}$, $E_g\times F_{2u}$	
		$A_{2u}\times E_u$, $A_{2u}\times F_{1g}$			$E_u\times F_{1g}$, $E_u\times F_{2g}$	
		$A_{2u}\times F_{1u}$, $A_{2u}\times F_{2u}$			$F_{1g}\times F_{1u}$, $F_{1g}\times F_{2u}$	
		$E_g\times E_g$, $E_g\times E_u$			$F_{1u}\times F_{2g}$, $F_{2g}\times F_{2u}$	
		$E_g\times F_{1g}$, $E_g\times F_{2g}$				
		$E_u\times E_u$, $E_u\times F_{1u}$				
		$E_u\times F_{2u}$, $F_{1g}\times F_{1g}$				
		$F_{1g}\times F_{2g}$, $F_{1u}\times F_{1u}$				
		$F_{1u}\times F_{2u}$, $F_{2g}\times F_{2g}$				
		$F_{2u}\times F_{2u}$				

Table 2-8B (*continued*)

Point group	Infrared			Raman		
	Fund.	Combinations	Overtones	Fund.	Combinations	Overtones
D_{6h}	A_{1g}	$A_{1g}\times A_{1g}$, $A_{1g}\times A_{2g}$	A_{1g}^n	A_{1u}	$A_{1g}\times A_{1u}$, $A_{1g}\times A_{2g}$	A_{1u}^n $(n = \text{odd})$
	A_{1u}	$A_{1g}\times B_{1g}$, $A_{1g}\times B_{1u}$	A_{1u}^n	A_{2g}	$A_{1g}\times A_{2u}$, $A_{1g}\times B_{1g}$	A_{2g}^n $(n = \text{odd})$
	A_{2g}	$A_{1g}\times B_{2g}$, $A_{1g}\times B_{2u}$	A_{2g}^n	A_{2u}	$A_{1g}\times B_{1u}$, $A_{1g}\times B_{2g}$	A_{2u}^n $(n = \text{odd})$
	B_{1g}	$A_{1g}\times E_{1g}$, $A_{1g}\times E_{1u}$	A_{2u}^n $(n = \text{even})$	B_{1g}	$A_{1g}\times B_{2u}$, $A_{1g}\times E_{1u}$	B_{1g}^n $(n = \text{odd})$
	B_{1u}	$A_{1g}\times E_{2g}$, $A_{1g}\times E_{2u}$	B_{1g}^n	B_{1u}	$A_{1g}\times E_{2u}$, $A_{1u}\times A_{2g}$	B_{1u}^n $(n = \text{odd})$
	B_{2g}	$A_{1u}\times A_{1u}$, $A_{1u}\times A_{2u}$	B_{1u}^n	B_{2g}	$A_{1u}\times A_{2u}$, $A_{1u}\times B_{1g}$	B_{2g}^n $(n = \text{odd})$
	B_{2u}	$A_{1u}\times B_{1g}$, $A_{1u}\times B_{1u}$	B_{2g}^n	B_{2u}	$A_{1u}\times B_{1u}$, $A_{1u}\times B_{2g}$	B_{2u}^n $(n = \text{odd})$
	E_{1g}	$A_{1u}\times B_{2g}$, $A_{1u}\times B_{2u}$	B_{2u}^n	E_{1u}	$A_{1u}\times B_{2u}$, $A_{1u}\times E_{1g}$	E_{1u}^n $(n = \text{odd})$
	E_{2g}	$A_{1u}\times E_{1u}$, $A_{1u}\times E_{2u}$	E_{1g}^n	E_{2u}	$A_{1u}\times E_{2g}$, $A_{2g}\times A_{2u}$	E_{2u}^n $(n = \text{odd})$
	E_{2u}	$A_{1u}\times E_{2g}$, $A_{2g}\times A_{2g}$	E_{1u}^n $(n = \text{even})$		$A_{2u}\times B_{1g}$, $A_{1g}\times B_{1u}$	
		$A_{2g}\times A_{2u}$, $A_{2g}\times B_{1g}$	E_{2g}^n		$A_{2g}\times B_{2g}$, $A_{2g}\times B_{2u}$	
		$A_{2g}\times B_{1u}$, $A_{2g}\times B_{2g}$	E_{2u}^n $(n = \text{even})$		$A_{2g}\times E_{1u}$, $A_{2g}\times E_{2u}$	
		$A_{2g}\times B_{2u}$, $A_{2g}\times E_{1g}$			$A_{2u}\times B_{1g}$, $A_{2u}\times B_{1u}$	
		$A_{2g}\times E_{2g}$, $A_{2g}\times E_{2u}$			$A_{2u}\times B_{2g}$, $A_{2u}\times B_{2u}$	
		$A_{2u}\times A_{2u}$, $A_{2u}\times B_{1g}$			$A_{2u}\times E_{1g}$, $A_{2u}\times E_{2g}$	
		$A_{2u}\times B_{1u}$, $A_{2u}\times B_{2g}$			$B_{1g}\times B_{1u}$, $B_{1g}\times B_{2g}$	
		$A_{2u}\times B_{2u}$, $A_{2u}E_{1u}$			$B_{1g}\times B_{2u}$, $B_{1g}\times E_{1u}$	
		$A_{2u}\times E_{2g}$, $A_{2u}\times E_{2u}$			$B_{1g}\times E_{2u}$, $B_{1u}\times B_{2g}$	
		$B_{1g}\times B_{1g}$, $B_{1g}\times B_{1u}$			$B_{1u}\times B_{2u}$, $B_{1u}\times E_{1g}$	
		$B_{1g}\times B_{2g}$, $B_{1g}\times E_{1g}$			$B_{1u}\times E_{2g}$, $B_{2g}\times B_{2u}$	
		$B_{1g}\times E_{1u}$, $B_{1g}\times E_{2g}$			$B_{2g}\times E_{1u}$, $B_{2g}\times E_{2u}$	
		$B_{1u}\times B_{1u}$, $B_{1u}\times B_{2u}$			$B_{2u}\times E_{1g}$, $B_{2u}\times E_{2g}$	

Table 2-8B *(continued)*

Point group	Infrared			Raman		
	Fund.	Combinations	Overtones	Fund.	Combinations	Overtones
		$B_{1u}\times E_{1g}$, $B_{1u}\times E_{1u}$ $B_{1u}\times E_{2u}$, $B_{2g}\times B_{2g}$ $B_{2g}\times B_{2u}$, $B_{2g}\times E_{1g}$ $B_{2g}\times E_{1u}$, $B_{2g}\times E_{2g}$ $B_{2u}\times B_{2u}$, $B_{2u}\times E_{1g}$ $B_{2u}\times E_{1u}$, $B_{2u}\times E_{2u}$ $E_{1g}\times E_{1g}$, $E_{1g}\times E_{2g}$ $E_{1u}\times E_{1u}$, $E_{1u}\times E_{2u}$ $E_{2g}\times E_{2g}$, $E_{2g}\times E_{2u}$ $E_{2u}\times E_{2u}$			$E_{1g}\times E_{1u}$, $E_{1g}\times E_{2u}$ $E_{1u}\times E_{2g}$, $E_{2g}\times E_{2u}$	

Table 2-9. Group Character $\chi_i(R)$ for Various Representations[4]*

Group character	Representation
$\chi_i(n_i) = \mu_R(\pm 1 + 2\cos\varphi)$	Total unit cell modes, internal and lattice
$\chi_i(T) = \pm 1 + 2\cos\varphi$	Acoustic modes (dipole moment vector)
$\chi_i(T') = (\mu_R(s) - 1](\pm 1 + 2\cos\varphi)$	Translatory lattice modes
$\chi_i(R') = [\mu_R(s-p)]\chi_i(P)$	Rotatory lattice modes

* Notation: μ_R is the number of atoms invariant under the symmetry operation R; $\mu_R(s)$ is the number of structural groups remaining invariant under symmetry operation R; $\mu_R(s-p)$ is the number of polyatomic groups remaining invariant under an operation R; p is the number of monatomic groups; φ is the angle of rotation corresponding to the symmetry operation R. Plus and minus signs stand, respectively, for proper and improper rotations; $\chi_i(P)$ is $(1 \pm 2\cos\varphi_R)$ for nonlinear polyatomic groups; $\pm 2\cos\varphi$ for operations $C(\varphi_R)$ and $S(\varphi_R)$ in a linear polyatomic group; and 0 for operations $C_2(\varphi)$ and σ_v in a linear polyatomic group.

molecules per (crystallographic) unit cell. It is helpful to have a model or figure showing the arrangement of the molecules in the unit cell, particularly since one has to perform the operation of a glide plane on the unit cell, and this may be difficult to do without a map of the unit cell. Figure 2-3 illustrates the unit cell for $NaNO_3$. In performing the various operations, the following transformation of the atoms occurs, where the numbers grouped together in parentheses indicate atoms which exchange positions for each set of operations:

$2S_6$	(1) (2) (3, 4) (5, 9, 7, 8, 6, 10) (1) (2) (3, 4) (5, 10, 6, 8, 7, 9)
$2C_3$	(1) (2) (3) (4) (8, 10, 9) (5, 7, 6) (1) (2) (3) (4) (8, 9, 10) (5, 6, 7)
i	(1) (2) (3, 4) (5, 8) (6, 9) (7, 10)
$3\sigma_v$(glide)	(1, 2) (3, 4) (5, 8) (6, 10) (7, 9) (1, 2) (3, 4) (5, 10) (6, 9) (7, 8) (1, 2) (3, 4) (5, 9) (6, 8) (7, 10)
$3C_2$	(1, 2) (3) (4) (5) (8) (6, 7) (9, 10) (1, 2) (3) (4) (6) (9) (5, 7) (8, 10) (1, 2) (3) (4) (7) (10) (5, 6) (8, 9)

Table 2-10 is the character table for D_{3d}^6 (isomorphic to the D_{3d} point group

Table 2-10. Character Table and Distribution of Unit Cell Modes in $NaNO_3$*

$D_{3d}^6(R\bar{3}c)$	E	$2C_2$	$3C_2$	i	$2S_6$	$3\sigma_d$ †	n_i	T	T'	R'	n_i'	Activity	
												IR	R
A_{1g}	1	1	1	1	1	1	1	0	0	0	1	ia	a
A_{1u}	1	1	1	−1	−1	−1	2	0	1	0	1	ia	ia
A_{2g}	1	1	−1	1	1	−1	3	0	1	1	1	ia	ia
A_{2u}	1	1	−1	−1	−1	1	4	1	1	1	1	a	ia
E_g	2	−1	0	2	−1	0	4	0	1	1	2	ia	a
E_u	2	−1	0	−2	1	0	6	1	2	1	2	a	ia
φ_R, deg	0	120	180	180	60	0							
$\cos\varphi_R$	1	−0.5	−1	−1	0.5	1							
$\pm 1 + 2\cos\varphi_R$	3	0	−1	−3	0	1							
μ_R	10	4	4	2	2	0							
$\mu_R(s)$	4	2	2	2	2	0							
$\mu_R(s-p)$	2	0	2	0	0	0							
$\chi_i(n_i)$	30	0	−4	−6	0	0							
$\chi_i(T)$	3	0	−1	−3	0	1							
$\chi_i(T')$	9	0	−1	−3	0	−1							
$\chi_i(R')$	6	0	−2	0	0	0							

* n_i = total number of unit cell modes (lattice plus internal).
T = acoustic modes.
T' = translatory lattice modes.
R' = rotatory lattice modes.
n_i' = internal modes of polyatomic unit.
$i \equiv S_2 \equiv S_6^3$.
† Glide plane.

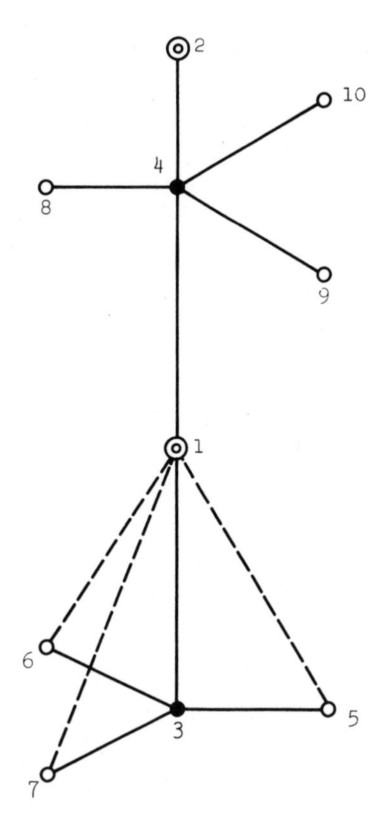

Fig. 2-3. Unit cell of $NaNO_3$. ● Nitrogen, ⊙ Sodium, ○ Oxygen.

except that the σ_v operation is a glide plane). The table has been developed to satisfy the equations for $\chi_i(n_i)$, $\chi_i(T)$, $\chi_i(T')$, and $\chi_i(R')$ corresponding to all of the symmetry operations, and which appear in Table 2-9.

Having obtained the characters of irreducible representations for $\chi_i(n_i)$, $\chi_i(T)$, $\chi_i(T')$, and $\chi_i(R')$, one is now ready to develop the right-hand part of the table and calculate n_i, T, T', R', and n_i' for each vibration species. To calculate n_i, use is made of the reduction formula

$$N_i = (1/N_G) \sum n_e \chi_i(R) \chi_i(n_i) \tag{2-38}$$

where N_i is the number of times an irreducible representation appears in a vibration species: N_G is the number of elements in the group, n_e is the number of elements in each operation class; $\chi_i(R)$ is the character of the vibration species for the space group involved; and $\chi_i(n_i)$ is the character for the total unit cell modes. For T one uses the same reduction formula but replaces $\chi_i(n_i)$ with $\chi_i(T)$, etc. Similar procedures were followed in the previous section for the calculation of the selection rules for the isolated molecule.

The development of Table 2-10 shows that of the 30 degrees of freedom ($3nZ'$) for $NaNO_3$, where n is the number of atoms and Z' is the number of molecules per Bravais unit cell, 18 are external or lattice modes and 12 correspond to the internal modes of the NO_3^-, $(3n - 6)Z'$. The different modes which correspond to the irreducible representations of the D_{3d}^6 space group are

$$\Gamma_{n_i} = 1A_{1g} + 2A_{1u} + 3A_{2g} + 4A_{2u} + 4E_g + 6E_u \tag{2-39}$$

All the gerade (g) modes, except for A_{2g}, are Raman active, and all ungerade (u) modes, except for A_{1u}, are infrared active. The A_{2g} and A_{1u} modes are forbidden. The breakdown of the vibrations between internal and external modes is as follows:

(acoustic) $$\Gamma_T = E_u + A_{2u} \tag{2-40}$$

(translatory) $$\Gamma_{T'} = A_{1u} + A_{2g} + A_{2u} + E_g + 2E_u \tag{2-41}$$

(rotatory) $$\Gamma_{R'} = A_{2g} + A_{2u} + E_g + E_u \tag{2-42}$$

(total internal) $$\Gamma_{n_i'} = A_{1g} + A_{1u} + A_{2g} + A_{2u} + 2E_g + 2E_u \tag{2-43}$$

B. The Halford–Hornig Site Group Method[6–8]

To illustrate the HH method or correlation method, we will again consider the $NaNO_3$ case. Whereas in the BV method a clear picture of the arrangement of the molecule(s) in the unit cell was necessary, the HH method avoids this complication.

In using this method, it is necessary to select an acceptable site group which must be a subgroup of both the space group and the molecular point group of the NO_3^- and the Na^+. In general, the site group will be of lower order than the molecular group. Further elimination can be made on the basis of the number of molecules contained in the unit cell.

For a $D_{3d}^6(R\bar{3}c)$ space group (No. 167) the following site symmetries are possible (see Appendix 3):

$$D_3(2);\quad C_{3i}(2);\quad C_3(4);\quad C_i(6);\quad C_2(6);\quad C_1(12)$$

Examining the correlation tables in Appendix 4, one observes that for the

D_{3h} molecular point group (NO_3^-) the following subgroups are possible:

$$C_{3h},\quad D_3,\quad C_{3v},\quad C_{2v},\quad C_3,\quad C_2,\quad C_s$$

Upon examination of the sites possible for the D_{3d}^6 factor group, a check of the number of molecules per unit cell or equivalent sites per set (the number in parentheses) indicates that for $NaNO_3$ with two molecules per Bravais unit cell (Z'), only the sites $D_3(2)$ and $C_{3i}(2)$ exist. Of the two, only the site symmetry of D_3 corresponds as a subgroup for both the D_{3d}^6 space group and the D_{3h} molecular point group for the NO_3^-. The other sites can be eliminated since the number of equivalent sites per set is greater than two. Thus, we may consider the NO_3^- to lie on a D_3 site, which in fact it does. The Na^+ can be considered to lie on a C_{3i} (equivalent to S_6) site. This method of selecting the site symmetry is ambiguous wherever there is a choice as to the axis of rotation or reflection plane involved. In these cases an unambiguous choice can be made by the procedure outlined in Section D. Having determined the site groups, one can set up a correlation table with help from Appendix 4. It is necessary to correlate the molecular point group D_{3h} and the factor group D_{3d}^6 with the site group D_3. Table 2-11A illustrates the correlation table for NO_3^-. For both the molecular point group and the factor group the vectors for the translations T and rotations R are included in Table 2-11A.

Using the methods for determining the selection rules for an isolated molecule, it can be shown that for the isolated NO_3^-, six internal ($3n - 6$) modes will be allowed (keeping in mind that E vibrations are doubly de-

Table 2-11A. Correlation Table for NO_3^- Internal Modes

Molecular point group D_{3h}	Site group D_3	Factor group D_{3d}^6
$1A_1'$	$1A_1$	$1A_{1g}$
$(R_z)\ 0A_2'$	$1A_2$	$1A_{2g}\ (R_z)$
$(T_x, T_y)\ 2E'$	$2E$	$2E_g\ (R_x, R_y)$
$0A_1''$	$1A_1$	$1A_{1u}$
$(T_z)\ 1A_2''$	$1A_2$	$1A_{2u}\ (T_z)$
$(R_x, R_y)\ 0E''$	$2E$	$2E_u\ (T_x, T_y)$

generate vibrations and are counted twice),

$$\Gamma_{n_i'} = 2E' + A_2'' + A_1' \tag{2-44}$$

Since two molecules are present per unit cell of solid $NaNO_3$, and if the interaction is sufficient between the two molecules, the doubling of vibrations could occur (called correlation field or factor group splitting). The latter type of splitting can split all types of vibrations, whereas a site group splitting will only split degenerate vibrations. Thus, 12 internal vibrations $(3n - 6)Z'$ of the NO_3^- in $NaNO_3$ may be allowed and these are

$$\Gamma_{n_i'} = 2E_u + 2E_g + A_{1g} + A_{2u} + A_{1u} + A_{2g}$$

Since a total $3nZ' = 3 \cdot 4 \cdot 2$ or 24 degrees of freedom are possible for the NO_3^- in $NaNO_3$, and 12 of these are internal vibrations, one would expect the other 12 modes to be external or lattice modes. These can be determined directly from Table 2-11B.

In determining the total number of lattice modes, it is necessary to consider the number of atoms or polyatomic groups lying on a site. It is recommended that the following procedure be followed:

(1) Determine the total degrees of freedom for each species in the site group. This is accomplished by multiplying the number of translations (T vectors) or rotations (R vectors) for each direction (x, y, z) appearing in the character table for the site group by the

Table 2-11B. Correlation Table for NO_3^- External Modes

Degrees of freedom		Site group	Factor group	Modes	
T	R	D_3	D_{3d}^6	T	R
			A_{1g}	0	0
0	0	A_1	A_{2g} (R_z)	1	1
			E_g (R_x, R_y)	1	1
2	2	(T_z, R_z) A_2	A_{1u}	0	0
			A_{2u} (T_z)	1	1
4	4	$(T_x, T_y), (R_x, R_y)$ E	E_u (T_x, T_y)	1	1

number of atoms (or polyatomic groups) present in the formula weight of the solid. This product is then multiplied by the number of molecules in the primitive cell.

(2) The T and R vectors are then traced from the site group to the factor group, and the correlations determined.

(3) The number of species correlating in the factor group must correspond to the total degrees of freedom determined in (1). The degeneracies of the species must be considered (e.g., $E = 2$, $F = 3$). Thus, it is necessary to determine the coefficients for each species of the factor group.

For the lattice modes involving the NO_3^-, $Z' = 2$, the polyatomic group NO_3^- in $NaNO_3$ is equal to one, and two translation vectors are involved (T_x, T_y) for the E species in the D_3 site group. The total number of degrees of freedom is $2 \cdot 2 \cdot 1 = 4$. The E species correlates with an $E_g + E_u$ species in the factor group. Since the total number of degrees of freedom for $E_g + E_u$ is equal to four (E degeneracies $= 2$), the coefficients of E_g and E_u must be one ($4/4 = 1$). For the A_2 species only the T_z vector is involved, and the total number of degrees of freedom is $1 \cdot 2 \cdot 1 = 2$. The A_2 species correlates with an $A_{2g} + A_{2u}$ species in the factor group. Since the number of degrees of freedom for $A_{2g} + A_{2u}$ is two, the coefficients of A_{2g} and A_{2u} are one. Therefore, the number of lattice translations is given by

$$\Gamma_{T'} = A_{2g} + A_{2u} + E_g + E_u \tag{2-45}$$

The lattice rotations are determined similarly, except that the R vectors are followed. Thus,

$$\Gamma_{R'} = A_{2g} + A_{2u} + E_g + E_u \tag{2-46}$$

Six lattice translations and six lattice rotations are allowed for the NO_3^-, and, as expected, a total of 12 external modes are possible.

For the Na^+ a similar procedure is followed. Since the Na^+ can be considered to be on a C_{3i} site (or S_6), a correlation table can be constructed, as shown in Table 2-12. A total of $3nZ' = 3 \cdot 1 \cdot 2$ or six modes is expected for the Na^+. Since the A_u species in the site group S_6 involves a T_z vector and since there are two Na atoms in the unit cell, then the number of degrees of freedom for this species is $1 \cdot 2 \cdot 1 = 2$. A_u correlates with $A_{1u} + A_{2u}$, giving a total of two degrees of freedom in the factor group. Thus the coefficients of A_{1u} and A_{2u} must be equal to one.

The species E_u of the site group has $2 \cdot 2 \cdot 1 = 4$ degrees of freedom. It correlates with an E_u species in the factor group which has only two degrees of freedom. Therefore the coefficient of E_u in the factor group must be equal to two ($4/2 = 2$). Thus

$$\Gamma_{T'+T} = A_{1u} + A_{2u} + 2E_u$$

Three of these will be acoustic modes and must be subtracted. These are determined by locating the T vectors in the factor group, and are found to be

$$\Gamma_T = E_u + A_{2u} \tag{2-47}$$

Subtracting these from the total lattice modes, one obtains

$$\Gamma_{T'} = E_u + A_{1u} \tag{2-48}$$

Thus six external modes have been determined. No rotational modes are possible with a monatomic ion.

Totaling $\Gamma_{T'}$ for the Na^+ and NO_3^-, one obtains for $NaNO_3$

$$\Gamma_{T'} = A_{1u} + A_{2g} + A_{2u} + E_g + 2E_u \tag{2-49}$$

and

$$\Gamma_{R'} = A_{2g} + A_{2u} + E_g + E_u \tag{2-50}$$

and for all modes

$$\Gamma_{n_i} = A_{1g} + 2A_{1u} + 3A_{2g} + 4A_{2u} + 4E_g + 6E_u \tag{2-51}$$

In summary, the HH treatment predicts results which are the same as those of the BV treatment. This is illustrated in Table 2-13, which is identical to the right-hand part of Table 2-10 obtained from the BV method.

Table 2-12. Correlation Table for Na^+

Degrees of freedom		$S_6(C_{3i})$	D_{3d}^6	Modes	
T	R			T	R
2	0	(T_z) A_u	A_{1u}	1	0
			A_{2u} (T_z)	1	0
4	0	(T_x, T_y) E_u	E_u (T_x, T_y)	2	0

Table 2-13. Results of Space Group Selection Rules for $NaNO_3$ as Obtained by the Halford and Hornig Method

Species	n_i	T	T'	R'	n_i'
A_{1g}	1	0	0	0	1
A_{1u}	2	0	1	0	1
A_{2g}	3	0	1	1	1
A_{2u}	4	1	1	1	1
E_g	4	0	1	1	2
E_u	6	1	2	1	2

C. Comparison of the BV and HH Methods

In the previous section, it was indicated that the BV and HH methods give the same results for $NaNO_3$, once the space group is known. However, Halford and Hornig have introduced some approximations which are helpful in correlating the classification of vibrational modes of an ion or molecule based on site symmetry (subgroup of molecular point group) in the cell with that of the unit cell group or space group. The latter method can be of considerable aid in solid-state chemistry and physics. For further reading on factor and site group analysis see the correlation bibliography at the end of this chapter.

D. Unambiguous Choice of Site Symmetry in the Bravais Unit Cell

For cases where an unambiguous choice of site symmetry cannot be made, the use of Wyckoff's tables[12] on crystallographic data can prove helpful. For example, consider orthorhombic $PuBr_3$, which has a D_{2h}^{17} (*Cmcm*) space group and a crystallographic unit cell of $Z = 4$ molecules. For a *C*-type lattice there are two repeat units in the cell (see Table 1-12) and therefore

$$Z' = \text{number of molecules in the Bravais cell}$$

$$= \frac{Z \text{ (number of molecules in crystallographic cell)}}{\text{repeat units in cell}} = \frac{4}{2} = 2$$

From Appendix 3 we can observe that for D_{2h}^{17} (space group 63) the following site symmetries are possible: $2C_{2h}(2)$; $C_{2v}(2)$; $C_i(4)$; $C_2(4)$; $2C_s(4)$; $C_1(8)$. With the number of molecules in the Bravais unit cell equal to two, we must place two M^{3+} ions on a site and six Br^- on another site. We observe that two site symmetries are available for the two M^{3+} ions—either C_{2h} or C_{2v}, each having two equivalent sites per set to place the metal ions. An unambiguous choice cannot be made with the data available. For the six Br^- no site symmetry has six equivalent sites available. Thus we must conclude that the six Br^- must be nonequivalent, and some are on one site and the others on another site. At this point one must consult the Wyckoff tables on published crystallographic data, and when this is done, we find the following notation:

Wyckoff's Table for $PuBr_3$

Ion	Site position
$2Pu^{3+}$	*c*
$2Br^-$	*c*
$4Br^-$	*f*

We can deduce the Wyckoff nomenclature of the site positions from the site symmetries by listing the site positions in alphabetical order as follows:

Site in Appendix 3	Alphabetical order	Wyckoff's alphabetical ordering of site position	Ion site
$2C_{2h}(2)$	$C_{2h}(2)$	*a*	
	$C_{2h}(2)$	*b*	
$C_{2v}(2)$	$C_{2v}(2)$	*c*	$2Pu^{3+}(c)$, $2Br^-(c)$
$C_i(4)$	$C_i(4)$	*d*	
$C_2(4)$	$C_2(4)$	*e*	
$2C_s(4)$	$C_s(4)$	*f*	$4Br^-(f)$
	$C_s(4)$	*g*	
$C_1(8)$	$C_1(8)$	*h*	

We can place the two Pu^{3+} ions on a *c* site (C_{2v}), two Br^- ions on a *c* site (C_{2v}), and four Br^- must be on an *f* site (C_s). If we examine the correlation tables in Appendix 4, we observe that three correlations are possible for a D_{2h} space group with a site symmetry of C_{2v}. Similarly, three correlations are possible for the site symmetry C_s. Each correlation is based on a different rotational axis or reflection plane being involved. For example:

	$C_2(Z)$	$C_2(Y)$	$C_2(X)$	$\sigma(xy)$	$\sigma(zx)$	$\sigma(yz)$
D_{2h}	C_{2v}	C_{2v}	C_{2v}	C_s	C_s	C_s

One must decide which site group to use. Table 2-14 can be used to determine the proper site. For each space group the correlation to go with each site is included. Knowing the site symmetry as given by the Wyckoff tables, one can determine which site correlation to use. For this example, the *c*-site position for a C_{2v} site is correlated with C_{2v} involving a C_2 rotation around the *y* axis, and the *f*-site position for a C_s site is correlated with C_s, involving a reflection plane in the *yz* plane. In this manner an unambiguous choice of the site symmetry for the Pu^{3+} and Br^- is made. This method of obtaining the proper site symmetry is possible whenever the Wyckoff tables contain the molecule of interest. If the information is not available in the Wyckoff tables, then one must resort to a study of the actual crystallographic structure of the molecule, if it is available.

Although only two equivalent sites per set are available for C_{2v} symmetry, it is possible to place the two Pu^{3+} and two Br^- ions in a C_{2v} site, since the number of such sites is infinite. When the site symmetry is C_p, C_{pv}, or C_s and $p = 1, 2, 3$, etc., the number of sites is infinite. This point should be kept in mind in using Appendix 3.

2-3. EXAMPLES OF THE HALFORD–HORNIG SITE GROUP METHOD

In the previous sections the space group selection rules were derived for $NaNO_3$ using the Bhagavantam–Venkatarayudu method and the Halford–Hornig site group approximation method (also called the correlation method). In this section, we shall attempt to further illustrate the HH method using several examples. Examples involving ionic crystals, covalent crystals, coordination compounds, linear groups in the crystal, and $LaCl_3$ are presented. The crystalline modifications represented in the examples given are monoclinic, orthorhombic, tetragonal, and hexagonal.

Table 2-14. Determination of the Proper Correlation Using Wyckoff's Tables

Space group number			Site correlation					
			$C_2(z)$	$C_2(y)$	$C_2(x)$	$\sigma(xy)$	$\sigma(zx)$	$\sigma(yz)$
D_2^i	16	D_2^1	*q, r, s, t*	*m, n, o, p*	*i, j, k, l*			
	17	D_2^2		*c, d*	*a, b*			
	18	D_2^3	*a, b*					
	20	D_2^5		*b*	*a*			
	21	D_2^6	*i, j, k*	*g, h*	*e, f*			
	22	D_2^7	*g, h*	*f, i*	*e, j*			
	23	D_2^8	*i, j*	*g, h*	*e, f*			
	24	D_2^9	*c*	*b*	*a*			
C_{2v}^i	25	C_{2v}^1					*e, f*	*g, h*
	26	C_{2v}^2						*a, b*
	28	C_{2v}^4						*c*
	31	C_{2v}^7						*a*
	35	C_{2v}^{11}					*d*	*e*
	36	C_{2v}^{12}						*a*
	38	C_{2v}^{14}					*c*	*d, e*
	39	C_{2v}^{15}					*c*	
	40	C_{2v}^{16}						*b*
	42	C_{2v}^{18}					*d*	*c*
	44	C_{2v}^{20}					*c*	*d*
	46	C_{2v}^{22}						*b*

D_{2h}^{i}	47	D_{2h}^{1}	*q, r, s, t*	*m, n, o, p*	*i, j, k, l*	*y, z*	*w, x*	*u, v*
	48	D_{2h}^{2}	*k, l*	*i, j*	*g, h*			
	49	D_{2h}^{3}	*a, b, c, d, m, n, o, p*	*k, l*	*i, j*	*q*		
	50	D_{2h}^{4}	*k, l*	*i, j*	*g, h*			
	51	D_{2h}^{5}	*e, f*	*a, b, c, d, g, h*			*i, j*	*k*
	52	D_{2h}^{6}	*c*		*d*			
	53	D_{2h}^{7}		*g*	*a, b, c, d, e, f*			*h*
	54	D_{2h}^{8}	*d, e*	*c*				
	55	D_{2h}^{9}	*a, b, c, d, e, f*			*g, h*		
	56	D_{2h}^{10}	*c, d*					
	57	D_{2h}^{11}			*c*	*d*		
	58	D_{2h}^{12}	*a, b, c, d, e, f*			*g*		
	59	D_{2h}^{13}	*a, b*				*f*	*e*
	60	D_{2h}^{14}		*c*				
	62	D_{2h}^{16}					*c*	
	63	D_{2h}^{17}		*c*	*a, b, e*	*g*		*f*
	64	D_{2h}^{18}		*e*	*a, b, d*			*f*
	65	D_{2h}^{19}	*e, f, k, l, m*	*i, j*	*g, h*	*p, q*	*o*	*n*
	66	D_{2h}^{20}	*c, d, e, f, i, j, k*	*h*	*g*	*l*		
	67	D_{2h}^{21}	*g, l*	*e, f, j, k*	*c, d, h, i*		*n*	*m*
	68	D_{2h}^{22}	*g, h*	*f*	*e*			
	69	D_{2h}^{23}	*e, i, j*	*d, h, k*	*c, g, l*	*o*	*n*	*m*
	70	D_{2h}^{24}	*g*	*f*	*e*			
	71	D_{2h}^{25}	*i, j*	*g, h*	*e, f*	*n*	*m*	*l*
	72	D_{2h}^{26}	*c, d, h, i*	*g*	*f*	*j*		
	73	D_{2h}^{27}	*e*	*d*	*c*			
	74	D_{2h}^{28}	*e*	*c, d, g*	*a, b, f*		*i*	*h*

Table 2-14 (*continued*)

Space group number			C_2	C_2'	C_2''	C_2 σ_v	C_2 σ_d	σ_h	σ_v	σ_d
D_4^i	89	D_4^1	i	e, f, l, m, n, o	j, k					
	90	D_4^2	d		e, f, a, b					
	91	D_4^3		a, b	c					
	92	D_4^4			a					
	93	D_4^5	g, h, i	a, b, c, d, j, k, l, m	e, f, n, o					
	94	D_4^6	c, d		a, b, e, f					
	95	D_4^7		a, b	c					
	96	D_4^8			a					
	97	D_4^9	f	c, h, i	d, g, j					
	98	D_0^{10}	c	f	a, b, d, e					
C_{4v}^i	99	C_{4v}^1				c, e, f	d			
	100	C_{4v}^2					b, c			
	101	C_{4v}^3					a, b, d			
	102	C_{4v}^4					a, c			
	105	C_{4v}^7				a, b, c, d, e				
	107	C_{4v}^9				b, d	c			
	108	C_{4v}^{10}					b, c			
	109	C_{4v}^{11}				a, b				
D_{2d}^i	111	D_{2d}^1	m	i, j, k, l						
	112	D_{2d}^2	k, l, m	g, h, i, j						

	113	D_{2d}^{3}	*d*							
	114	D_{2d}^{4}	*c, d*							
	115	D_{2d}^{5}		*h, i*						
	116	D_{2d}^{6}	*g, h, i*	*e, f*						
	117	D_{2d}^{7}	*e, f*	*g, h*						
	118	D_{2d}^{8}	*e, h*	*f, g*						
	119	D_{2d}^{9}		*g, h*						
	120	D_{2d}^{10}	*f, g*	*e, h*						
	121	D_{2d}^{11}	*h*	*f, g*						
	122	D_{2d}^{12}	*c*	*d*						
D_{4h}^{i}	123	D_{4h}^{1}		*e, f, l, m, n, o*	*j, k*	*i*		*p, q*	*s, t*	*r*
	124	D_{4h}^{2}	*i, e*	*f, k, l*	*j*			*m*		
	125	D_{4h}^{3}		*c, d, k, l*	*e, f, i, j*		*h*			*m*
	126	D_{4h}^{4}	*g*	*c, i, j*	*h*					
	127	D_{4h}^{5}			*c, d, g, h*		*f*	*i, j*		*k*
	128	D_{4h}^{6}	*c, f*		*d, g*			*h*		
	129	D_{4h}^{7}			*a, b, d, e, g, h*	*f*			*i*	*j*
	130	D_{4h}^{8}	*e*		*a, f*					
	131	D_{4h}^{9}		*a, b, c, d, j, k, l, m*	*e, f, n*	*g, h, i*		*q*	*o, p*	
	132	D_{4h}^{10}	*f, k*	*b, d, e, l, m*	*a, c, i, j*		*g, h*	*n*		*o*
	133	D_{4h}^{11}	*f, g*	*a, b, h, i*	*c, j*					
	134	D_{4h}^{12}	*h*	*a, b, c, i, j*	*d, e, f, k, l*		*g*			*m*
	135	D_{4h}^{13}	*a, c, e, f*		*d, g*			*h*		
	136	D_{4h}^{14}	*c, h*		*a, b, f, g*		*e*	*i*		*j*
	137	D_{4h}^{15}			*a, b, f*	*c, d*			*g*	
	138	D_{4h}^{16}	*f*		*a, c, d, g, h*		*e*			*i*
	139	D_{4h}^{17}		*c, i, j*	*d, f, h, k*	*g*		*l*	*n*	*m*

Table 2-14 (*continued*)

Space group number	C_2	C_2'	C_2''	C_2 σ_v	C_2 σ_d	σ_h	σ_v	σ_d
D_{4h}^i 140 D_{4h}^{18}		b, j	d, e, h, i		g	k		l
141 D_{4h}^{19}		c, d, f	a, b, g	e			h	
142 D_{4h}^{20}	d	e	b, f					

Space group number	C_2	C_2'	C_2''	Space group number	σ_v	σ_d	Space group number	σ_h	σ_v
D_6^i 177 D_6^1	i	j, k	c, d, l, m	C_{6v}^i 183 C_{6v}^1	b, e	d	D_{3h}^i 187 D_{3h}^1	l, m	n
178 D_6^2		a	b	185 C_{6v}^3		a, c	188 D_{3h}^2	k	
179 D_6^3		a	b	186 C_{6v}^4	a, b, c		189 D_{3h}^3	j, k	i
180 D_6^4	e, f	g, h	i, j				190 D_{3h}^4	h	
181 D_6^5	e, f	g, h	i, j						
182 D_6^6		a, g	b, c, d, h						

Space group number	C_2	C_2'	C_2''	σ_h	σ_d	σ_v
D_{6h}^i 191 D_{6h}^1	i	j, k	l, m, c, d	p, q	h, o	n
192 D_{6h}^2	g, i	j	c, k	l		
193 D_{6h}^3		a, g	b, d, f, i	j		k
194 D_{6h}^4		a, g, i	b, c, d, h	j	k, e, f	

Space group number	C_2	C'_2	$3C_2$	$C_2, 2C'_2$
O^i 207 O^1	h	i, j		
208 O^2	h, i, j	k, l	d	e, f
209 O^3	i	g, h		d
210 O^4	f	g		
211 O^5	g	h, i		d
212 O^6		d		
213 O^7		d		
214 O^8	f	g, h		c, d

Space group number	C_2	$3C_2$	$C_2, 2C'_2$	C'_2, σ_h	C_2, σ_h	C_2, σ_d	σ_h	σ_d
O_h^i 221 O_h^1				i, j	h		k, l	m
222 O_h^2	g			h				
223 O_h^3		b		c, d, j	f, g, h		k	
224 O_h^4	d, h		f	i, j		g		k
225 O_h^5			d	h, i		g	j	k
226 O_h^6				c, h	e		i	
227 O_h^7				h		f		g
228 O_h^8	f			g				
229 O_h^9				i, d, h	g		j	k
230 O_h^{10}	f		c	g				

* Taken from W. G. Fateley, F. R. Dollish, N. T. McDevitt, and F. F. Bentley, *Infrared and Raman Selection Rules for Molecular and Lattice Vibrations: The Correlation Method,* Wiley-Interscience, New York (1972), courtesy of Wiley-Interscience.
Correlation tables are listed in Appendix 4. Note that only space groups where a choice in C_2 axis and reflection plane are possible are recorded in this table. In the case of orthorhombic crystals particular care must be exercised in the proper choice of the principal axis and/or mirror planes (see reference 23 in Appendix 11).

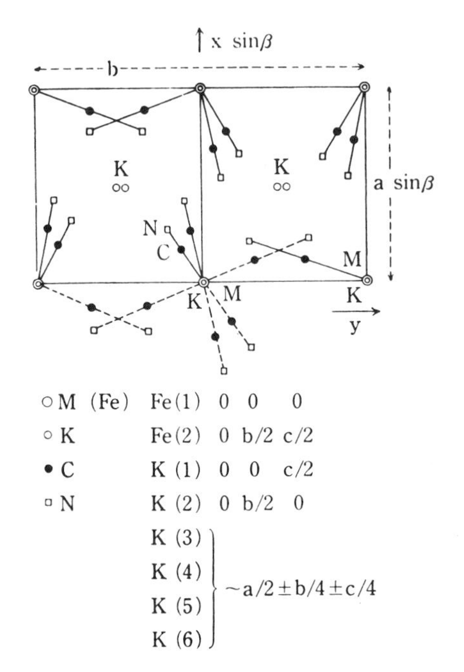

Fig. 2-4. Structure of $K_3[Fe(CN)_6]$ with the space group $C_{2h}^5(P2_1/c$; $Z' = 2)$. [From I. Nakagawa, *Coord. Chem. Rev.* **4**:423 (1969). Courtesy of *Coord. Chem. Rev.*]

Example 1

$K_3[Fe(CN)_6]$ (Monoclinic)

$K_3[Fe(CN)_6]$ has a space group of C_{2h}^5–$P2_1/c$ (No. 14), with $Z' = 2$, and the $Fe(CN)_6^{3-}$ ion has a molecular point group of O_h.[13] Figure 2-4 shows the structure of $K_3[Fe(CN)_6]$. C_i is a subgroup of the C_{2h}^5 space group and can be correlated with the O_h through a T_h subgroup. For $[Fe(CN)_6]^{3-}$ in $K_3[Fe(CN)_6]$, $3nZ' = 3 \cdot 13 \cdot 2 = 78$ modes are expected and of these $(3n - 6)Z' = 66$ will be internal modes. Thus 12 external modes are predicted. The correlation tables for the internal and external modes are given in Tables 2-15A and 2-15B.

Internal Modes for $[Fe(CN)_6]^{3-}$:

$$\Gamma_{n_i'} = 15A_g + 15B_g + 18A_u + 18B_u$$

External Modes for $[Fe(CN)_6]^{3-}$:

$$\Gamma_{T'} = 3A_u + 3B_u, \qquad \Gamma_{R'} = 3A_g + 3B_g$$

External Modes for K^+. Two types of K^+ sites exist. One K^+ is on a C_i site (see Table 2-16). Since the total number of degrees of freedom is

Table 2-15A. Correlation Table for $[Fe(CN)_6]^{3-}$ in $K_3[Fe(CN)_6]$
Internal Modes

Molecular point group O_h	Site group C_i	Factor group C_{2h}^5
$2A_{1g}$		
$0A_{2g}$		$15A_g$
$2E_g$	$15A_g$	
(R_x, R_y, R_z) $1F_{1g}$		$15B_g$
$2F_{2g}$		
$0A_{1u}$		
$0A_{2u}$		$18A_u$ (T_z)
$0E_u$	$18A_u$	
(T_x, T_y, T_z) $4F_{1u}$		$18B_u$ (T_x, T_y)
$2F_{2u}$		

equal to six for the translations, the coefficients of A_u and B_u must be three:

$$\Gamma_T = A_u + 2B_u$$
$$\Gamma_{T+T'} = 3A_u + 3B_u$$
$$\Gamma_{T'} = 2A_u + B_u$$

Table 2-15B. Correlation Table for $[Fe(CN)_6]^{3-}$ in $K_3[Fe(CN)_6]$
External Modes

Degrees of freedom		Site group C_i	Factor group C_{2h}^5	Modes	
T	R			T	R
0	6	(R_x, R_y, R_z) A_g	A_g (R_z)	0	3
			B_g (R_x, R_y)	0	3
6	0	(T_x, T_y, T_z) A_u	A_u (T_z)	3	0
			B_u (T_x, T_y)	3	0

Table 2-16. Correlation Table of One K^+ on a C_i Site

Degrees of freedom		Site group	Factor group	Modes	
T	R	C_i	C_{2h}^5	T	R
0	0	$(R_x, R_y, R_z)\ A_g$	$A_g\ (R_z)$	0	0
			$B_g\ (R_x, R_y)$	0	0
6	0	$(T_x, T_y, T_z)\ A_u$	$A_u\ (T_z)$	3	0
			$B_u\ (T_x, T_y)$	3	0

Two K^+ are on a C_1 site (see Table 2-17):

$$\Gamma_{T'} = 3A_g + 3B_g + 3A_u + 3B_u$$

Since the total number of degrees of freedom is 12 for T, the sum of the coefficients of $A_g + B_g + A_u + B_u$ must be 12, and each coefficient is three ($12/4 = 3$).

Summary for $K_3Fe(CN)_6$

A total of 96 modes are allowed.

$Fe(CN)_6^{3-}$:

$$\Gamma_{n_i'} = 15A_g + 15B_g + 18A_u + 18B_u$$

Table 2-17. Correlation Table of Two $2K^+$ on a C_1 Site

Degrees of freedom		Site group	Factor group	Modes	
T	R	C_1	C_{2h}^5	T	R
			$A_g\ (R_z)$	3	0
12	0	$(T_x, T_y, T_z)\ A_1$	$B_g\ (R_x, R_y)$	3	0
			$A_u\ (T_z)$	3	0
			$B_u\ (T_x, T_y)$	3	0

$$\Gamma_{T'} = \qquad\qquad + \; 3A_u + \; 3B_u$$
$$\Gamma_{R'} = \; 3A_g + \; 3B_g$$

3K⁺:

$$\left.\begin{aligned} \Gamma_T &= \qquad\qquad + \; A_u + \; 2B_u \\ \Gamma_{T'} &= \qquad\qquad + \; 2A_u + \; B_u \end{aligned}\right\} \text{one K}^+ \text{ on a } C_i \text{ site}$$
$$\Gamma_{T'} = \; 3A_g + \; 3B_g + \; 3A_u + \; 3B_u \qquad \text{two K}^+ \text{ on a } C_1 \text{ site}$$

$$\Gamma_{n_i} = 21A_g + 21B_g + 27A_u + 27B_u$$

42 of these will be Raman active and 51 infrared active ($A_u + 2B_u$ are acoustic modes).

Example 2

Naphthalene, $C_{10}H_8$ (Monoclinic)

Naphthalene crystallizes in the monoclinic system, space group C_{2h}^5–$P2_1/c$ (No. 14) with $Z' = 2$.[14] Figure 2-5 shows the unit cell of naphthalene. The isolated molecules belong to the D_{2h} point group. The site C_i is a subgroup of C_{2h}^5 and the D_{2h} point group. Tables 2-18A and 2-18B are the correlation tables. Naphthalene has $3nZ' = 3 \cdot 18 \cdot 2 = 108$ total modes of which $(3n - 6)Z' = 96$ are internal modes.

Internal Modes for Naphthalene:

$$\Gamma_{n_i'} = 24A_g + 24A_u + 24B_g + 24B_u$$

External Modes for Naphthalene:

$$\Gamma_{T+T'} = 3A_u + 3B_u$$

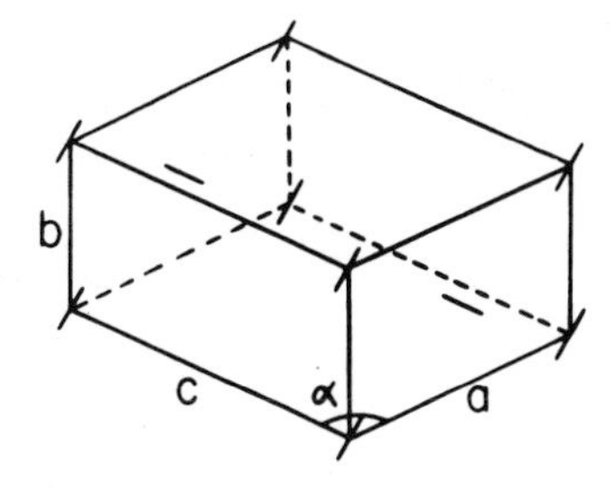

Fig. 2-5. Unit cell of anthracene and naphthalene. [From S. S. Mitra, *Optical Properties of Solids*, S. Nudelman and S. S. Mitra (eds.), Plenum Press, New York (1969). Courtesy of Plenum Press.]

Table 2-18A. Correlation Table for Naphthalene Internal Modes

Molecular point group D_{2h}	Site group C_i	Factor group C_{2h}^5
$9A_g$		$24A_g$
$4A_u$		
(R_z) $8B_{1g}$	$24A_g$	$24A_u$ (T_z)
(T_z) $4B_{1u}$		
(R_y) $4B_{2g}$		$24B_g$
(T_y) $8B_{2u}$	$24A_u$	
(R_x) $3B_{3g}$		$24B_u$ (T_x, T_y)
(T_x) $8B_{3u}$		

$$\Gamma_T = A_u + 2B_u$$
$$\Gamma_{T'} = 2A_u + B_u$$
$$\Gamma_{R'} = 3A_g + 3B_g$$

Summary for Naphthalene:

$$\Gamma_{n_i'} = 24A_g + 24A_u + 24B_g + 24B_u$$

Table 2-18B. Correlation Table for Naphthalene External Modes

Degrees of freedom		Site group C_i	Factor group C_{2h}^5	Modes	
T	R			T	R
0	6	(R_x, R_y, R_z) A_g	A_g (R_z)	0	3
			A_u (T_z)	3	0
6	0	(T_x, T_y, T_z) A_u	B_g (R_x, R_y)	0	3
			B_u (T_x, T_y)	3	0

$$\begin{aligned}
\Gamma_T &= \quad\quad + A_u \quad\quad + 2B_u \\
\Gamma_{T'} &= \quad\quad + 2A_u \quad\quad + B_u \\
\Gamma_{R'} &= 3A_g \quad\quad + 3B_g \\
\hline
\Gamma_{n_i} &= 27A_g + 27A_u + 27B_g + 27B_u
\end{aligned}$$

Since all species will be active, there will be 51 infrared-active modes and 54 Raman-active modes ($A_u + 2B_u$ are acoustic modes).

Example 3

Cyclopropane, C_3H_6

Cyclopropane belongs to the C_{2v}^7–$Pmn2_1$ space group (No. 31), with $Z' = 2$.[15] Figure 2-6 shows the structure of cyclopropane. The molecular point group is D_{3h}. The site group C_s is a subgroup of both the C_{2v}^7 and D_{3h} groups. The proper choice of C_s is obtained from Table 2-14, and is

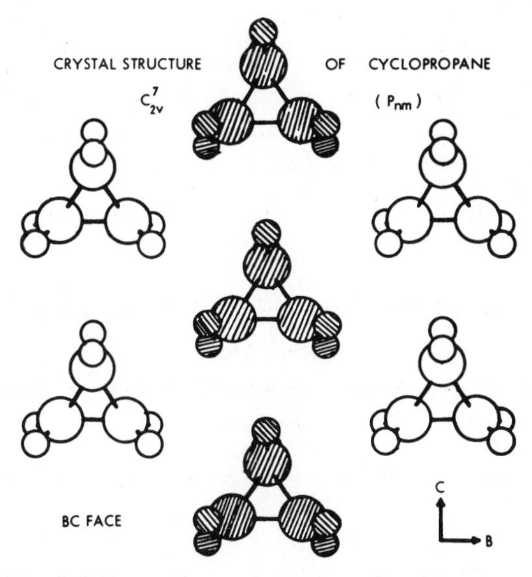

Fig. 2-6. Proposed crystal structure of cyclopropane. The shaded molecules are not in the same plane as the unshaded ones and are inclined oppositely. [From S. S. Mitra, *Optical Properties of Solids*, S. Nudelman and S. S. Mitra (eds.), Plenum Press, New York, (1969). Courtesy of Plenum Press.]

found to be $C_s(\sigma_{yz})$. Tables 2-19A and 2-19B show the correlation charts for cyclopropane. A total of $3n \cdot Z' = 3 \cdot 9 \cdot 2 = 54$ modes are expected, of which $(3n - 6)Z' = 42$ are internal modes.

Internal Modes for C_3H_6:

$$\Gamma_{n_i'} = 12A_1 + 9A_2 + 9B_1 + 12B_2$$

External Modes for C_3H_6:

$$\begin{aligned} \Gamma_{T+T'} &= A_2 + B_1 + 2A_1 + 2B_2 \\ \Gamma_T &= A_1 + B_1 + B_2 \\ \Gamma_{T'} &= A_2 + A_1 + B_2 \\ \Gamma_{R'} &= A_1 + 2B_1 + 2A_2 + B_2 \end{aligned}$$

Summary for C_3H_6

A total of $3nZ' = 3 \cdot 9 \cdot 2 = 54$ modes are expected:

$$\begin{array}{lcccccccc} \Gamma_{n_i} & = & 12A_1 & + & 9A_2 & + & 9B_1 & + & 12B_2 \\ \Gamma_T & = & A_1 & + & & & B_1 & + & B_2 \\ \Gamma_{T'} & = & A_1 & + & A_2 & + & & & B_2 \\ \Gamma_{R'} & = & A_1 & + & 2A_2 & + & 2B_1 & + & B_2 \\ \hline \Gamma_{n_i} & = & 15A_1 & + & 12A_2 & + & 12B_1 & + & 15B_2 \\ \text{Activity} & & (\text{IR, R}) & & (\text{R}) & & (\text{IR, R}) & & (\text{IR, R}) \end{array}$$

Table 2-19A. Correlation Table for Cyclopropane

Internal Modes

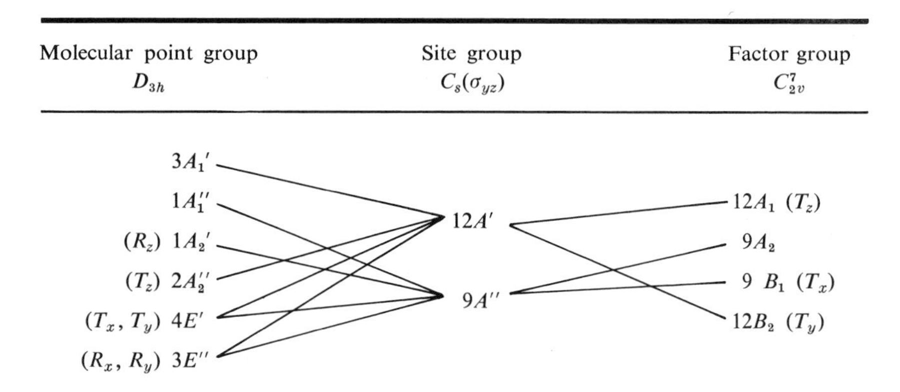

Table 2-19B. Correlation Table for Cyclopropane External Modes

Degrees of freedom T	Degrees of freedom R	Site group $C_s(\sigma_{yz})$	Factor group C_{2v}^7	Modes T	Modes R
4	2	$(T_x, T_y)(R_z)$ A'	A_1 (T_z)	2	1
			A_2 (R_z)	1	2
2	4	$(T_z)(R_x, R_y)$ A''	B_1 $(T_x)(R_y)$	1	2
			B_2 $(T_y)(R_x)$	2	1

Of these, 39 are infrared active and 51 are Raman active, and $A_1 + B_1 + B_2$ are acoustic modes and not observed.

Example 4

Ammonium Nitrate, NH_4NO_3 (Orthorhombic)

Ammonium nitrate at room temperature is found to have the space group D_{2h}^{13}–*Pmmn* (No. 59) with two molecules per unit cell.[16] The crystallographic cell is equal to the Bravais unit cell, since this is a primitive space group. First, one can consider the nitrate ion in NH_4NO_3. The molecular point group for the NO_3^- is D_{3h}. Consulting Appendix 3, it can be observed that for a space group of D_{2h}^{13} the following site symmetries are possible:

$$2C_{2v}(2);\quad 2C_i(4);\quad 2C_s(4);\quad C_1(8)$$

From Appendix 4 it can be found that for the D_{3h} group the following subgroups are possible:

$$C_{3h};\quad D_3;\quad C_{3v};\quad C_{2v};\quad C_3;\quad C_2;\quad C_s$$

Both the C_s and the C_i site groups can be eliminated since four equivalent sites per set are available and only two are needed ($Z' = 2$). Only the C_{2v} site group with two molecules per unit cell is a subgroup for both the D_{2h}^{13} and the D_{3h} groups, and this is selected as the site group for the NO_3^-. Referring to Table 2-14, it can be determined that the C_{2v} site involves the $C_2(z)$ axis (Wyckoff notation for site of N atoms is $2a$, and the NO_3^- can be considered to sit on a $2a$ site as well). Having determined the site group for the NO_3^-, the correlation table for the NO_3^- can be formulated. Table 2-20 illustrates the correlation table for the NO_3^- in a molecular point group D_{3h}, a site group $C_{2v}[C_2(z)]$, and the factor group D_{2h}^{13}, and lists the number of internal modes.

For a factor group of D_{2h}^{13} and $Z' = 2$, it can be shown that the NO_3^- should have $(3n - 6)Z' = 12$ internal modes (since $n = 4$ and $Z' = 2$). The six internal modes for the molecular point group D_{3h} consist of the species

$$\Gamma_{n'} = 2E' + A_1' + A_2''$$

This can be determined from the procedures described previously. It can be shown that the six internal modes of the molecular point group correlate with the 12 internal modes of the factor group, and consist of the following species:

$$\Gamma_{n_i'} = 3A_g + 3B_{1u} + 2B_{3g} + 2B_{2u} + B_{2g} + B_{3u}$$

Since a total of $3nZ' = 24$ total modes ($n = 4$, $Z' = 2$) are expected, 12 of these must be external modes. Following the procedures given in the previous section and tracing the T vectors from site group to space group in Table 2-21, we can show that

$$\Gamma_{T'} = A_g + B_{2u} + B_{2g} + B_{3u} + B_{3g} + B_{1u}$$

Tracing the R vectors, we obtain

$$\Gamma_{R'} = B_{1g} + B_{2u} + B_{2g} + A_u + B_{3g} + B_{3u}$$

Table 2-20. Correlation Table for the NO_3^- in NH_4NO_3 Internal Modes

Molecular point group D_{3h}	Site group $C_{2v}[C_2(z)]$	Factor group D_{2h}^{13}
$1A_1'$	$3A_1$	$3A_g$
		$1B_{2g}$
		$2B_{3g}$
(T_x, T_y) $2E'$	$2B_2$	$3B_{1u}$ (T_z)
		$2B_{2u}$ (T_y)
(T_z) $1A_2''$	$1B_1$	$1B_{3u}$ (T_x)

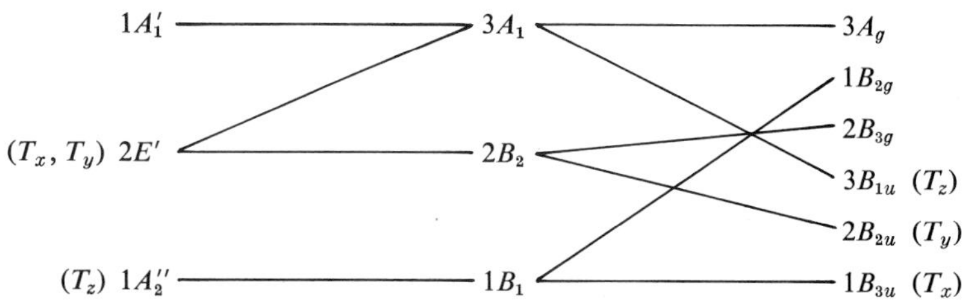

Table 2-21. Correlation Table for NO_3^- in NH_4NO_3

External Modes

Degrees of freedom		Site group $C_{2v}[C_2(z)]$	Factor group D_{2h}^{13}	Modes	
T	R			T	R
			A_g	1	0
2	0	(T_z) A_1	B_{1g} (R_z)	0	1
			B_{2g} (R_y)	1	1
2	2	$(T_y,\ R_x)$ B_2	B_{3g} (R_x)	1	1
			A_u	0	1
0	2	(R_z) A_2	B_{1u} (T_z)	1	0
			B_{2u} (T_y)	1	1
2	2	$(T_x,\ R_y)$ B_1	B_{3u} (T_x)	1	1

We have determined that 12 external modes are possible, belonging to the above species.

Let us next consider the NH_4^+, which is in a T_d molecular point group. The NH_4^+ is also on a site group where the C_2 axis is on the z axis (Wyckoff notation for the NH_4^+ site is $2b$). By consulting Appendices 3 and 4, the correlations for the species for an NH_4^+ which lies on a $C_{2v}[C_2(z)]$ site group can be found. The correlation table can then be constructed (Table 2-22). It can be determined that $(3n - 6)Z' = 18$ internal modes and $3nZ' = 30$ total modes corresponding to the NH_4^+ are allowed. The 18 internal modes relating to the NH_4^+ can be determined as was done previously for the NO_3^-. The number of each species which will be allowed in the molecular point group can be determined by the procedures outlined earlier in this chapter. Thus 12 external modes must be allowed. Since we did not determine the three acoustic modes when determining the external modes for the NO_3^-, we must do so for the NH_4^+. It does not matter in the case of ionic crystals which ion is used to determine the acoustic modes. As a result, the only T vectors in the space group D_{2h}^{13} correspond to the species B_{1u}, B_{2u}, and B_{3u}. Consequently,

$$\Gamma_T = B_{1u} + B_{2u} + B_{3u}$$

Tracing the T vectors across the correlation table for NH_4^+, we can find the

Table 2-22. Correlation Table for the NH_4^+ in NH_4NO_3 Internal Modes

Molecular point group T_d	Site group $C_{2v}[C_2(z)]$	Factor group D_{2h}^{13}
		$4A_g$
	$4A_1$	$1B_{1g}$
$1A_1$		$2B_{2g}$
	$1A_2$	$2B_{3g}$
$1E$		$1A_u$
	$2B_1$	$4B_{1u}$ (T_z)
$2F_2$		$2B_{2u}$ (T_y)
	$2B_2$	$2B_{3u}$ (T_x)

total number of translations (optical and acoustic)

$$\Gamma_{T+T'} = B_{2g} + B_{3u} + A_g + B_{1u} + B_{3g} + B_{2u}$$

and for the lattice modes

$$\Gamma_{T'} = \Gamma_{T+T'} - \Gamma_T = A_g + B_{2g} + B_{3g}$$

For the rotational lattice modes (and tracing the R vectors across the table)

$$\Gamma_{R'} = B_{1g} + A_u + B_{2g} + B_{3u} + B_{3g} + B_{2u}$$

A total of 12 external modes relating to the NH_4^+ are found (see Table 2-23).

Summary for NH_4NO_3

For NH_4NO_3 a total of $3nZ' = 3 \cdot 9 \cdot 2 = 54$ modes are expected.

NH_4^+:

$$\begin{aligned}
\Gamma_{n_i'} &= 4A_g + 4B_{1u} + 2B_{3g} + 2B_{2u} + 2B_{2g} + 2B_{3u} + B_{1g} + A_u \\
\Gamma_T &= \quad + B_{1u} \quad + B_{2u} \quad + B_{3u} \\
\Gamma_{T'} &= A_g \quad + B_{3g} \quad + B_{2g} \\
\Gamma_{R'} &= \quad + B_{3g} + B_{2u} + B_{2g} + B_{3u} + B_{1g} + A_u
\end{aligned}$$

NO_3^-:

$$\Gamma_{n_i'} = 3A_g + 3B_{1u} + 2B_{3g} + 2B_{2u} + B_{2g} + B_{3u}$$
$$\Gamma_{T'} = A_g + B_{1u} + B_{3g} + B_{2u} + B_{2g} + B_{3u}$$
$$\Gamma_{R'} = \qquad + B_{3g} + B_{2u} + B_{2g} + B_{3u} + B_{1g} + A_u$$

$$\Gamma_{n_i} = 9A_g + 9B_{1u} + 8B_{3g} + 8B_{2u} + 7B_{2g} + 7B_{3u} + 3B_{1g} + 3A_u$$

A total of 54 modes of the above species are accounted for as expected.

The internal modes consist of the following 30 species:

$$\Gamma_{n_i'} = 7A_{1g} + 7B_{1u} + 4B_{3g} + 4B_{2u} + 3B_{2g} + 3B_{3u} + B_{1g} + A_u$$

and the external modes are of the following 24 species:

$$\Gamma_{T+T'+R} = 2A_{1g} + 2B_{1u} + 4B_{3g} + 4B_{2u} + 4B_{2g} + 4B_{3u} + 2B_{1g} + 2A_u$$

All g modes are Raman active and all u modes are infrared active, with the exception of the A_u mode, which is inactive. Thus, for the internal modes, 14 are infrared active and 15 are Raman active. For the external modes, 7 are infrared active and 12 are Raman active. The acoustic modes B_{1u}, B_{2u}, and B_{3u} are not observed.

Table 2-23. Correlation Table for NH_4^+ in NH_4NO_3

External Modes

Degrees of freedom		Site group $C_{2v}[C_2(z)]$	Factor group D_{2h}^{13}	Modes	
T	R			T	R
			A_g	1	0
2	0	(T_z) A_1	B_{1g} (R_z)	0	1
			B_{2g} (R_y)	1	1
0	2	(R_z) A_2	B_{3g} (R_x)	1	1
			A_u	0	1
2	2	(T_x, R_y) B_1	B_{1u} (T_z)	1	0
			B_{2u} (T_y)	1	1
2	2	(T_y, R_x) B_2	B_{3u} (T_x)	1	1

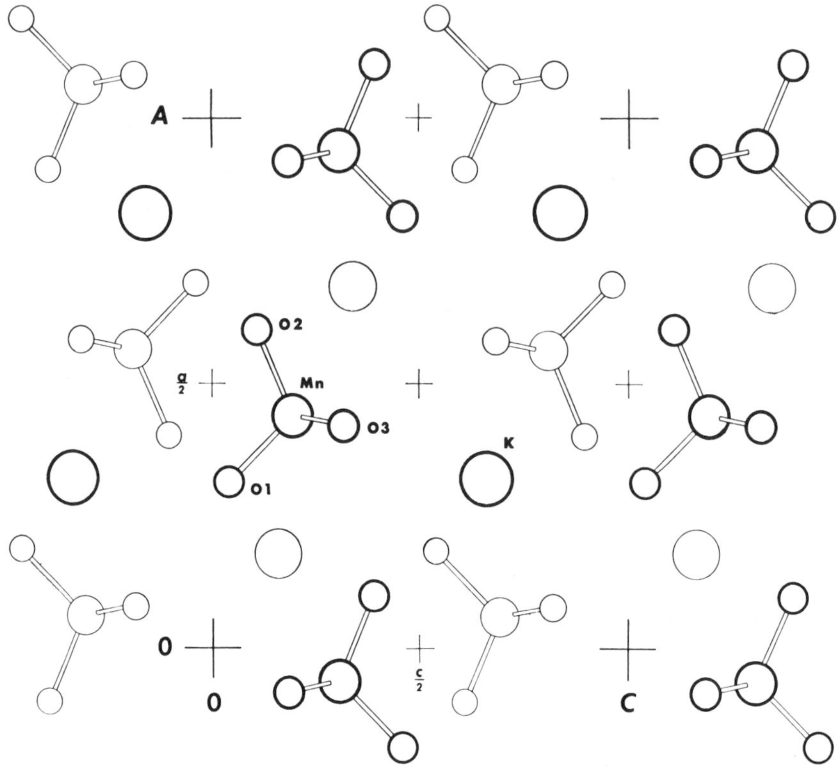

Fig. 2-7. Structure of $KMnO_4$. A(010) projection, illustrating the ionic packing. [From G. J. Palenik, *Inorg. Chem.* **6**:503 (1967). Courtesy of Inorganic Chemistry.]

Example 5

$KMnO_4$ (Orthorhombic)

Potassium permanganate $KMnO_4$, has a space group of D_{2h}^{16}–*Pnma* (No. 62), with $Z' = 4$, and the MnO_4^- has a molecular point group of T_d.[17] Figure 2-7 shows the structure of $KMnO_4$. C_s is a subgroup of both T_d and D_{2h}^{16}. The selection of the proper C_s site for the MnO_4^- and the K^+ ions is facilitated by use of Table 2-14 and Wyckoff's tables.[12] For MnO_4^- in $KMnO_4$, $3nZ' = 3 \cdot 5 \cdot 4 = 60$ total modes are expected and of these $(3n - 6)Z' = 36$ will be internal modes. Thus, 24 external modes are predicted. The correlation charts for MnO_4^- are shown in Tables 2-24A and 2-24B. Table 2-25 shows the correlation chart for the K^+ in $KMnO_4$. We obtain the following results.

Internal Modes for MnO_4^-:

$$\Gamma_{n_i'} = 6A_g + 6B_{2g} + 6B_{1u} + 6B_{3u} + 3B_{1g} + 3B_{3g} + 3A_u + 3B_{2u}$$

External Modes for MnO_4^-:

$$\Gamma_{T'} = 2A_g + 2B_{2g} + 2B_{1u} + 2B_{3u} + B_{1g} + B_{3g} + A_u + B_{2u}$$

$$\Gamma_{R'} = A_g + B_{2g} + B_{1u} + B_{3u} + 2B_{1g} + 2B_{3g} + 2A_u + 2B_{2u}$$

The coefficients for the A' species correlating with $A_g + B_{2g} + B_{1u} + B_{2u}$ for the translations are equal to two since the total number of degrees of freedom equals eight. Similarly, for the A'' species correlating with $B_{1g} + B_{3g} + A_u + B_{3u}$ for the rotations the coefficients are also equal to two.

The K^+ can be considered also to lie on a C_s site and the correlation table can be prepared as shown in Table 2-25. For the K^+ one should expect $3nZ' = 3 \cdot 1 \cdot 4 = 12$ external modes. Using the previous procedures, the following results are obtained.

External Modes for K^+:

$$\Gamma_T = B_{1u} + B_{2u} + B_{3u}$$

$$\Gamma_{T+T'} = 2A_g + 2B_{2g} + 2B_{1u} + 2B_{3u} + B_{1g} + B_{3g} + A_u + B_{2u}$$

$$\Gamma_{T'} = \Gamma_{T+T'} - \Gamma_T = 2A_g + 2B_{2g} + B_{1g} + B_{3g} + A_u + B_{1u} + B_{3u}$$

Thus a total of 12 external modes are accounted for.

Table 2-24A. Correlation Table for MnO_4^- in $KMnO_4$

Internal Modes

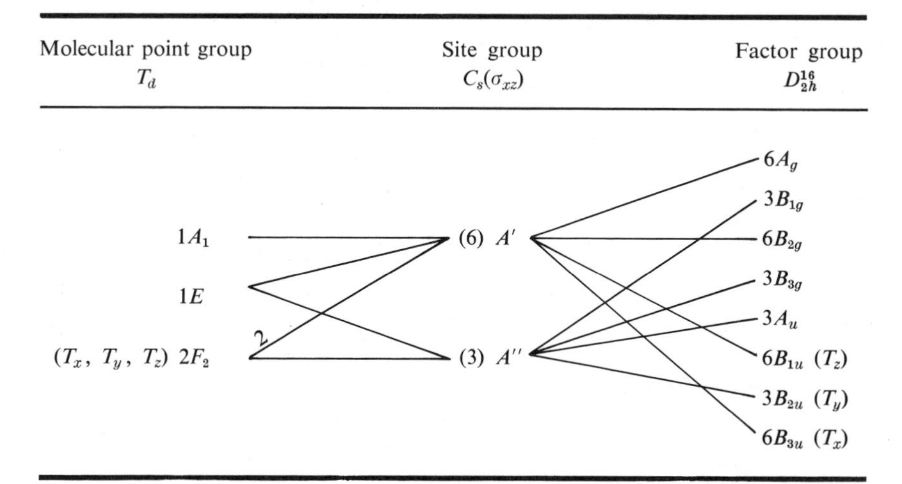

Table 2-24B. Correlation Table for MnO_4^- in $KMnO_4$

External Modes

Degrees of freedom		Site group $C_s(\sigma_{xz})$	Factor group D_{2h}^{16}	Modes	
T	R			T	R
			A_g	2	1
			B_{1g} (R_z)	1	2
8	4	$(T_x,\ T_y)(R_z)$ A'	B_{2g} (R_y)	2	1
			B_{3g} (R_x)	1	2
			A_u	1	2
4	8	$(T_z)(R_x,\ R_y)$ A''	B_{1u} (T_z)	2	1
			B_{2u} (T_y)	1	2
			B_{3u} (T_x)	2	1

Table 2-25. Correlation Table for K^+ in $KMnO_4$

Degrees of freedom		Site group $C_s(\sigma_{xz})$	Factor group D_{2h}^{16}	Modes	
T	R			T	R
			A_g	2	0
			B_{1g} (R_z)	1	0
8	0	$(T_x,\ T_y,\ R_z)$ A'	B_{2g} (R_y)	2	0
			B_{3g} (R_x)	1	0
			A_u	1	0
4	0	$(T_z,\ R_x,\ R_y)$ A''	B_{1u} (T_z)	2	0
			B_{2u} (T_y)	1	0
			B_{3u} (T_x)	2	0

Summary for $KMnO_4$

For $KMnO_4$ a total of $3nZ' = 3 \cdot 6 \cdot 4 = 72$ modes are expected.

MnO_4^-:

$$\Gamma_{n_i'} = 6A_g + 6B_{2g} + 6B_{1u} + 6B_{3u} + 3B_{1g} + 3B_{3g} + 3A_u + 3B_{2u}$$
$$\Gamma_{T'} = 2A_g + 2B_{2g} + 2B_{1u} + 2B_{3u} + B_{1g} + B_{3g} + A_u + B_{2u}$$
$$\Gamma_{R'} = A_g + B_{2g} + B_{1u} + B_{3u} + 2B_{1g} + 2B_{3g} + 2A_u + 2B_{2u}$$

K^+:

$$\Gamma_T = + B_{1u} + B_{3u} + B_{2u}$$
$$\Gamma_{T'} = 2A_g + 2B_{2g} + B_{1u} + B_{3u} + B_{1g} + B_{3g} + A_u$$

$$\Gamma_{n_i} = 11A_g + 11B_{2g} + 11B_{1u} + 11B_{3u} + 7B_{1g} + 7B_{3g} + 7A_u + 7B_{2u}$$

Of the 72 modes expected, the $7A_u$ modes are inactive. Therefore 36 modes are Raman active and 26 are infrared active, with the three acoustic modes $(B_{1u} + B_{2u} + B_{3u})$ not observed.

Example 6

PH_4I (Tetragonal)

Phosphonium iodide, PH_4I, crystallizes in tetragonal crystals of D_{4h}^7–P_4/nmm space group (No. 129), with $Z' = 2$.[18] Figure 2-8 shows the unit cell of PH_4I. The PH_4^+ can be considered to be in a T_d point group. Only the D_{2d} site group is a subgroup of both the T_d and D_{4h}^7 groups. The $2P$ atoms are located on a D_{2d} site from Wyckoff's tables[12] (site $2a$). The central atom of a polyatomic ion will generally determine the configuration of the ion. Therefore we can infer that PH_4^+ is on a D_{2d} site as well. Examining the correlation tables in Appendix 4, it can be observed that two D_{2d} sites are possible. For a site $2a$ (from Wyckoff's tables) and from Table 2-14 for space group 129, the axis of interest is C_2''. Tables 2-26A and 2-26B show the correlation tables for the PH_4^+ for the internal and external modes, respectively. For PH_4^+ in PH_4I $3nZ' = 3 \cdot 5 \cdot 2 = 30$ total modes are possible. Of these, 18 are internal modes, and 12 external modes should be found. Using the aforementioned procedures, the following results are obtained.

Internal Modes for PH_4^+:

$$\Gamma_{n_i'} = 2A_{1g} + 2B_{2u} + B_{2g} + A_{1u} + 2E_g + 2E_u + 2B_{1g} + 2A_{2u}$$

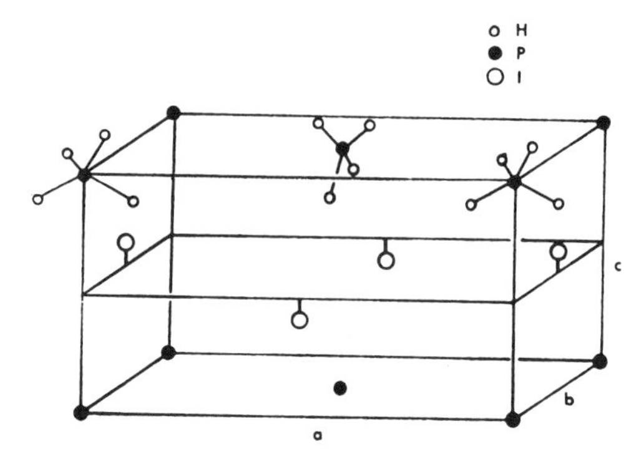

Fig. 2-8. Unit cell of PH_4I. The dimensions a and c have been expanded for reasons of clarity: $a = b \neq c$. [From J. R. Durig, D. J. Antion, and F. G. Baglin, *J. Chem. Phys.* **49**:671 (1968). Courtesy of American Institute of Physics.].

Table 2-26A. Correlation Table for the PH_4^+ in PH_4I Internal Modes

Molecular point group T_d	Site group $D_{2d}(C_2'')$	Factor group D_{4h}^7
		$2A_{1g}$
$1A_1$	$2A_1$	
		$2B_{2u}$
		$1B_{2g}$
$0A_2$	$1B_1$	
		$1A_{1u}$
		$0A_{2g}$
$1E$	$0A_2$	
		$0B_{1u}$
		$2E_g$
$(R_x, R_y, R_z)\ 0F_1$	$2E$	
		$2E_u\ (T_x, T_y)$
		$2B_{1g}$
$(T_x, T_y, T_z)\ 2F_2$	$2B_2$	
		$2A_{2u}\ (T_z)$

Table 2-26B. Correlation Table for PH_4^+ in PH_4I

External Modes

Degrees of freedom		Site group $D_{2d}(C_2'')$	Factor group D_{4h}^7	Modes	
T	R			T	R
0	0	A_1	A_{1g}	0	0
			B_{2u}	0	0
0	0	B_1	B_{2g}	0	0
			A_{1u}	0	0
0	2	(R_z) A_2	A_{2g} (R_z)	0	1
			B_{1u}	0	1
4	4	$(T_x, T_y)(R_x, R_y)$ E	E_g (R_x, R_y)	1	1
			E_u (T_x, T_y)	1	1
2	0	(T_z) B_2	B_{1g}	1	0
			A_{2u} (T_z)	1	0

External Modes for PH_4^+:

$$\Gamma_{T'} = E_g + E_u + B_{1g} + A_{2u}$$

$$\Gamma_{R'} = A_{2g} + B_{1u} + E_g + E_u$$

A total of 12 external modes are found.

The iodide ion can be considered to be on a C_{4v} site (see Wyckoff[12]). This is the other site possible for a D_{4h}^7 space group. The correlation table can be formulated as in Table 2-27.

A total of $3nZ' = 3 \cdot 1 \cdot 2 = 6$ modes should be accounted for and these will all be external translation modes.

External Modes for I^-:

$$\Gamma_T = A_{2u} + E_u$$

$$\Gamma_{T+T'} = A_{1g} + A_{2u} + E_g + E_u$$

$$\Gamma_{T'} = \Gamma_{T+T'} - \Gamma_T = A_{1g} + E_g$$

A total of six external translational modes are accounted for. No rotations are possible with a monatomic cation.

Table 2-27. Correlation Table for I^- in PH_4I
External Modes

Degrees of freedom		Site group	Factor group	Modes	
T	R	C_{4v}	D_{4h}^{7}	T	R
2	0	$(T_z)\ A_1$	A_{1g}	1	0
			B_{1u}	0	0
0	0	$(R_z)\ A_2$	B_{1g}	0	0
			A_{1u}	0	0
0	0	B_1	$A_{2g}\ (R_z)$	0	0
			B_{2u}	0	0
0	0	B_2	$E_g\ (R_x, R_y)$	1	0
			$E_u\ (T_x, T_y)$	1	0
4	0	$(T_x, T_y, R_x, R_y)\ E$	B_{2g}	0	0
			$A_{2u}\ (T_z)$	1	0

Summary for PH_4I

For PH_4I a total of $3nZ' = 3 \cdot 6 \cdot 2 = 36$ modes are expected.

PH_4^+:

$$\begin{aligned}
\Gamma_{n_i'} &= 2A_{1g} + 2B_{2u} + B_{2g} + A_{1u} + 2E_g + 2E_u + 2B_{1g} + 2A_{2u} \\
\Gamma_{T'} &= E_g + E_u + B_{1g} + A_{2u} \\
\Gamma_{R'} &= A_{2g} + B_{1u} + E_g + E_u
\end{aligned}$$

I^-:

$$\begin{aligned}
\Gamma_{T'} &= A_{1g} + E_g \\
\Gamma_{T} &= E_u + A_{2u}
\end{aligned}$$

$$\Gamma_{n_i} = 3A_{1g} + 2B_{2u} + B_{2g} + A_{1u} + A_{2g} + B_{1u} + 5E_g + 5E_u + 3B_{1g} + 4A_{2u}$$

Of the 36 modes expected, 18 are Raman active and 11 infrared active, with the A_{1u}, B_{1u}, and B_{2u} modes inactive, the three acoustic modes $(E_u + A_{2u})$ not being observed.

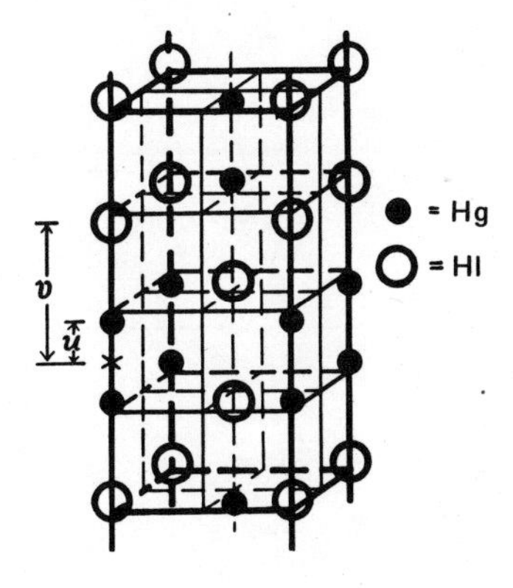

Fig. 2-9. Structure of mercurous halide. [From R. J. Havighurst, *J. Amer. Chem. Soc.* **48**:2113 (1926). Courtesy of *J. Amer. Chem. Soc.*]

Example 7

Hg_2Cl_2 (Tetragonal)

One of the possible space groups for Hg_2Cl_2 is D_{4h}^{17}–$I4/mmm$ (No. 139), with $Z' = 1$, with a molecular point group of $D_{\infty h}$ and a site group of D_{4h}.[19] Figure 2-9 shows the structure of Hg_2Cl_2. The correlation tables for Hg_2Cl_2 are shown in Tables 2-28A and 2-28B. This example is illustrative of a linear molecule and is handled very easily by the HH method.

A total of $3nZ' = 3 \cdot 4 \cdot 1 = 12$ total modes should be allowed for Hg_2Cl_2. A total of $(3n - 5)Z' = (3 \cdot 4 - 5)1 = 7$ internal modes are allowed and five external modes. In tracing the R vectors for the linear case, only those in the x and y directions are considered, since rotation along the z axis has a zero moment of inertia.

Table 2-28A. Correlation Table for Hg_2Cl_2

Internal Modes

Molecular point group $D_{\infty h}$	Site group D_{4h}	Factor group D_{4h}^{17}
$2A_{1g}(\Sigma_g^+)$	$2A_{1g}$	$2A_{1g}$
(T_z) $1A_{2u}(\Sigma_u^+)$	$1A_{2u}$	$1A_{2u}$ (T_z)
(R_x, R_y) $1E_g(\pi_g)$	$1E_g$	$1E_g$
(T_x, T_y) $1E_u(\pi_u)$	$1E_u$	$1E_u$ (T_x, T_y)

Table 2-28B. Correlation Table for Hg_2Cl_2
External Modes

Degrees of freedom		Site group		Factor group	Modes	
T	R	D_{4h}		D_{4h}^7	T	R
0	0	A_{1g}	———	A_{1g}	0	0
0	0	(R_z) A_{2g}	———	A_{2g} (R_z)	0	0
1	0	(T_z) A_{2u}	———	A_{2u} (T_z)	1	0
0	2	(R_x, R_y) E_g	———	E_g (R_x, R_y)	0	1
2	0	(T_x, T_y) E_u	———	E_u (T_x, T_y)	1	0

Internal Modes for Hg_2Cl_2:

$$\Gamma_{n_i'} = 2A_{1g} + A_{2u} + E_g + E_u$$

External Modes for Hg_2Cl_2:

$$\Gamma_T = A_{2u} + E_u$$

$$\Gamma_{T+T'} = A_{2u} + E_u$$

$$\Gamma_{T'} = 0$$

$$\Gamma_{R'} = E_g$$

Summary for Hg_2Cl_2:

$$\begin{aligned} \Gamma_{n_i'} &= 2A_{1g} + A_{2u} + E_g + E_u \\ \Gamma_T &= \quad + A_{2u} \quad + E_u \\ \Gamma_{R'} &= \quad + E_g \\ \hline \Gamma_{n_i} &= 2A_{1g} + 2A_{2u} + 2E_g + 2E_u \end{aligned}$$

Six Raman-active modes and three infrared-active modes are to be expected, with A_{2u} and E_u the acoustic modes.

Example 8

$LaCl_3$ (Hexagonal)

Lanthanum chloride crystallizes in a C_{6h}^2–$P6_3/m$ space group (No. 176), with $Z' = 2$. The lanthanum ions can be considered to be on a C_{3h} site and

Table 2-29. Correlation Table for La^{3+} in $LaCl_3$

Degrees of freedom		Site group	Factor group	Modes	
T	R	C_{3h}	C_{6h}^2	T	R
0	0	A'	A_g (R_z)	0	0
			B_u	0	0
4	0	(T_x, T_y) E'	E_{2g}	1	0
			E_{1u} (T_x, T_y)	1	0
2	0	(T_z) A''	A_u (T_z)	1	0
			B_g	1	0
0	0	E''	E_{2u}	0	0
			E_{1g} (R_x, R_y)	0	0

the chlorines on C_s sites.[20] The correlation table for the La^{3+} is given in Table 2-29 and that for the Cl^- in Table 2-30.

A total of $3nZ' = 3 \cdot 4 \cdot 2 = 24$ total modes should be allowed for $LaCl_3$, and all of these can be considered as external modes.

Table 2-30. Correlation Table for Cl^- in $LaCl_3$

Degrees of freedom		Site group	Factor group	Modes	
T	R	C_s	C_{6h}^2	T	R
12	0	(T_x, T_y) A'	A_g (R_z)	2	0
			B_u	2	0
			E_{2g}	2	0
			E_{1u} (T_x, T_y)	2	0
6	0	(T_z) A''	A_u (T_z)	1	0
			B_g	1	0
			E_{2u}	1	0
			E_{1g} (R_x, R_y)	1	0

Modes for LaCl₃:

For La^{3+}: $\Gamma_{T'} = E_{2g} + E_{1u} + A_u + B_g$

For $6Cl^-$: $\Gamma_{T'+T} = 2A_g + 2B_u + 2E_{2g} + 2E_{1u} + A_u + B_g + E_{2u} + E_{1g}$

$\Gamma_T = A_u + E_{1u}$

$\Gamma_{T'} = 2A_g + 2B_u + 2E_g + E_{1u} + B_g + E_{2u} + E_{1g}$

Summary for LaCl₃

A total of 24 modes are allowed.

La^{3+}:

$$\Gamma_{T'} = E_{2g} + E_{1u} + A_u + B_g$$

$6Cl^-$:

$$\Gamma_T = \quad + E_{1u} + A_u$$

$$\Gamma_{T'} = 2E_g + E_{1u} \quad + B_g + 2A_g + 2B_u + E_{2u} + E_{1g}$$

$$\Gamma_{n_i} = 3E_g + 3E_{1u} + 2A_u + 2B_g + 2A_g + 2B_u + E_{2u} + E_{1g}$$

The B_g, B_u, and E_{2u} modes are inactive. There will be ten Raman-active modes and five infrared-active modes ($E_{1u} + A_u$ are acoustic modes).

Tables for factor group or point group analyses have been prepared by Adams and Newton.[21,22] One can read the number and type of species allowed directly from the tables. Although useful, this approach does not provide the reader with the procedures needed to derive the tables.

PROBLEMS

1. Derive the selection rules for the fundamentals, combinations, and overtones for the D_{2h}, D_{3h}, D_{4h}, D_{5h}, C_{2v}, C_{3v}, C_{4v}, D_{2d}, and O_h point groups.

2. For the gaseous molecule $(AlCl_3)_2$, D_{2h} symmetry, determine the following:
   a) Indicate the total number of fundamentals and to which species they belong.
   b) Indicate infrared-active vibrations.
   c) Indicate Raman-active vibrations.
   d) Number of polarized bands.

3. Repeat Problem 2 with $N_2O_4(g)$, D_{2h} symmetry.

4. Repeat Problem 2 with $PCl_5(g)$, D_{3h} symmetry.

5. Repeat Problem 2 with $[SbCl_5]^{2-}$, C_{4v} symmetry.

6. Determine the number of fundamentals (which species) and infrared-active and Raman-active modes for C_{3v}, C_{2v}, and C_s symmetry for the NSF_3 molecule.

7. Verify the classification of the normal modes for diacetylene. Also determine the selection rules for Raman displacements, etc.

8. Derive the selection rules for monochlorodiacetylene.

9. Derive the selection rules for Hg $(C{\equiv}CCl)_2$ and HgCl $(C{\equiv}CCl)$.

10. Derive selection rules for: a) $N^{15}N^{14}O^{16}$; b) $N^{14}N^{15}O^{16}$; c) $N^{15}N^{15}O^{16}$; d) $N^{14}N^{14}O^{16}$.

11. Assume that your instrument provides sufficient resolution. Show how you would distinguish one species from the other on the basis of the results in Problem 10.

12. Determine the selection rules for solid CuF_2, space group C_{2h}^5–$P2_1/c$ (No. 14) with $Z' = 2$. Cu is on a $2a$ site and F on a $4e$ site.

13. Determine the selection rules for solid ZrS, space group D_{4h}^7–$P4/nmm$ (No. 129) and $Z' = 2$. Zr is on a $2c$ site, S on a $2a$ site.

14. Determine the selection rules for solid AlF_3, space group D_3^7–$R32$ (No. 155) with $Z' = 2$. Al^{3+} is on a $2c$ site, one F^- is on a $3d$ site, and two F^- are on a $3e$ site.

15. Determine the selection rules for solid CuO, space group C_{2h}^6–$C2/c$ (No. 15) and a crystallographic unit cell $Z = 4$. Cu^{2+} has a $4c$ site and O^{2-} a $4e$ site.

16. Determine the selection rules for solid paraelectric SbSI with a C_{2h}^2–$P2_1/m$ (No. 11) space group, $Z' = 2$, and all atoms on a $4e$ site. Upon cooling, the solid transforms into a ferroelectric material having a space group C_2^2–$P2_1$ (No. 4) with all atoms on a C_1 site and $Z' = 2$. Calculate the changes in selection rules that occur.

17. Determine the selection rules for solid K_2CrO_4, space group D_{2h}^{16}–$Pnma$ (No. 62) and $Z' = 4$. Consider the K^+ and CrO_4^- on a C_s site and the CrO_4 in a T_d molecular point group.

18. α-HgS crystallizes in a D_3^4–$P3_121$ (No. 152) space group with $Z' = 3$. The Hg atoms are on a $3a$ site and the sulfur atoms on a $3b$ site. Calculate the solid-state selection rules.

19. Cubic β-HgS crystallizes in a T_d^2–$F\bar{4}3/m$ (No. 216) space group with the crystallographic unit cell $Z = 4$. The Hg and S atoms are on a T_d site. Calculate selection rules.

20. Zircon ($ZrSiO_4$) crystallizes in a D_{4h}^{19}–$14_1/amd$ (No. 141) with the crystallographic unit cell equal to 4. The four Zr atoms are on a $4a$ site and the four SiO_4^{2-} ions can be considered to be on a $4b$ site since the four Si atoms are on a $4b$ site. Calculate the selection rules.

REFERENCES

1. A. G. Meister, F. F. Cleveland, and M. J. Murray, *Am. J. Phys.*, **11**:239 (1943).
2. J. E. Rosenthal and G. M. Murphy, *Rev. Mod. Phys.*, **8**:317 (1936).
3. S. M. Ferigle and A. G. Meister, *Am. J. Phys.*, **20**:421 (1952).
4. S. Bhagavantam and T. Venkatarayudu, *Proc. Ind. Acad. Sci.*, **A9**:224 (1939); **A13**:543 (1941).
5. S. S. Mitra, *Z. Kristallogr.*, **116S**:149 (1961).
6. R. S. Halford, *J. Chem. Phys.*, **14**:8 (1946).
7. H. Winston and R. S. Halford, *J. Chem. Phys.*, **17**:607 (1949).
8. D. F. Hornig, *J. Chem. Phys.*, **16**:1063 (1948).
9. M. H. Brooker and D. E. Irish, *Can. J. Chem.*, **48**:1183 (1970).
10. D. W. James and W. H. Leong, *J. Chem. Phys.*, **49**:5089 (1968).
11. I. Nakagawa and J. L. Walter, *J. Chem. Phys.*, **51**:1389 (1969).
12. R. W. C. Wyckoff, *Crystal Structures*, Vol. 2, Wiley–Interscience, New York (1964).
13. I. Nakagawa, *Coord. Chem. Rev.*, **4**:423 (1969).
14. V. C. Sinclair, J. M. Robertson, and A. Mathieson, *Acta Cryst.* **3**:251 (1950); S. S. Mitra, Air Force Materials Laboratory Report, AFML-TR-66-203, Wright-Patterson AFB, Ohio (1966).
15. S. S. Mitra, "Infrared and Raman Spectra Due to Lattice Vibrations," in *Optical Properties of Solids*, S. Nudelman and S. S. Mitra, Eds., Plenum Press. New York (1969), pp. 333–341.
16. J. W. Menary, *Acta Cryst.*, **8**:840 (1955).
17. G. J. Palenik, *Inorg. Chem.*, **6**:503 (1967).
18. R. G. Dickinson, *J. Am. Chem. Soc.*, **44**:1489 (1922).
19. J. R. Durig, K. K. Lau, G. Nagarajan, M. Walker, and J. Bragen, *J. Chem. Phys.*, **50**:2130 (1969); R. J. Havighurst, *J. Am. Chem. Soc.*, **48**:2113 (1926).
20. N. F. M. Henry and K. Lonsdale (eds.), *International Tables for X-Ray Crystallography*, Vol. I, The Kynoch Press, Birmingham, England (1965).
21. D. M. Adams and D. C. Newton, "Tables for Factor Group and Point Group Analysis," Beckman-RIIC Limited, Surley House, England, *J. Chem. Soc.*, **1970**:2822.
22. D. M. Adams, *Coord. Chem. Rev.* **10**:183 (1973).

RECENT CORRELATION BIBLIOGRAPHY

1. R. K. Khanna and C. W. Reimann, "Spectra-Physics," Raman Technical Bulletin No. 3 (1970).
2. W. G. Fateley, N. T. McDevitt, and F. F. Bentley, *Appl. Spectry.*, **25**:155 (1971).
3. W. G. Fateley, F. R. Dollish, N. T. McDevitt, and F. F. Bentley, *Infrared and Raman Selection Rules for Molecular and Lattice Vibrations: The Correlation Method*, Wiley–Interscience, New York (1972).
4. J. R. Ferraro and P. LaBonville, "Space Group Selection Rules," Argonne National Laboratory, Argonne, Illinois (1971).
5. R. Kopelman, *J. Chem. Phys.*, **47**:2631 (1967).
6. D. M. Adams and D. C. Newton, *J. Chem. Soc.*, **1970**:2822.
7. J. E. Bertie and J. W. Bell, *J. Chem. Phys.*, **54**:160 (1971).
8. B. DeAngelis, R. E. Newnham, and W. B. White, *Am. Mineralogist*, **57**:255 (1972).

Chapter 3

POTENTIAL FORCE FIELDS

3-1. INTRODUCTION

One of the most important problems in the interpretation of Raman and infrared spectra of a polyatomic molecule is the determination of the molecular potential function. Its importance lies in the fact that it furnishes information on the forces between the atoms in a molecule, and it serves as a starting point for making perturbation calculations that could lead to the finer details of the spectrum.

Any theoretical work in the interpretation or prediction of properties or reactions of matter has as a fundamental basis the *properties of molecules* and their interactions. The properties of a molecule depend upon the *masses* and *relative positions of* and *forces between* the atoms constituting the molecule. One of the most practical properties of a molecule for use in attempting to obtain a physical understanding of the molecule is the set of its fundamental vibrational frequencies. A vibration represents a departure from the equilibrium configuration of the molecule. This departure from equilibrium may be due to the stretching of a bond or the bending of the molecule due to a change in the angle between two bonds. The motion involved in a vibration is associated with energy: kinetic as the atoms move away from and return to their equilibrium positions, and potential as restoring forces come into action to bring the atoms back to their equilibrium positions. A basic assumption which we make about these displacements is that they are small. This permits one to use Taylor expansions in calculating change in functions related to them and to disregard cubic and quartic terms in the potential energy function.

The vibrational frequencies are dependent on the masses of the atoms and the restoring forces. The former are constants in the kinetic energy terms; the latter are related to the potential energy function. The coordinate system used to express the kinetic energy and potential energy functions must represent the geometry of the atoms in the molecule as well as meet certain symmetry requirements to simplify solution of the secular equation.

It is known that the force bringing about a displacement in a potential energy field V is the negative of the first derivative of the energy with respect to the displacement, i.e.,

$$F_i = -\left(\frac{\partial V}{\partial q_i}\right)_{q_1, q_{i-1}, \ldots, q_{i+1}, \ldots, q_n} \tag{3-1}$$

where $V = (V(q_1, q_2, \ldots))$ and $q_1, q_2, \ldots$ are displacements. The general term for the potential energy is, where q_i represents all types of displacements,

$$V = V_0 + \sum_i A_i q_i + \frac{1}{2!}\sum_{i,j} a_{ij} q_i q_j + \frac{1}{3!}\sum_{i,j,k} b_{ijk} q_i q_j q_k + \cdots \tag{3-2}$$

and the restoring force F is

$$\begin{aligned} F_i &= -\left(\frac{\partial V}{\partial q_i}\right)_{q_1, q_{i-1}, \ldots, q_{i+1}, \ldots, q_n} \\ &= -\sum_i A_i - \frac{1}{2!}\sum_{i,j} a_{ij} q_i - \frac{1}{3!}\sum_{i,j,k} b_{ijk} q_j q_k - \cdots \end{aligned} \tag{3-3}$$

Now we must examine the form which the potential energy function takes for the molecule at equilibrium (i.e., all $q_i \simeq 0$).

1. Define $V_0 = 0$.
2. $\partial V/\partial q_i = 0$ (minimum in potential energy curve) is a necessary condition; therefore $\sum A_i = 0$.
3. The potential energy has to satisfy the following sufficiency conditions in order to have a minimum at the equilibrium position $\mathbf{r}_0$:

$$V(\mathbf{r}_0 + \delta\mathbf{r}) - V(\mathbf{r}_0) > 0$$

and

$$V(\mathbf{r}_0 - \delta\mathbf{r}) - V(\mathbf{r}_0) > 0$$

where $\delta\mathbf{r}$ represents a displacement from the equilibrium position $\mathbf{r}_0$. More specifically

$$V(\mathbf{r}) - V(\mathbf{r}_0) > 0 \qquad \text{for} \quad \mathbf{r} \in N(\mathbf{r}_0) \quad \text{except } \mathbf{r}_0$$

where $\mathbf{r}$ is a position in the neighborhood $N(\mathbf{r}_0)$. As a result, the potential energy expression becomes

$$V = \frac{1}{2!}\sum_{i,j} a_{ij} q_i a_j + \frac{1}{3!}\sum_{i,j,k} b_{ijk} q_i q_j q_k + \frac{1}{4!}\sum_{i,j,k,l} c_{ijkl} q_i q_j q_k q_l + \cdots > 0 \tag{3-4}$$

where

$$a_{ij} = \left(\frac{\partial^2 V}{\partial q_i\, \partial q_j}\right)_0$$

$$b_{ijk} = \left(\frac{\partial^3 V}{\partial q_i\, \partial q_j\, \partial q_k}\right)_0$$

$$c_{ijkl} = \left(\frac{\partial^4 V}{\partial q_i\, \partial q_j\, \partial q_k\, \partial q_l}\right)_0$$

evaluated at $\mathbf{r}_0$.

Also,

(a) $\sum_{i,j} a_{ij} q_i q_j$ is the quadratic portion,

(b) $\sum_{i,j,k} b_{ijk} q_i q_j q_k$ is the cubic portion,

(c) $\sum_{i,j,k,l} c_{ijkl} q_i q_j q_k q_l$ is the quartic portion,

etc., of potential energy.

4. If the potential function is taken with terms in the second degree only, it becomes

$$V = \tfrac{1}{2} \sum_{i,j} a_{ij} q_i q_j \tag{3-5}$$

and since it has to be positive for all q_i's in the neighborhood of the equilibrium position, it is thus a positive-definite form. This latter requirement can be expressed in terms of the following relations among a_{ij}'s:

(a) $a_{ii} > 0$ for $i = 1, 2, \ldots, 3n - 6$.

(b) Two-rowed principal minors

$$\begin{vmatrix} a_{ii} a_{ij} \\ a_{ji} a_{jj} \end{vmatrix} > 0 \quad \text{for every } i \text{ and } j,\ i > j.$$

(c) Three-rowed principal minors

$$\begin{vmatrix} a_{ii} a_{ij} a_{ik} \\ a_{ji} a_{jj} a_{jk} \\ a_{ki} a_{kj} a_{kk} \end{vmatrix} > 0 \quad \text{for all } i, j, k \rightarrow i > j > k$$

and all n-rowed principal minors must be greater than zero up to and including

$$\det \begin{vmatrix} a_{11} \cdots a_{1(3n-6)} \\ \cdots \\ a_{3n-6,1} \cdots a_{3n-6,3n-6} \end{vmatrix} > 0$$

Thus this represents the lowest order of approximation of the potential function. This latter requirement that the potential function be a positive definite is only for the quadratic form. Since the potential function is quadratic in the displacement coordinates, then in this lowest order of approximation the molecule can be regarded as a set of noninteracting harmonic oscillators, $3n - 6$ for a polyatomic molecule and $3n - 5$ for a linear molecule, where n is the number of atoms in the molecule. This process of analysis corresponds to transforming to normal coordinate and normal frequencies. The molecular system can be examined using quantum mechanics but the results to this order of approximation are identical with the ones obtained from the use of classical techniques.

The expression for V now reduces to

$$V = \tfrac{1}{2} \sum_{ij} a_{ij} q_i q_j \tag{3-6}$$

which includes the squared term when $i = j$. Furthermore, this expression can be transformed to the sum of the squares on the basis that this molecular system, within the degree of approximation, can be transformed to an equivalent molecular system whose potential energy is given by

$$\tfrac{1}{2} \sum A_{ii} Q_i^2 \tag{3-7a}$$

and whose kinetic energy is given by

$$\tfrac{1}{2} \sum m_i \dot{Q}_i^2 \tag{3-7b}$$

and the restoring force F_i is given by the quadratic force constant

$$F_i = -(\partial V / \partial q_i)_0 = -2 \times \tfrac{1}{2} A_{ii} \phi_i = -A_{ii} \phi_i \tag{3-8}$$

Therefore the restoring force is proportional to the displacement and the proportionality constant A_{ii} is known as the quadratic force constant for these displacements. This is the condition for simple harmonic motion. For the diatomic molecule it is represented by (see Fig. 3-1)

$$f_i = \mathrm{FC}_i \text{ (force constant)} = (\partial^2 V / \partial q_i^2)_0$$

In general we now have the expressions for kinetic energy

$$T = \tfrac{1}{2} \sum_i m_i \dot{q}_i^{\,2} \tag{3-9}$$

and for potential energy V

$$V = \tfrac{1}{2} \sum_{i,j} a_{ij} q_i q_j \tag{3-10}$$

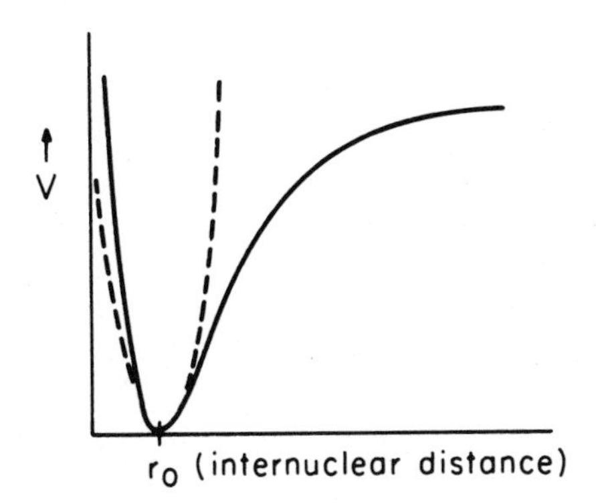

Fig. 3-1. Potential energy curve for a diatomic molecule at equilibrium. Solid line, actual potential. Dashed line, harmonic potential.

A coordinate system is required which can express the complex motions of a polyatomic molecule as superpositions of fundamental modes (normal coordinates). This selection of coordinates is a major consideration in solving force field problems.

3-2. OF WHAT VALUE TO CHEMISTRY IS THE SOLUTION OF FORCE FIELD PROBLEMS?

1. The potential energy expression for a diatomic molecule is a combination of an attractive term due to the valence electrons and a term representing repulsion due to the nuclei and to electrons localized at the nuclei, $V(r) = V^{\mathrm{R}}(r) - V^{\mathrm{A}}(r)$. (See Fig. 3-2.) Therefore, force constants, which are the coefficients in the potential energy expression, should afford insights with regard to bond order and delocalization of electrons.

As examples of the utility of the normal coordinate treatment, calculations of Shimanouchi[(1)] are cited in Table 3-1.

2. If we know the force constant, we can, in principle, calculate the vibrational frequency for the molecule as a harmonic oscillator and thus make assignments in the spectra of molecules.

3. One can calculate the structures of rotational isomers since the coupling of two skeletal deformation vibrations or of one skeletal deformation and one skeletal stretching vibration is very closely related to angle of internal rotation. Thus, one can calculate frequencies for various rotational isomers and compare them with the experimental spectra.

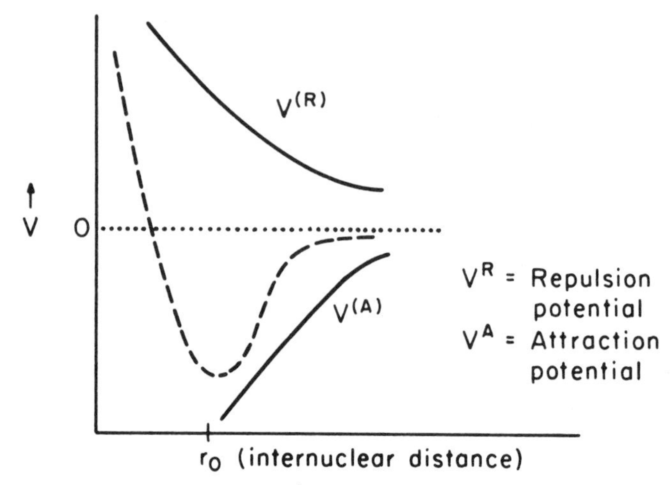

Fig. 3-2. The resultant potential energy curve for a diatomic molecule; V^R, repulsion potential; V^A, attraction potential.

4. Force constants have been used in calculations for *cis* and *trans* deuterated ethylene isomers to determine whether the *cis* or *trans* double bond has opened in polymerization.

5. Off-diagonal force constants, and potential constants for interaction between two different displacements, give information on forces between nonbonded atoms.

Table 3-1. LSFF and MUBFF Constants for the CC Stretching Vibrations[1,2]*

Stretching vibration	LSFF constant f_{CC}, mdyn/Å	MUBFF constant K_{CC}, mdyn/Å
C—C (CH_3—CH_3)	4.92	2.54
(CH_2=CH—CH=CH_2)	4.78	3.82
C—C (C_6H_6)	5.78	5.23
C=C (CH_2=CH_2)	8.56	7.33
(CH_2=CH—CH=CH_2)	8.10	6.64
C≡C (CH≡CH)	15.68	—

* LSFF = local symmetry force field; MUBFF = modified Urey–Bradley force field.

3-3. GENERAL PROCEDURE IN DETERMINING FREQUENCIES FROM FORCE CONSTANTS AND VICE VERSA

1. Choose a model for the molecule, e.g., the MX_4 tetrahedral structure.
2. Obtain the F matrix related to the potential energy of the molecule.

(a) Choose internal coordinates based on the model (r_k).

(b) Form linear combinations of these which transform as the symmetry operations of the molecule (R_j):

$$R_j = \sum_k U_{jk} r_k \tag{3-11}$$

where U_{jk} is the coefficient of the kth internal coordinate. In matrix notation, $R = Ur$. Symmetry coordinates must be orthonormal:

$$\sum_k (U_{jk})^2 = 1, \qquad \sum_k (U_{jk} U_{lk}) = 0 \tag{3-12}$$

where j and l refer to two different symmetry coordinates.

(c) Derive the f matrix, i.e., the matrix of force constants. Then the F matrix is calculated by

$$F = UfU' \tag{3-13}$$

where U' is the transpose of U.

3. Obtain the G matrix—the g matrix is similar to f; then

$$G = UgU' \tag{3-14}$$

The secular determinant is

$$\begin{vmatrix} \sum_i G_{1i}F_{i1} - \lambda & \sum_j G_{1j}F_{j2} & \cdots & \sum_k G_{1k}F_{kn} \\ \sum_i F_{2i}F_{i1} & \sum_j G_{2j}F_{j2} - \lambda & \cdots & \\ \sum_i G_{ni}F_{in} & & \cdots & \sum_k G_{nk}F_{kn} - \lambda \end{vmatrix} = 0 = | GF - \lambda E |$$

and since

$$\lambda = 4\pi^2 c^2 \nu^2 \tag{3-15}$$

the frequency ν (cm^{-1}) can be calculated.

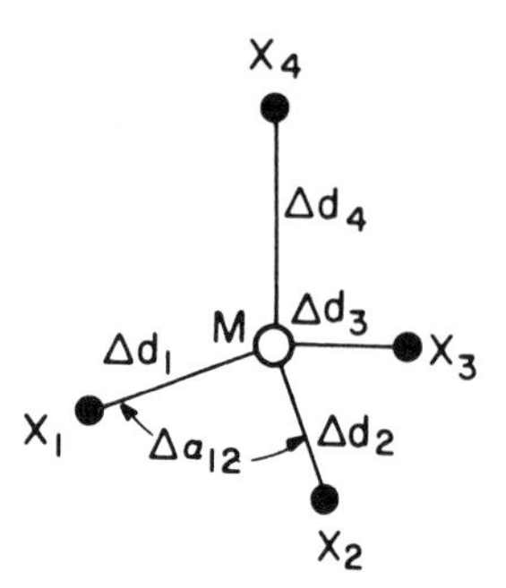

Fig. 3-3. Internal coordinates for MX_4 tetrahedral molecules.

If the molecule has symmetry elements, there will be some duplication in values of λ (degeneracy), and therefore the determinant can be factored into smaller determinants and hence into equations of lower degree. For a tetrahedral molecule MX_4 two first degree equations and two second degree equations are obtained.

Let us apply this method to the tetrahedral MX_4 molecule. The internal coordinates (see Fig. 3-3) are

changes in bond lengths: Δd_1, Δd_2, Δd_3, Δd_4

changes in bond angle: $\Delta\alpha_{12}$, $\Delta\alpha_{23}$, $\Delta\alpha_{31}$, $\Delta\alpha_{14}$, $\Delta\alpha_{24}$, $\Delta\alpha_{34}$

Note that there are ten of these internal, coordinates and there are only $3n - 6 = 3(5) - 6 = 9$ fundamental modes. Therefore there is a redundancy in these coordinates, i.e., they are not all independent of each other.

As we shall see later, the use of symmetry coordinates simplifies the solution of force field problems for molecules having several symmetry elements such as MX_4. Sets of symmetry coordinates are not unique. An appropriate set for MX_4 is as follows:

For the A_1 vibration:

$$R_1 = \tfrac{1}{2}(\Delta d_1 + \Delta d_2 + \Delta d_3 + \Delta d_4)$$
$$R_2 = (1/\sqrt{6})(\Delta\alpha_{12} + \Delta\alpha_{23} + \Delta\alpha_{31} + \Delta\alpha_{14} + \Delta\alpha_{24} + \Delta\alpha_{34}) = 0 \quad \text{(3-16A)}$$

For the E vibration:

$$R_{3a} = (1/\sqrt{12})(2\,\Delta\alpha_{12} - \Delta\alpha_{23} - \Delta\alpha_{31} - \Delta\alpha_{14} - \Delta\alpha_{24} + 2\,\Delta\alpha_{34})$$
$$R_{3b} = \tfrac{1}{2}(\Delta\alpha_{14} - \Delta\alpha_{31} + \Delta\alpha_{23} - \Delta\alpha_{24}) \quad \text{(3-16}B\text{)}$$

For the F vibrations:

$$
\begin{aligned}
R_{4a} &= (1/\sqrt{12})(2\,\Delta\alpha_{12} - \Delta\alpha_{23} - \Delta\alpha_{31} + \Delta\alpha_{14} + \Delta\alpha_{24} - 2\,\Delta\alpha_{34}) \\
R_{4b} &= (1/\sqrt{6})(\Delta\alpha_{12} + \Delta\alpha_{23} + \Delta\alpha_{31} - \Delta\alpha_{14} - \Delta\alpha_{24} - \Delta\alpha_{34}) \\
R_{4c} &= \tfrac{1}{2}(\Delta\alpha_{23} - \Delta\alpha_{31} - \Delta\alpha_{14} + \Delta\alpha_{24}) \\
R_{5a} &= (1/\sqrt{6})(\Delta d_1 + \Delta d_2 - 2\,\Delta d_3) \\
R_{5b} &= (1/\sqrt{12})(\Delta d_1 + \Delta d_2 + \Delta d_3 - 3\,\Delta d_4) \\
R_{5c} &= (1/\sqrt{2})(\Delta d_2 - \Delta d_1)
\end{aligned}
\tag{3-16C}
$$

The table of force constants for MX_4 is constructed using a general force field for a molecule which has n atoms and $3n - 6$ internal displacement coordinates to represent the $3n - 6$ fundamental modes of vibration into which all of the internal motions of this molecule are to be resolved.

The force constants are the coefficients of the $q_i q_i$ terms of the potential energy function of the general force field for MX_4:

$$
\begin{aligned}
2V = {} & f_d \sum_{i}^{4} (\Delta d_i)^2 + d_0^2 f_\alpha \sum_{i,j}^{6} \Delta\alpha_{ij}^2 + 2f_{dd} \sum_{i,j}^{6} \Delta d_i\,\Delta d_j \\
& + 2\,d_0 f_{d\alpha} \sum_{i,j}^{12} \Delta d_i\,\Delta\alpha_{ij} + 2\,d_0\,f_{d\alpha'} \sum_{i,j,k}^{12} \Delta d_i\,\Delta\alpha_{jk} \\
& + 2\,d_0^2 f_{\alpha\alpha} \sum_{i,j,k}^{12} \Delta\alpha_{ij}\,\Delta\alpha_{jk} + 2\,d_0^2 f_{\alpha\alpha'} \sum_{i,j,k,l}^{3} \Delta\alpha_{ij}\,\Delta\alpha_{kl}
\end{aligned}
\tag{3-17}
$$

This set of coefficients constitutes the f matrix of order 10×10 (Table 3-2). The potential energy expression for the general force field (GFF) using the ten internal coordinates listed is

$$
\begin{aligned}
2V = {} & f_d[(\Delta d_1)^2 + (\Delta d_2)^2 + (\Delta d_3)^2 + (\Delta d_4)^2] \\
& + 2f_{dd}[(\Delta d_1)(\Delta d_2 + \Delta d_3 + \Delta d_4) + (\Delta d_2)(\Delta d_3 + \Delta d_4) + (\Delta d_3)(\Delta d_4)] \\
& + f_\alpha[(d\Delta\alpha_{12})^2 + (d\Delta\alpha_{23})^2 + (d\Delta\alpha_{31})^2 + (d\Delta\alpha_{14})^2 + (d\Delta\alpha_{24})^2 + (d\Delta\alpha_{34})^2] \\
& + 2f_{\alpha\alpha}[(d\Delta\alpha_{12})(d\Delta\alpha_{31} + d\Delta\alpha_{14} + d\Delta\alpha_{23} + d\Delta\alpha_{24}) \\
& + (d\Delta\alpha_{23})(d\Delta\alpha_{24} + d\Delta\alpha_{34} + d\Delta\alpha_{31}) + (d\Delta\alpha_{31})(d\Delta\alpha_{14} + d\Delta\alpha_{34}) \\
& + (d\Delta\alpha_{14})(d\Delta\alpha_{24} + d\Delta\alpha_{34}) + (d\Delta\alpha_{24})(d\Delta\alpha_{34})] \\
& + 2f_{\alpha\alpha'}[(d\Delta\alpha_{12})(d\Delta\alpha_{34}) + (d\Delta\alpha_{23})(d\Delta\alpha_{14}) + (d\Delta\alpha_{31})(d\Delta\alpha_{24})] \\
& + 2f_{d\alpha}[\Delta d_1(d\Delta\alpha_{12} + d\Delta\alpha_{31} + d\Delta\alpha_{14}) + \Delta d_2(d\Delta\alpha_{12} + d\Delta\alpha_{23} + d\Delta\alpha_{24})
\end{aligned}
$$

Table 3-2. The f Matrix for MX_4-Type Molecules

f	Δd_1	Δd_2	Δd_3	Δd_4	$\Delta\alpha_{12}$	$\Delta\alpha_{23}$	$\Delta\alpha_{31}$	$\Delta\alpha_{14}$	$\Delta\alpha_{24}$	$\Delta\alpha_{34}$
Δd_1	f_d	f_{dd}	f_{dd}	f_{dd}	$f_{d\alpha}$	$f_{d\alpha'}$	$f_{d\alpha}$	$f_{d\alpha}$	$f_{d\alpha'}$	$f_{d\alpha'}$
Δd_2		f_d	f_{dd}	f_{dd}	$f_{d\alpha}$	$f_{d\alpha}$	$f_{d\alpha'}$	$f_{d\alpha'}$	$f_{d\alpha}$	$f_{d\alpha'}$
Δd_3			f_d	f_{dd}	$f_{d\alpha'}$	$f_{d\alpha}$	$f_{d\alpha}$	$f_{d\alpha'}$	$f_{d\alpha'}$	$f_{d\alpha}$
Δd_4				f_d	$f_{d\alpha'}$	$f_{d\alpha'}$	$f_{d\alpha'}$	$f_{d\alpha}$	$f_{d\alpha}$	$f_{d\alpha}$
Δd_{12}					f_α	$f_{\alpha\alpha}$	$f_{\alpha\alpha}$	$f_{\alpha\alpha}$	$f_{\alpha\alpha}$	$f_{\alpha\alpha'}$
$\Delta\alpha_{23}$						f_α	$f_{\alpha\alpha}$	$f_{\alpha\alpha'}$	$f_{\alpha\alpha}$	$f_{\alpha\alpha}$
$\Delta\alpha_{31}$							f_α	$f_{\alpha\alpha}$	$f_{\alpha\alpha'}$	$f_{\alpha\alpha}$
$\Delta\alpha_{14}$		$f_{ij} = f_{ji}$						f_α	$f_{\alpha\alpha}$	$f_{\alpha\alpha}$
$\Delta\alpha_{24}$									f_α	$f_{\alpha\alpha}$
$\Delta\alpha_{34}$										f_α

$$
\begin{aligned}
&+ \Delta d_3(d\Delta\alpha_{31} + d\Delta\alpha_{23} + d\Delta\alpha_{34}) + \Delta d_4(d\Delta\alpha_{14} + d\Delta\alpha_{24} + d\Delta\alpha_{34})] \\
&+ 2f_{d\alpha'}[\Delta d_1(d\Delta\alpha_{23} + d\Delta\alpha_{24} + d\Delta\alpha_{34}) + \Delta d_2(d\Delta\alpha_{31} + d\Delta\alpha_{14} + d\Delta\alpha_{34}) \\
&+ \Delta d_3(d\Delta\alpha_{12} + d\Delta\alpha_{14} + d\Delta\alpha_{24}) + \Delta d_4(d\Delta\alpha_{12} + d\Delta\alpha_{23} + d\Delta\alpha_{31})]
\end{aligned}
\tag{3-18}
$$

There are seven force constants in this field even when the assumptions due to the identity of the four X atoms and tetrahedral symmetry have been made:

f_d = distortion along one bond interacting with itself
f_{dd} = interaction between two stretches
$f_{d\alpha}$ = interaction between stretch and distortion of angle of which it is a side
$f_{d\alpha'}$ = interaction between stretch and distortion of angle of which it is not a side
f_α = interaction of angle bend with itself
$f_{\alpha\alpha}$ = interaction of angle bend with angle having common side
$f_{\alpha\alpha'}$ = interaction of angle bend with angle not having common side

Thus, in this case there are seven force constants and only four vibrational frequencies. This is equivalent to four equations in seven unknowns and the problem is not solvable for a unique set of seven force constants.

There are, in principle, several alternatives.

1. One can form at least two such compounds as MX_3X', MX_4', MX_2X_2', where X′ is the isotope of X, and use the four frequencies of each molecule. The frequencies and g elements for these will differ but the force constants will be the same. Now there are more frequencies than force constants. This is known as the "*small isotope shift in frequencies*" approach. If the X's are hydrogens, deuterium could be used with much more effect.

2. Derive the Coriolis constants for the infrared spectra to get relationships between frequencies, or determine centrifugal distortion constants or use mean amplitudes of vibration to determine relationships between frequencies.

3. Set force constants associated with a stretch–bend or bend–bend interaction equal to zero if the two internal coordinates do not share a common bond.

In practice, the data for these methods are rarely available, so the alternative is to turn to a model force field, i.e., a modified f matrix with fewer force constants.

There are several such force fields: e.g., central force field (CFF), simple valence force field (SVFF), general valence force field (GVFF). Urey–Bradley force field (UBFF), orbital valence force field (OVFF), local symmetry force field (LSFF). All have as their objective the reduction of force constants from the seven of the general force field (GFF). The last four also endeavor to better represent the properties (force constants or frequencies) of the molecule than the oversimplified CFF or SVFF are able to do.

We shall use the Urey–Bradley model force field[3,4] now to follow through the procedure of calculating force constants from observed frequencies of the tetrahedral MX_4 molecule. The Urey–Bradley model keeps the interaction of stretching and bending vibrations of the SVFF and adds repulsions between nonbonded atoms. The Urey–Bradley potential energy is given by

$$2V = K \sum_{i}^{4} \Delta r_i^2 + r_0^2 H \sum_{i,j}^{6} \Delta \alpha_{ij}^2 + F \sum_{i,j}^{6} \Delta q_{ij}^2 + 2q_0 F' \sum_{i,j}^{6} \Delta q_{ij} + 2r_0 K' \sum_{i}^{4} \Delta r_i \tag{3-19}$$

where q_{ij} is the distance between nonbonded atoms X_i and X_j and $q_{ij}^2 = r_i^2 + r_j^2 - 2r_i r_j \cos \alpha_{ij}$. Using a Taylor expansion in the neighborhood of the equilibrium configuration ($q = q_0$) to obtain Δq_{ij} as a function of

Δr_i, Δr_j, and $\Delta\alpha_{ij}$ and substituting in Eq. (3-19), we obtain

$$\begin{aligned} V = {} & \tfrac{1}{2}\sum_i \Big[K_i + \sum_{j(\neq i)} (t_{ij}^2 F'_{ij} + s_{ij}^2 F_{ij})\Big] (\Delta r_i)^2 \\ & + \tfrac{1}{2}\sum_{i<j} [H_{ij} - s_{ij}s_{ji}F'_{ij} + t_{ij}t_{ji}F_{ij} + (3\varkappa/\sqrt{8}\, r_{ij}^2)](r_{ij}\,\Delta\alpha_{ij})^2 \\ & + \sum_{i<j} [-t_{ij}t_{ji}F'_{ij} + s_{ij}s_{ji}F_{ij}](\Delta r_i)(\Delta r_j) \\ & + \sum_{i\neq j} [t_{ij}s_{ji}F'_{ij} + t_{ji}s_{ij}F_{ij}](r_j/r_i)^{1/2}(\Delta r_i)(r_{ij}\,\Delta\alpha_{ij}) \\ & + \sum_{i\neq j\neq k} (\varkappa/\sqrt{2}\, r_{ij} r_{ik})(r_{ij}\,\Delta\alpha_{ij})(r_{ik}\,\Delta\alpha_{ik}) \end{aligned} \tag{3-20}$$

where $s_{ij} = s_{ij} = r/q_{ij}(1 - \cos\alpha_{ij})$, $t_{ij} = t_{ij} = (r \sin\alpha_{ij})/q_{ij}$ (where $r = r_i = r_j$), and $\varkappa$ is the intramolecular tension.

Using the redundancy relationship (symmetry coordinates exceed nine in number) the potential energy can be expressed in terms of four force constants K, H, F, and F' as

$$\begin{aligned} 2V = {} & (K + 2F + F')\sum_i^4 \Delta r_i^2 + r_0^2(H + \tfrac{1}{3}F - \tfrac{5}{3}F')\sum_{ij}^6 \Delta\alpha_{ij}^2 \\ & + 2(\tfrac{2}{3}F - \tfrac{1}{3}F')\sum_{ij}^6 \Delta r_i\,\Delta r_j + 2r_0[(\tfrac{1}{3}\sqrt{2}\,F + \tfrac{1}{3}\sqrt{2}\,F')\sum_{iij}^{12} \Delta r_i\,\Delta\alpha_{ij}] \\ & + 2r_0^2(-\tfrac{2}{3}F')\sum_{ijjk}^{12} \Delta\alpha_{ij}\,\Delta\alpha_{jk} \end{aligned} \tag{3-21}$$

The equivalence between the seven force constants of the GVFF and the form of the Urey–Bradley model field for MX_4 is demonstrated as follows:

GVFF	UBFF
f_r	$K + 2F + F'$
f_α	$H + \frac{1}{3}F - \frac{5}{3}F'$
f_{rr}	$\frac{2}{3}F - \frac{1}{3}F'$
$f_{r\alpha}$	$\frac{1}{3}\sqrt{2}\,F + \frac{1}{3}\sqrt{2}\,F'$
$f_{r\alpha'}$	0
$f_{\alpha\alpha}$	$-\frac{2}{3}F'$
$f_{\alpha\alpha'}$	0

The U matrix of the symmetry coordinates is written out along with the f and U' matrices and the calculation $UfU' = F$ is shown in Tables 3-3 to 3-7.

Table 3-3. The U Matrix for MX_4 Molecules (T_d)

	U	Δd_1	Δd_2	Δd_3	Δd_4	$\Delta\alpha_{12}$	$\Delta\alpha_{23}$	$\Delta\alpha_{31}$	$\Delta\alpha_{14}$	$\Delta\alpha_{24}$	$\Delta\alpha_{34}$
A_1 —	R_1	$\frac{1}{2}$	$\frac{1}{2}$	$\frac{1}{2}$	$\frac{1}{2}$	0	0	0	0	0	0
E	R_{3a}	0	0	0	0	$2/\sqrt{12}$	$-1/\sqrt{12}$	$-1/\sqrt{12}$	$-1/\sqrt{12}$	$-1/\sqrt{12}$	$2/\sqrt{12}$
	R_{3b}	0	0	0	0	0	$\frac{1}{2}$	$-\frac{1}{2}$	$\frac{1}{2}$	$-\frac{1}{2}$	0
$2F$	R_{5a}	$1/\sqrt{6}$	$1/\sqrt{6}$	$-2/\sqrt{6}$	0	0	0	0	0	0	0
	R_{4a}	0	0	0	0	$2/\sqrt{12}$	$-1/\sqrt{12}$	$-1/\sqrt{12}$	$1/\sqrt{12}$	$1/\sqrt{12}$	$-2/\sqrt{12}$
	R_{5b}	$1/\sqrt{12}$	$1/\sqrt{12}$	$1/\sqrt{12}$	$-3/\sqrt{12}$	0	0	0	0	0	0
	R_{4b}	0	0	0	0	$1/\sqrt{6}$	$1/\sqrt{6}$	$1/\sqrt{6}$	$-1/\sqrt{6}$	$-1/\sqrt{6}$	$-1/\sqrt{6}$
	R_{5c}	$-1/\sqrt{2}$	$1/\sqrt{2}$	0	0	0	0	0	0	0	0
	R_{4c}	0	0	0	0	0	$\frac{1}{2}$	$-\frac{1}{2}$	$-\frac{1}{2}$	$\frac{1}{2}$	0
Redundant coordinate	R_2	0	0	0	0	$1/\sqrt{6}$	$1/\sqrt{6}$	$1/\sqrt{6}$	$1/\sqrt{6}$	$1/\sqrt{6}$	$1/\sqrt{6}$

Table 3-4. The f Matrix for MX_4 Molecules (T_d)*

Q	M	M	M	W	0	W	W	0	0
M	Q	M	M	W	W	0	0	W	0
M	M	Q	M	0	W	W	0	0	W
M	M	M	Q	0	0	0	W	W	W
W	W	0	0	P	Z	Z	Z	Z	0
0	W	W	0	Z	P	Z	0	Z	Z
W	0	W	0	Z	Z	P	Z	0	Z
W	0	0	W	Z	0	Z	P	Z	Z
0	W	0	W	Z	Z	0	Z	P	Z
0	0	W	W	0	Z	Z	Z	Z	P

* $Q = K + F' + 2F$.
$M = -\frac{1}{3}F' + \frac{2}{3}F$.
$P = H - \frac{5}{3}F' + \frac{1}{3}F$.
$W = \frac{1}{3}\sqrt{2}(F' + F)$.
$Z = -\frac{2}{3}F'$.

Table 3-5. The U' Matrix for MX_4 Molecules (T_d)

$\frac{1}{2}$	0	0	$1/\sqrt{6}$	0	$1/\sqrt{12}$	0	$-1/\sqrt{2}$	0	0
$\frac{1}{2}$	0	0	$1/\sqrt{6}$	0	$1/\sqrt{12}$	0	$1/\sqrt{2}$	0	0
$\frac{1}{2}$	0	0	$-2/\sqrt{6}$	0	$1/\sqrt{12}$	0	0	0	0
$\frac{1}{2}$	0	0	0	0	$-3/\sqrt{12}$	0	0	0	0
0	$2/\sqrt{12}$	0	0	$2/\sqrt{12}$	0	$1/\sqrt{6}$	0	0	$1/\sqrt{6}$
0	$-1/\sqrt{12}$	$\frac{1}{2}$	0	$-1/\sqrt{12}$	0	$1/\sqrt{6}$	0	$\frac{1}{2}$	$1/\sqrt{6}$
0	$-1/\sqrt{12}$	$-\frac{1}{2}$	0	$-1/\sqrt{12}$	0	$1/\sqrt{6}$	0	$-\frac{1}{2}$	$1/\sqrt{6}$
0	$-1/\sqrt{12}$	$\frac{1}{2}$	0	$1/\sqrt{12}$	0	$-1/\sqrt{6}$	0	$-\frac{1}{2}$	$1/\sqrt{6}$
0	$-1/\sqrt{12}$	$-\frac{1}{2}$	0	$1/\sqrt{12}$	0	$-1/\sqrt{6}$	0	$\frac{1}{2}$	$1/\sqrt{6}$
0	$2/\sqrt{12}$	0	0	$-2/\sqrt{12}$	0	$-1/\sqrt{6}$	0	0	$1/\sqrt{6}$

Table 3-6. The Partitioned *F* Matrix for MX_4 Molecules (T_d) in Terms of *Q*, *M*, *P*, *W*, and *Z*

A_1	E	E	F		F		F		R
$Q + 3M$	0	0	0	0	0	0	0	0	$\sqrt{6}\,W$
0	$P - 2Z$	0	0	0	0	0	0	0	0
0	0	$P - 2E$	0	0	0	0	0	0	0
0	0	0	$Q - M$	$\sqrt{2}\,W$	0	0	0	0	0
0	0	0	$\sqrt{2}\,W$	P	0	0	0	0	0
0	0	0	0	0	$Q - M$	$\sqrt{2}\,W$	0	0	0
0	0	0	0	0	$\sqrt{2}\,W$	P	0	0	0
0	0	0	0	0	0	0	$Q - M$	$\sqrt{2}\,W$	0
0	0	0	0	0	0	0	$\sqrt{2}\,W$	P	0
$\sqrt{6}\,W$	0	0	0	0	0	0	0	0	$P + 4Z$

Table 3-7. The Partitioned F Matrix for MX_4 Molecules (T_d)*

A_1	F_{11} $K + 4F$								
E		F_{22} $H +$ $\frac{1}{2}(F - F')$							
			F_{22} $H +$ $\frac{1}{3}(F - F')$						
$2F$				F_{33} $K +$ $\frac{4}{3}(F + F')$	F_{34} $\frac{2}{3}(F + F')$				
				F_{43} $\frac{2}{3}(F + F')$	F_{44} $H +$ $\frac{1}{3}F - \frac{5}{3}F'$				
						F_{33} $K +$ $\frac{4}{3}(F + F')$	F_{34} $\frac{2}{3}(F + F')$		
						F_{43} $\frac{2}{3}(F + F')$	F_{44} $H +$ $\frac{1}{3}F - \frac{5}{8}F'$		
								F_{33} $K +$ $\frac{4}{3}(F + F')$	F_{34} $\frac{2}{3}(F + F')$
								F_{43} $\frac{2}{3}(F + F')$	F_{44} $H +$ $\frac{1}{3}F - \frac{5}{3}F'$

* $F_{11} = K + 4F$, $F_{22} = H + \frac{1}{3}F - \frac{1}{3}F'$, $F_{33} = K + \frac{4}{3}F + \frac{4}{3}F'$, $F_{34} = F_{43} = \frac{2}{3}(F + F')$, $F_{44} = H + \frac{1}{3}F - \frac{5}{3}F'$,

Table 3-8. The g Matrix for MX_4 Molecules (T_d)

$$g = \begin{bmatrix} g_d & g_{dd} & g_{dd} & g_{dd} & g_{d\alpha} & g_{d\alpha'} & g_{d\alpha} & g_{d\alpha} & g_{d\alpha'} & g_{d\alpha'} \\ & g_d & g_{dd} & g_{dd} & g_{d\alpha} & g_{d\alpha} & g_{d\alpha'} & g_{d\alpha'} & g_{d\alpha} & g_{d\alpha'} \\ & & g_d & g_{dd} & g_{d\alpha'} & g_{d\alpha} & g_{d\alpha} & g_{d\alpha'} & g_{d\alpha'} & g_{d\alpha} \\ & & & g_d & g_{d\alpha'} & g_{d\alpha'} & g_{d\alpha'} & g_{d\alpha} & g_{d\alpha} & g_{d\alpha} \\ & & & & g_\alpha & g_{\alpha\alpha} & g_{\alpha\alpha} & g_{\alpha\alpha} & g_{\alpha\alpha} & g_{\alpha\alpha'} \\ & & & & & g_\alpha & g_{\alpha\alpha} & g_{\alpha\alpha'} & g_{\alpha\alpha} & g_{\alpha\alpha} \\ & & & & & & g_\alpha & g_{\alpha\alpha} & g_{\alpha\alpha'} & g_{\alpha\alpha} \\ & g_{ij} = g_{ji} & & & & & & g_\alpha & g_{\alpha\alpha} & g_{\alpha\alpha} \\ & & & & & & & & g_\alpha & g_{\alpha\alpha} \\ & & & & & & & & & g_\alpha \end{bmatrix}$$

The G matrix has a similar form:

$$G = UgU'$$

The G matrix (for kinetic energy) depends on the masses of the elements involved. The g matrix is analogous to the f matrix (see Table 3-8). Similarly, the $UgU' = G$ matrix blocks out just as the F matrix does. There is a straightforward method for setting up the G matrix devised by Wilson, Decius, and Cross[5] (Table 3-9). Thus, the F and G matrices each factors into three

Table 3-9. The Partitioned G Matrix for MX_4 Molecules (T_d)

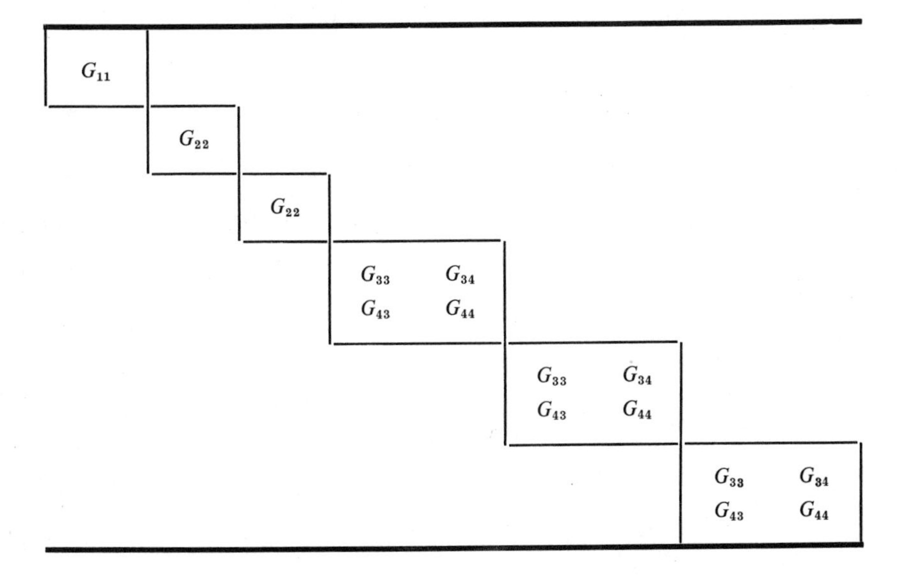

Table 3-10. The F and G Submatrices for T_d Molecules

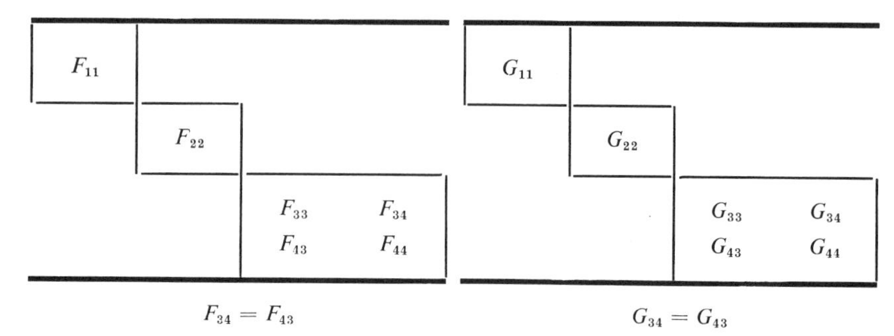

F_{11}			
	F_{22}		
		F_{33}	F_{34}
		F_{43}	F_{44}

$F_{34} = F_{43}$

G_{11}			
	G_{22}		
		G_{33}	G_{34}
		G_{43}	G_{44}

$G_{34} = G_{43}$

distinct submatrices: two 1×1 and one 2×2 (Table 3-10). Hence, the secular equation $G_i F_i - E\lambda_i = 0$ factors into

$$\text{for } A_1 \text{ species:} \quad G_{11}F_{11} - \lambda_1 = 0, \qquad G_{11} = \mu_M = 1/m_M \tag{3-22}$$

$$\text{for } E \text{ species:} \quad G_{22}F_{22} - \lambda_2 = 0, \qquad G_{22} = 3\mu_M \tag{3-23}$$

$$\text{for } 2F \text{ species:} \quad \begin{vmatrix} G_{33}F_{33} - \lambda & G_{34}F_{34} \\ G_{43}F_{43} & G_{44}F_{44} - \lambda \end{vmatrix} = 0 \tag{3-24}$$

where

$$G_{33} = \mu_X + (4/3)\mu_M$$

$$G_{34} = G_{43} = -(8/3)\mu_M$$

$$G_{44} = (16/3)(\mu_M + 2\mu_X)$$

Making the substitution, we have

$$\lambda_1 = \mu_M(K + 4F) \tag{3-25}$$

$$\lambda_2 = 3\mu_M(H + \tfrac{5}{3} - \tfrac{1}{3}F') \tag{3-26}$$

$$G_{33}F_{33}G_{44}F_{44} - (G_{33}F_{33} + G_{44}F_{44})\lambda + \lambda^2 - (G_{34}F_{34})^2 = 0 \tag{3-27}$$

$$\lambda^2 - (G_{33}F_{33} + G_{44}F_{44})\lambda + (G_{33}F_{33}G_{44}F_{44} - G_{34}^2F_{34}^2) = 0 \tag{3-28}$$

$$\begin{aligned} \lambda_3 + \lambda_4 &= G_{33}F_{33} + G_{44}F_{44} \\ \lambda_3\lambda_4 &= G_{33}F_{33}G_{44}F_{34} - G_{34}F_{34} \end{aligned} \tag{3-29}$$

Making substitutions, we have

$$\lambda_3 + \lambda_4 = (K + 2H + 2F - 2F')\mu_X + [(4/3)K + (16/3)M - (32/3)F']\mu_M$$

$$\begin{aligned} \lambda_3\lambda_4 = {} & [2KH + (2/3)KF - (10/3)KF' + (8/3)HF + (8/3)HF' \\ & -(16/3)FF' - (16/3)(F)^2](\mu_X{}^2 + 4\mu_X\mu_M) \end{aligned} \tag{3-30}$$

From Eqs. (3-25), (3-26), and (3-30) we can see that if we know the force constants, we can, in principle, calculate the frequencies and vice versa.

We have seen how to calculate force constants for a given set of frequencies for the simple Urey–Bradley force field. Once one changes the force field, the potential energy function changes and so will the F matrix. In the procedure used in deriving the F matrix for a given force field, the following steps are taken:

(1) The potential energy function is written in terms of the Δr_i and $\Delta\alpha_{ij}$ (after expressing the Δq_{ij} as functions of the Δr_i and $\Delta\alpha_{ij}$).
(2) The coefficients of the different terms (Δr_i^2, $\Delta\alpha_{ij}^2$, etc.) are calculated as functions of the force constants being used for the given force field.
(3) The f matrix is set up from the results of step 2.
(4) The U matrix is written from the symmetry coordinates, placing them in the proper order with the redundant coordinate last, and its transverse U' is written.
(5) $UfU' = F$ is calculated, where F is a diagonalized matrix if the symmetry coordinates are taken in the suitable order.

The relationship between the general valence force field and the Urey–Bradley force field for MX_4 molecules (T_d) is as follows:

Symmetrized	GVFF	UBFF
F_{11}	$f_r + 3f_{rr}$	$K + 4F$
F_{22}	$f_\alpha - 2f_{\alpha\alpha} + f_{\alpha\alpha'}$	$H + \frac{1}{3}F - \frac{1}{3}F'$
F_{33}	$f_r - f_{rr}$	$K + \frac{4}{3}F + \frac{4}{3}F'$
F_{34}	$-\sqrt{2}(f_{r\alpha} - f_{r\alpha'})$	$-\frac{2}{3}(F + F')$
F_{44}	$f_\alpha - f_{\alpha\alpha'}$	$H + \frac{1}{3}F - \frac{5}{3}F'$

One can test to see which of several force fields best describes the molecule by using a computer to calculate the set of force constants which will give frequencies closest to the observed frequencies. The computer program starts with a trial set of force constants and calculates the frequencies. The force constants are then varied in an attempt to minimize the weighted sum of the squares of the deviations of the frequencies so calculated from the experimental values:

$$S = (\nu_{\text{obs}} - \nu_{\text{calc}})W(\nu'_{\text{obs}} - \nu'_{\text{calc}}) \tag{3-31}$$

where W is a matrix each element of which gives the weight assigned to each ν_{obs}.

Assume a linear relationship between the variation in the frequencies and the variation of the force constants

$$\Delta\nu = J\,\Delta f \qquad (J = \text{Jacobian matrix}) \tag{3-32}$$

$$J_{ik} = \partial\nu_i/\partial f_k \tag{3-33}$$

$$\Delta\nu_i = \sum_k (\partial\nu_i/\partial f_k)\,\Delta f_k \tag{3-34}$$

By an iterative process a new set of force constants modified by Δf_k calculates a new set of frequencies differing from the last by $\Delta\nu_i$ and a new S is calculated. When there is no further change in S the system has converged to a set of force constants corresponding to a set of calculated frequencies. When the sets of frequencies calculated with each force field are compared with the set of frequencies observed, the one showing the smallest differences is the most effective force for the given molecule.

There are a number of sources of error in this procedure. The vibrations may depart considerably from being simple harmonic, in which case cubic terms, etc. need to be included. There are definite procedures for anharmonicity corrections.

When the number of force constants calculated is equal to the number of frequencies from which they are obtained, the set of force constants may not be unique. Sometimes when there are two sets, one can be ruled out because K, F, or H has a negative value. Most of the computer programs utilize least square fits to the observed frequencies by adjusting the potential constants through a linear relation between $\Delta\nu$ and Δf.[6–8] For other programs see Hart.[9]

Now we discuss briefly some other model force fields and then show from our work how some of these compare in solving the octahedral and tetrahedral problems. All of the other model force fields decrease the number of force constants associated with a general force field by making certain assumptions about the off-diagonal elements.

1. *Central Force Field* (*CFF*)

This field includes only those force constants associated with changes in interatomic distances between bonded (Δr_{ij}) or nonbonded (Δq_{ij}) atoms

$$V = \tfrac{1}{2}\sum_{i,j} K_{ij}(\Delta r_i)^2 + \tfrac{1}{2}\sum_{i,j} F_{ij}(\Delta q_{ij})^2 \tag{3-35}$$

Hence it does not reflect the contribution of the bending modes to the potential energy of the molecule.

2. *Valence Force Field* (*VFF*)

This field includes force constants associated with stretching along bonds, changes in bond angles, and internal rotation (K, H, Y):

$$V = \tfrac{1}{2}\sum_{i,j} K_{ij}(\Delta r_{ij})^2 + \tfrac{1}{2}\sum_{i,j,k} H_{ijk}(\Delta\alpha_{ijk})^2 + \tfrac{1}{2}\sum_{i,j,k,l} Y_{ijkl}(\Delta t_{ijkl})^2 \qquad (3\text{-}36)$$

where r, α, and t are bond length, bond angle, and angle of internal rotation. This field takes no account of interactions between nonbonded atoms or between stretches and bends.

3. *General Valence Force Field* (*GVFF*)

This field is identical with the general force field in simple molecules where there is no redundancy in the internal coordinates. It includes all interaction terms. For the MX_4 molecule, the potential force field is given in Eq. (3-18). Its greatest limitation is that it usually involves calculation of more force constants than there are observed frequencies.

4. *Orbital Valence Force Field* (*OVFF*)

This field, developed by Heath and Linnett,[10] differs from the Urey-Bradley field primarily in its treatment of angle bending. Instead of considering the changes in the angle between the bonds as a result of bending ($\Delta\alpha_{ij}$), Heath and Linnett used the angle between the equilibrium position of the line through the axis of the hybrid orbital and the position of this imaginary line in the deformed configuration (β_{ij}). So the potential energy function for the OVFF is

$$\begin{aligned}2V = {} & (K + 2F + F')\sum_{i}^{4} \Delta r_i^2 + r_0^2(\tfrac{1}{2}D + \tfrac{1}{3}F - \tfrac{5}{3}F')\sum_{ij}^{6} \Delta\beta_{ij}^2 \\ & + 2(\tfrac{2}{3}F - \tfrac{1}{3}F')\sum_{ij}^{6} \Delta r_i\,\Delta r_j + 2r_0(\tfrac{1}{3}\sqrt{2}\,F + \tfrac{1}{3}\sqrt{2}\,F')\sum_{iij}^{12} \Delta r_i\,\Delta\beta_{ij} \\ & + 2r_0^2(\tfrac{1}{2}D - \tfrac{2}{3}F')\sum_{ijjk}^{12} \Delta\beta_{ij}\,\Delta\beta_{jk}\end{aligned} \qquad (3\text{-}37)$$

where K, F, and F' correspond to the force constants of the Urey–Bradley field and D corresponds to H in the Urey–Bradley model. As noted before, $\Delta\beta_{ij}$ corresponds to but represents a different quantity from the Urey–Bradley $\Delta\alpha_{ij}$. As we shall see in Chapter 4, the OVFF in some cases gives a better account of bending vibrations than does the Urey–Bradley force field.

Table 3-11. The Partitioned *F* Matrix for *A′* Species of $CH_3CH_2CH_2Cl$

$F(CH_3)$		
	$F(CH_2)$	
		$F(CH_2Cl)$

5. *Local Symmetry Force Field* (*LSFF*)

This force field was developed to provide a consistent set of force constants for basic organic molecules which could be transferred to more complex organic molecules to explain their frequencies, and to analogous molecules containing deuterium. The force constants for the LSFF are based on various models but must conform to the following concepts.

1. CH_3 has C_{3v} symmetry and CH_2 has C_{2v} symmetry and the symmetry coordinates for these can be used as basis coordinates for organic molecules containing these groups.

2. In general, CH_3 (or CH_2) groups are uninfluenced by next to nearest neighbor groups or those farther removed.

3. The general basis coordinates of a complex organic molecule are constituted by the symmetry coordinates for each CH_3, CH_2, and functional group in the molecule. The force constants outside the submatrices are, in general, set equal to zero so the *F* matrix becomes blocked out as given by Shimanouchi[1,2] for the in-plane *A′* species of $CH_3CH_2CH_2Cl$ and shown in Table 3-11.

REFERENCES

1. T. Shimanouchi, National Standard Reference Data Series, National Bureau of Standards, US NSRD-NBS, No. 6 (1967).
2. I. Harada and T. Shimanouchi, National Standard Reference Data Series, National Bureau of Standards, US NSRD-NBS, No. 11, (1968).
3. H. C. Urey and C. A. Bradley, *Phys. Rev.*, **38**:1969 (1931).
4. T. Shimanouchi, *J. Chem. Phys.*, **17**:245, 734, 848 (1949).
5. E. B. Wilson, J. C. Decius, and P. C. Cross, *Molecular Vibrations*, McGraw-Hill, New York (1955).
6. J. Overend and J. R. Scherer, *J. Chem. Phys.*, **32**:1289, 1296, 1720 (1160); **33**:446 (1960); **34**:547 (1961); **36**:3308 (1962).
7. J. H. Schactschneider and R. G. Snyder, *Spec. Acta*, **19**:117 (1963).

8. W. A. Yeranos, *Bull. Soc. Chim. Belg.*, **74**, 414 (1965).
9. R. R. Hart, in *Developments in Applied Spectroscopy*, Vol. 4, E. N. Davis (ed.), Plenum Press, New York (1965).
10. D. F. Heath and J. W. Linnett, *Trans. Faraday Soc.*, **44**:556, 873, 878, 884 (1948); **45**:264 (1949).

Chapter 4

THE NORMAL COORDINATE TREATMENT FOR MOLECULES WITH C_{2v}, C_{3v}, AND O_h SYMMETRY

The selection rules for a molecule, based on a hypothetical model, were derived in Chapter 2. The normal modes of this model can be determined using the normal coordinate treatment (NCT) method, which also serves to check the frequency assignments for the model. This method, which illustrates a useful application of group theory, depends on the bond lengths and bond angles of the model. Knowledge of force constants (the restoring force between two atoms) is also necessary, and the molecule must possess a high degree of symmetry.

This chapter attempts to illustrate the NCT method by presenting the detailed procedure necessary to calculate the theoretical frequencies for H_2O, NH_3, and UF_6. The method of Wilson *et al.*[(1)] will be used.

4-1. PROCEDURE NECESSARY IN THE NCT METHOD

The following steps are necessary in carrying out the NCT of a molecule:

(1) Determine the most probable model or models based on the largest amount of data available, including spectroscopic data.

(2) Choose a set of internal coordinates, and convert them to symmetry coordinates. For a general method of obtaining symmetry coordinates see Appendix 7. The symmetry coordinates are of the form

$$R_j = \sum_k U_{jk} r_k \qquad (4\text{-}1)$$

where R_j is the jth symmetry coordinate ($j = 1, 2, \ldots 3n - 6$), U_{jk} is the coefficient of the kth internal coordinate r_k, and the summation is over all of the equivalent internal coordinates. The symmetry coordinates can be written in matrix notation as

$$R = Ur, \qquad r = U^T R = U'R, \qquad r' = R'U \qquad (4\text{-}2)$$

The selection of symmetry coordinates is not necessarily unique, but must satisfy the following conditions:

A. Normalization of the symmetry coordinates. This is accomplished by satisfying Eq. (4-3):

$$\sum_k (U_{jk})^2 = 1 \tag{4-3}$$

B. Orthogonalization of the symmetry coordinates. This is possible when Eq. (4-4) is satisfied:

$$\sum_k (U_{jk})(U_{lk}) = 0 \tag{4-4}$$

where j and l refer to two different symmetry coordinates.

C. Transformation of the symmetry coordinates. A symmetry coordinate belongs to an irreducible representation if it transforms accordingly. For a nondegenerate vibration the coordinate transforms symmetrically or antisymmetrically into itself. Symbolically,

$$RS_a = +S_a \quad \text{(symmetric)}$$

$$RS_a = -S_a \quad \text{(antisymmetric)}$$

For a degenerate vibration, a pair of coordinates transforms into itself. Symbolically,

$$RS_a = \alpha S_a + \beta S_b$$

$$RS_b = \gamma S_a + \delta S_b$$

or

$$R(S_a, S_b) = (S_a, S_b)$$

Next, derive the f matrix, i.e., the matrix formed by the force constants. Finally, obtain the U and U' matrices, where U is the matrix formed by the coefficients of the symmetry coordinates, and U' is the transpose of U. Then

$$F = UfU' \tag{4-5}$$

(3) *Obtain the G Matrix* (G). This G can be defined in terms of elements of a modified form of the kinetic energy matrix.[1,2] For nondegenerate vibrations it can be expressed as

$$G_{jl} = \sum p\mu_p g_p S_j^{(t)} \cdot S_l^{(t)} \tag{4-6}$$

where jl refers to the symmetry coordinates used to determine the S vectors, p is the set of equivalent atoms, μ_p is the reciprocal of the mass of the typical atom t, and g_p is the number of equivalent atoms in the pth set. The summation is over all equivalent atoms. The dot indicates the scalar product of the S vectors. The S vectors are defined as

$$S_j^{(t)} = \sum U_{jk} s_{kt} \tag{4-7}$$

where j refers to the jth symmetry coordinate and t to the typical atom, U_{jk} is the coefficient of the kth internal coordinate, s_{kt} are vectors expressed in terms of unit vectors along chemical bonds, and the summation extends over all internal coordinates. For a more thorough presentation of this subject, see Wilson *et al.*[1] and Meister and Cleveland.[3]

The G matrix can be determined more simply by using the notation of Appendix 6.[1,2,4] The G matrix can be expressed as

$$G = UgU' \tag{4-8}$$

where U and U' are defined exactly as they were for the F matrix, and g is similar to the f matrix. The elements of the g matrix are given in terms of internal coordinates, such as bond distances (d) and deformation angles (α).

(4) *Once the F and G matrices are known, the secular determinant can be written.* This has the form

$$\begin{vmatrix} \sum_i G_{1i}F_{i1} - \lambda & \sum_j G_{1j}F_{j2} & \cdots & \sum_k G_{1k}F_{kn} \\ \sum_i G_{2i}F_{i1} & \sum_j G_{2j}F_{j2} - \lambda & \cdots & \sum_k G_{2k}F_{kn} \\ \cdots & \cdots & \cdots & \cdots \\ \sum_i G_{ni}F_{i1} & \sum_j G_{nj}F_{j2} & \cdots & \sum_k G_{nk}F_{kn} - \lambda \end{vmatrix} = | GF - E\lambda | = 0 \tag{4-9}$$

where E is the unit matrix and λ is related to the frequency in wave number units. After substitution of the proper values for G and F, the secular determinant may be expanded into an equation in terms of λ. The equation can then be solved for λ, and since

$$\lambda = 4\pi^2 c^2 \nu^2 \tag{4-10}$$

ν in wave numbers can be obtained.

4-2. NORMAL COORDINATE TREATMENT OF H_2O (C_{2v} SYMMETRY)

Since water is a nonlinear molecule of C_{2v} symmetry, it will have $3n - 6$ vibrational degrees of freedom. Since $n = 3$, three coordinates are necessary to describe the molecule. Following the steps outlined above, the internal coordinates can be determined on the basis of a C_{2v} model.

Figure 4-1 shows the bond distances and bond angle for H_2O. The changes in these quantities are the three internal coordinates Δd_1, Δd_2, and $\Delta\alpha$ of the water molecule.

Decius[5] has devised a method of determining the types and numbers of internal coordinates in a molecule. For the general case, there are four types:

(1) Bond stretching r (change in the interatomic distance); the number of such coordinates is given by

$$n_r = b \tag{4-11}$$

where b is the number of bonds (disregarding type).

(2) Angle deformation φ (change in interbond angle); the number of such coordinates is given by

$$n_\varphi = 4b - 3a + a_1 \tag{4-12}$$

where a is the number of atoms, and a_1 is the number of atoms of multiplicity one (where multiplicity indicates the number of bonds meeting at the central atom).

(3) Torsional coordinates τ, whose number is given by

$$n_\tau = b - a_1 \tag{4-13}$$

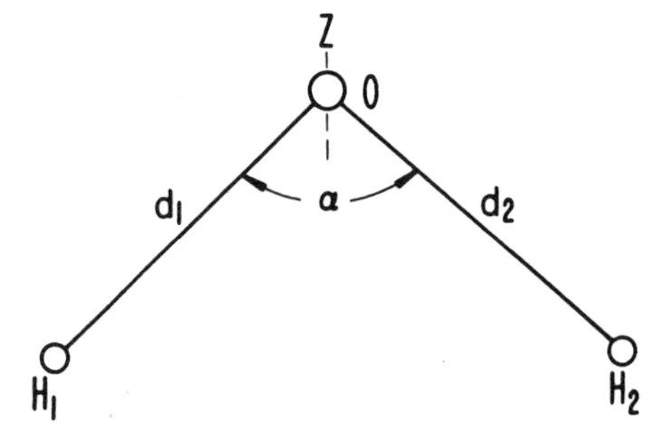

Fig. 4-1. Bond distances and bond angle for H_2O.

(4) Out-of-plane bending φ', which occurs only for linear molecules or for a linear portion within a molecule.

Decius also discusses the cases of linear and planar molecules.[5]

We choose H_2O to illustrate the method. Here $b = 2$, $a = 3$, and $a_1 = 2$, and therefore

$$n_r = 2$$

$$n_\varphi = 1$$

$$n_\tau = 0$$

giving H_2O two types of internal coordinates, bond stretching and angle deformation.

We now proceed to the selection of symmetry coordinates, which are defined as linear combinations of internal coordinates and are designated as R. Three are necessary to describe H_2O. A suitable set of symmetry coordinates would be as follows. For the two A_1 vibrations found for H_2O

$$R_1 = \Delta d_1 + \Delta d_2 \tag{4-14}$$

$$R_2 = \Delta\alpha \tag{4-15}$$

and for the B_2 vibration

$$R_3 = \Delta d_1 - \Delta d_2 \tag{4-16}$$

The selection of symmetry coordinates is facilitated by experience, and the choice of such coordinates is not unique. (See Appendix 7 for a method which can be used to obtain symmetry coordinates.)

The next step is the normalization of the symmetry coordinates. The condition for normalization is that Eq. (4-3) hold. Since from Eq. (4-14) the U coefficients for $\Delta d_1 + \Delta d_2$ are 1, and $1^2 + 1^2 \neq 1$, the R_1 symmetry coordinate is not normalized as it stands. However, if we set each of the coefficients equal to $1/\sqrt{2}$, Eq. (4-14) becomes

$$R_1 = (1/\sqrt{2})\Delta d_1 + (1/\sqrt{2})\Delta d_2 \tag{4-17}$$

and $(1/\sqrt{2})^2 + (1/\sqrt{2})^2 = 1$. Thus, the R_1 symmetry coordinate is now normalized. The R_2 symmetry coordinate, Eq. (4-15), is normalized as it stands, since $(1)^2 = 1$. Finally, R_3 can be normalized by making the coefficients $1/\sqrt{2}$; then $(1/\sqrt{2})^2 + (-1/\sqrt{2})^2 = 1$. Thus

$$R_3 = (1/\sqrt{2})\Delta d_1 - (1/\sqrt{2})\Delta d_2 \tag{4-18}$$

and R_1, R_2, and R_3 are now all normalized.

For orthogonalization of the symmetry coordinates, it is necessary that Eq. (4-4) hold, i.e., that

$$\sum_k U_{jk}U_{lk} = 0$$

For R_1 and R_2, the coefficients of Δd_1 and Δd_2 are multiplied by zero since Δd_1 and Δd_2 are not involved in R_2. Similarly, the coefficient of R_2 is multiplied by zero since no $\Delta\alpha$ is involved in R_1. For example,

$$(1/\sqrt{2})(0) + (1/\sqrt{2})(0) + (0)1 = 0$$

For R_1 and R_3

$$(1/\sqrt{2})(1/\sqrt{2}) + 1/\sqrt{2}(-1/\sqrt{2}) + 0(0) = 0$$

For R_2 and R_3

$$(0)(1/\sqrt{2}) + 0(-1/\sqrt{2}) + 1(0) = 0$$

Hence, R_1, R_2, and R_3 are orthogonal. Thus, the symmetry coordinates have both been normalized and orthogonalized.

It is now necessary to see if the symmetry coordinates transform properly. Table 4-1 restates the character table for C_{2v}. If the covering operations are applied to the internal coordinates, they will transform as indicated in Table 4-2.

If the symmetry coordinates are now considered for each covering operation, the following results are obtained:

$$E(R_1) = (1)R_1$$
$$C_2(R_1) = (1/\sqrt{2})\Delta d_2 + (1/\sqrt{2})\Delta d_1 = (1)R_1$$
$$\sigma_v(R_1) = (1/\sqrt{2})\Delta d_2 + (1/\sqrt{2})\Delta d_1 = (1)R_1$$
$$\sigma_{v'}(R_1) = (1/\sqrt{2})\Delta d_2 + (1/\sqrt{2})\Delta d_1 = (1)R_1$$

Table 4-1. Character Table for C_{2v} Symmetry

C_{2v}	E	C_2	$\sigma_v(xz)$	$\sigma_v'(yz)$
A_1	1	1	1	1
A_2	1	1	−1	−1
B_1	1	−1	1	−1
B_2	1	−1	−1	1

Table 4-2. Transformation of the Internal Coordinates for H_2O

	E	C_2	$\sigma_v(xz)$	$\sigma_v'(yz)$
Δd_1	Δd_1	Δd_2	Δd_2	Δd_1
Δd_2	Δd_2	Δd_1	Δd_1	Δd_2
$\Delta\alpha$	$\Delta\alpha$	$\Delta\alpha$	$\Delta\alpha$	$\Delta\alpha$

also

$$E(R_2) = 1(R_2)$$
$$C_2(R_2) = 1(R_2)$$
$$\sigma_v(R_2) = 1(R_2)$$
$$\sigma_{v'}(R_2) = 1(R_2)$$

and

$$E(R_3) = 1(R_3)$$
$$C_2(R_3) = (1/\sqrt{2})\Delta d_2 - (1/\sqrt{2})\Delta d_1 = (-1)R_3$$
$$\sigma_v(R_3) = (1/\sqrt{2})\Delta d_2 - (1/\sqrt{2})\Delta d_1 = (-1)R_3$$
$$\sigma_{v'}(R_3) = (1/\sqrt{2})\Delta d_1 - (1/\sqrt{2})\Delta d_2 = (1)R_3$$

Table 4-3 summarizes these transformations.

Since only nondegenerate vibrations are involved, the results show that in all the transformations the symmetry coordinate goes either into itself (positive) or into its negative.

A comparison of Tables 4-1 and 4-3 shows that R_3 transforms according to the character of the type B_2 vibration, while R_1 and R_2 transform according to the character of the two A_1 vibrations. Hence, the transformation is proper, and all the requirements for the symmetry coordinates are fulfilled.

We now proceed to the derivation of the F matrix. To obtain equations from which vibrational frequencies can be calculated, it is necessary to

Table 4-3. Transformation of Symmetry Coordinates for H_2O

	E	C_2	$\sigma_v(xz)$	$\sigma_v'(yz)$
R_1	1	1	1	1
R_2	1	1	1	1
R_3	1	−1	−1	1

derive the F matrix. The F matrix is related to the potential energy of the molecule,[6,7] which for a harmonic oscillator can be expressed as

$$2V = \sum_{i,k} f_{ik} r_i r_k \tag{4-19}$$

where $f_{ik} = f_{ki}$ and are force constants in units of dynes/cm, and i, k extend over all the internal coordinates r. The potential energy can also be expressed in terms of the symmetry coordinates:

$$2V = \sum F_{jl} R_j R_l \tag{4-20}$$

where $F_{jl} = F_{lj}$, and j and l extend over symmetry coordinates. For H_2O this becomes[8]

$$2V = f_d[(\Delta d_1)^2 + (\Delta d_2)^2] + f_\alpha(d\Delta\alpha)^2 + 2f_{d\alpha}(\Delta d_1 + \Delta d_2)(d\Delta\alpha) + 2f_{dd}(\Delta d_1)(\Delta d_2) \tag{4-21}$$

where $d = d_1 = d_2$ is the equilibrium value of the O–H bond distance. In matrix notation Eq. (4-20) takes the form

$$2V = r'fr \tag{4-22}$$

or

$$2V = R'FR \tag{4-23}$$

where r' and R' are the transposes of the r and R matrices. Combining Eqs. (4-22) and (4-23), one obtains

$$r'fr = R'FR \tag{4-24}$$

Since from Eq. (4-2) $R = Ur$ or $r = RU^{-1}$, where U^{-1} is the inverse of U, and $U^{-1} = U'$,

$$r = U'R \tag{4-25}$$

and

$$r' = (U'R)' = R'U \tag{4-26}$$

From Eq. (4-24), where $r'fr = R'FR$, then

$$(R'U)f(U'R) = R'FR \tag{4-27}$$

and it follows that

$$R'(UfU')R = R'FR \tag{4-28}$$

and

$$UfU' = F \tag{4-29}$$

Table 4-4. The f Matrix for H_2O

f	Δd_1	Δd_2	$\Delta\alpha$
Δd_1	$f_{d_1d_1}$	$f_{d_1d_2}$	$f_{d_1\alpha}$
Δd_2	$f_{d_2d_1}$	$f_{d_2d_2}$	$f_{d_2\alpha}$
$\Delta\alpha$	$f_{\alpha d_1}$	$f_{\alpha d_2}$	$f_{\alpha\alpha}$

Thus, F can be determined as the product of the U, U', and f matrices, and we now proceed to derive these.

The f matrix for H_2O is illustrated in Table 4-4, and corresponds to the potential function for H_2O written in Eq. (4-21). In order to express all force constants in units of dynes/cm, all changes in angle must be multiplied by a distance d in centimeter. When this is done, and since $f_{d_1d_1} = f_d = f_{d_2d_2}$ and $f_{\alpha\alpha} = f_\alpha$, Table 4-4 can be simplified and the f matrix becomes

$$f_{kl} = \begin{bmatrix} f_d & f_{dd} & df_{d\alpha} \\ f_{dd} & f_d & df_{d\alpha} \\ df_{d\alpha} & df_{d\alpha} & d^2 f_\alpha \end{bmatrix} \qquad k = \text{row} \downarrow,\ l = \text{column} \rightarrow$$

If a generalized valence force field is used, four force constants are necessary, and since only three vibrations exist for H_2O, it is impossible to determine all four simultaneously. In the case of H_2O, one can obtain additional frequencies by using D_2O and assuming the same set of force constants; one thus arrives at values for all four force constants.

The U matrix is derived as follows. For the two A_1 and the B_2 vibrations the U coefficients are obtained from Eqs. (4-15), (4-17), and (4-18). Thus, the U matrix is

	Δd_1	Δd_2	$\Delta\alpha$
A_1 R_1	$1/\sqrt{2}$	$1/\sqrt{2}$	0
A_1 R_2	0	0	1
$B_2 \rightarrow R_3$	$1/\sqrt{2}$	$-1/\sqrt{2}$	0

$$= U_{ik} \qquad i = \text{row} \downarrow,\ k = \text{column} \rightarrow$$

The U' matrix is simply the transpose of U and is, therefore,

$$U'_{lj} = \begin{bmatrix} 1/\sqrt{2} & 0 & 1/\sqrt{2} \\ 1/\sqrt{2} & 0 & -1/\sqrt{2} \\ 0 & 1 & 0 \end{bmatrix} \qquad l = \text{row} \downarrow,\ j = \text{column} \rightarrow$$

Now that we have obtained f, U, and U', we can solve Eq. (4-29) for the F matrix. This solution involves a triple multiplication of matrices. If the f matrix is multiplied by the U' matrix following the procedure outlined in Chapter 3, and using the U' matrix for the A_1 vibrations, we obtain by multiplication of a 3×3 matrix and a 3×2 matrix

$$fU' = \begin{bmatrix} (1/\sqrt{2})(f_d + f_{dd}) & df_{d\alpha} \\ (1/\sqrt{2})(f_d + f_{dd}) & df_{d\alpha} \\ \sqrt{2}\, df_{d\alpha} & d^2 f_\alpha \end{bmatrix}$$

If this is then multiplied by the U matrix for the A_1 vibrations, we obtain

$$\begin{bmatrix} F_{11} & F_{12} \\ F_{21} & F_{22} \end{bmatrix} = \begin{bmatrix} f_d + f_{dd} & \sqrt{2}\, df_{d\alpha} \\ \sqrt{2}\, df_{d\alpha} & d^2 f_\alpha \end{bmatrix}$$

Four matrix elements are involved since two A_1 vibrations exist.

Alternatively, the following procedure can be followed, which has an advantage in cases when larger matrices are to be multiplied. From Eq. (4-29)

$$F_{ij} = \sum_{k,l} (U_{ik} f_{kl} U'_{lj}) \tag{4-30}$$

where i refers to the row in the matrices F and U, j to the column in the matrices F and U', and k to the column in the U matrix and to the row in the f matrix. These definitions are illustrated above for the f, U, and U' matrices.

For $i = j = 1$

$$F_{11} = \sum_{k,l} (U_{1k} f_{kl} U'_{l1}) \tag{4-31}$$

Summing over k values from 1 to 3, one obtains

$$F_{11} = \sum_{l} (U_{11} f_{1l} U'_{l1} + U_{12} f_{2l} U'_{l1} + U_{13} f_{3l} U'_{l1}) \tag{4-32}$$

Since $U_{13} = 0$ and U_{11} and $U_{12} = 1/\sqrt{2}$ for H_2O,

$$F_{11} = (1/\sqrt{2}) \sum_{l} (f_{1l} + f_{2l}) U'_{l1} \tag{4-33}$$

Summing over values of l from 1 to 3, one obtains

$$F_{11} = (1/\sqrt{2})\ [(f_{11} + f_{21}) U'_{11} + (f_{12} + f_{22}) U'_{21} + (f_{13} + f_{23}) U'_{31}] \tag{4-34}$$

Since $U'_{11} = U'_{21} = 1/\sqrt{2}$, $U'_{31} = 0$, and substituting for the f_{kl} from the f

matrix, one obtains

$$F_{11} = (1/\sqrt{2})(f_d + f_{dd})(1/\sqrt{2}) + (1/\sqrt{2})(f_{dd} + f_d)(1/\sqrt{2})$$

or

$$F_{11} = f_d + f_{dd} \tag{4-35}$$

For F_{12}, $i = 1$, $j = 2$, one obtains from Eq. (4-30)

$$F_{12} = \sum_{k,l} (U_{1k} f_{kl} U'_{l2}) \tag{4-36}$$

summing over k values from 1 to 3, one obtains

$$F_{12} = \sum_{l} (U_{11} f_{1l} U'_{l2} + U_{12} f_{2l} U'_{l2} - U_{13} f_{3l} U'_{l2}) \tag{4-37}$$

Since $U_{11} = U_{12} = 1/\sqrt{2}$ and $U_{13} = 0$,

$$F_{12} = (1/\sqrt{2}) \sum_{l} (f_{1l} + f_{2l}) U'_{l2} \tag{4-38}$$

Summing over values of l from 1 to 3, one obtains

$$F_{12} = (1/\sqrt{2})\ [(f_{11} + f_{21})U'_{12} + (f_{12} + f_{22})U'_{22} + (f_{13} + f_{23})U'_{32}] \tag{4-39}$$

By substituting for the f and U' values from their respective matrices, the following result is obtained:

$$F_{12} = (1/\sqrt{2})(df_{d\alpha} + df_{d\alpha}) = \sqrt{2}\, df_{d\alpha} \tag{4-40}$$

For F_{21}, $i = 2$, $j = 1$,

$$F_{21} = \sum_{k,l} (U_{2k} f_{kl} U'_{l1}) \tag{4-41}$$

Summing over k values from 1 to 3, one obtains

$$F_{21} = \sum_{l} (U_{21} f_{1l} U'_{l1} + U_{22} f_{2l} U'_{l1} + U_{23} f_{3l} U'_{l1}) \tag{4-42}$$

Since $U_{21} = U_{22} = 0$ and $U_{23} = 1$

$$F_{21} = \sum_{l} (f_{3l}) U'_{l1} \tag{4-43}$$

Summing next over values of l from 1 to 3, one obtains

$$F_{21} = f_{31} U'_{11} + f_{32} U'_{21} + f_{33} U'_{31} \tag{4-44}$$

By substituting for the f and U' values, F_{21} becomes

$$F_{21} = (1/\sqrt{2})(df_{d\alpha} + df_{d\alpha}) = \sqrt{2}\, df_{d\alpha} \tag{4-45}$$

For F_{22}, $i = 2$, $j = 2$,

$$F_{22} = \sum_{k,l} (U_{2k} f_{kl} U'_{l2}) \tag{4-46}$$

and following similar procedures, after summing for k and l values from 1 to 3, the result is

$$F_{22} = d^2 f_\alpha \tag{4-47}$$

For the B_2 type vibration only one matrix element exists—F_{33}. Using the U and U' matrix for the B_2 type vibration and first multiplying the f matrix by the U' matrix, we obtain

$$fU' = \begin{bmatrix} (1/\sqrt{2})(f_d - f_{dd}) \\ (1/\sqrt{2})(f_{dd} - f_d) \\ 0 \end{bmatrix}$$

This is then multiplied by U and, therefore,

$$F = (F_{33}) = f_d - f_{dd} \tag{4-48}$$

The same result for F_{33} can be obtained by solving Eq. (4-30).

We next determine the G matrix, which is expressed by Eq. (4-8). It is first necessary to write a g matrix for H_2O. When this is done and the result is simplified, as was done for the f matrix, one obtains the matrix in Table 4-5 and

$$g = \begin{bmatrix} g_d & g_{dd} & g_{d\alpha} \\ g_{dd} & g_d & g_{d\alpha} \\ g_{d\alpha} & g_{d\alpha} & g_\alpha \end{bmatrix}$$

Having obtained a g matrix and using the U and U' values for the two

Table 4-5. The g Matrix for H_2O

g	Δd_1	Δd_2	$\Delta\alpha$
Δd_1	$g_{d_1 d_1}$	$g_{d_1 d_2}$	$g_{d_1 \alpha}$
Δd_2	$g_{d_2 d_1}$	$g_{d_2 d_2}$	$g_{d_2 \alpha}$
$\Delta\alpha$	$g_{\alpha d_1}$	$g_{\alpha d_2}$	$g_{\alpha\alpha}$

A_1 vibrations in H_2O, we can now obtain the G matrix. The product gU' equals

$$gU' = \begin{bmatrix} (1/\sqrt{2})(g_d + g_{dd}) & g_{d\alpha} \\ (1/\sqrt{2})(g_d + g_{dd}) & g_{d\alpha} \\ \sqrt{2}\, g_{d\alpha} & g_\alpha \end{bmatrix}$$

When this is multiplied by the U matrix, we obtain

$$\begin{bmatrix} G_{11} & G_{12} \\ G_{21} & G_{22} \end{bmatrix} = \begin{bmatrix} g_d + g_{dd} & \sqrt{2}\, g_{d\alpha} \\ \sqrt{2}\, g_{d\alpha} & g_\alpha \end{bmatrix}$$

Note the similarity with the F matrix.

Alternatively, the same results could have been obtained using a modified form of Eq. (4-30), such as

$$G_{ij} = \sum_{k,l} U_{ik} g_{kl} U'_{lj} \tag{4-49}$$

Using the notation of Appendix 6, the elements of the G matrix for H_2O can be shown to be

$$g_d = G^2_{rr} = \mu_x + \mu_y \tag{4-50}$$

$$g_{dd} = G^1_{rr} = \mu_x \cos\alpha \tag{4-51}$$

$$g_{d\alpha} = G^2_{r\varphi} = -(1/d)\mu_x \sin\alpha \tag{4-52}$$

$$g_\alpha = G^3_{\varphi\varphi} = (2/d^2)[\mu_y + \mu_x(1 - \cos\alpha)] \tag{4-53}$$

Therefore

$$G_{11} = g_d + g_{dd} = \mu_x + \mu_y + \mu_x \cos\alpha \tag{4-54}$$

$$G_{11} = \mu_x(1 + \cos\alpha) + \mu_y \tag{4-55}$$

Since x = oxygen and y = hydrogen,

$$G_{11} = \mu_0(1 + \cos\alpha) + \mu_H \tag{4-56}$$

For $G_{21} = G_{12}$

$$G_{21} = \sqrt{2}\, g_{d\alpha} = -(\sqrt{2}/d)\mu_x \sin\alpha \tag{4-57}$$

and

$$G_{21} = -(\sqrt{2}/d)\mu_0 \sin\alpha \tag{4-58}$$

For G_{22}

$$G_{22} = g_\alpha = (2/d^2)[\mu_y + \mu_x(1 - \cos\alpha)] \tag{4-59}$$

and

$$G_{22} = (2/d^2)[\mu_H + \mu_0(1 - \cos\alpha)] \tag{4-60}$$

The G matrix for the type A_1 vibrations in H_2O is then

$$\begin{bmatrix} G_{11} & G_{12} \\ G_{21} & G_{22} \end{bmatrix} = \begin{bmatrix} \mu_0(1 + \cos\alpha) + \mu_H & -(\sqrt{2}/d)\mu_0 \sin\alpha \\ -(\sqrt{2}/d)\mu_0 \sin\alpha & (2/d^2)[\mu_H + \mu_0(1 - \cos\alpha)] \end{bmatrix}$$

The G matrix for the type B_2 vibrations is obtained in the same manner. The U matrix is

$$[1/\sqrt{2} \quad -1/\sqrt{2} \quad 0]$$

and the U' matrix is

$$\begin{bmatrix} 1/\sqrt{2} \\ -1/\sqrt{2} \\ 0 \end{bmatrix}$$

Therefore, gU' will be

$$gU' = \begin{bmatrix} (1/\sqrt{2})(g_d - g_{dd}) \\ (1/\sqrt{2})(g_{dd} - g_d) \\ 0 \end{bmatrix}$$

Multiplying this by U, one obtains

$$G_{33} = (g_d - g_{dd}) \tag{4-61}$$

and

$$G_{33} = \mu_x + \mu_y - \mu_x \cos\alpha \tag{4-62}$$

$$G_{33} = \mu_0(1 - \cos\alpha) + \mu_H \tag{4-63}$$

The F and G matrices for H_2O have now been obtained. It remains to solve the secular equation. Wilson[6,7] has shown that the secular equation for type A_1 fundamentals can be expanded from the secular determinant as follows:

$$(-\lambda)^2 + (F_{11}G_{11} + 2F_{12}G_{12} + F_{22}G_{22})(-\lambda) + \begin{vmatrix} F_{11} & F_{12} \\ F_{21} & F_{22} \end{vmatrix} \cdot \begin{vmatrix} G_{11} & G_{12} \\ G_{21} & G_{22} \end{vmatrix} = 0 \tag{4-64}$$

or

$$\lambda^2 - (F_{11}G_{11}+2F_{12}G_{12}+F_{22}G_{22})\lambda+(F_{11}F_{22} - F_{12}^2)(G_{11}G_{22} - G_{12}^2) = 0$$

The constant term is the product of the determinants of the F and G matrices, and the coefficient of the λ term is the sum of the products of the one-rowed minors of the determinants of the F and G matrices. It is simpler at this stage to substitute the values for force constant, bond distance, and bond angle in the F and G matrices. The secular equation can then be solved for λ. For example, Eq. (4-64) becomes

$$\lambda^2 - (6.159 \times 10^{29} \text{ sec}^{-2})\lambda + 5.032 \times 10^{58} \text{ sec}^{-4} = 0 \tag{4-65}$$

By substituting the λ values into

$$\lambda = 4\pi^2c^2\nu^2 \tag{4-66}$$

where c is the velocity of light in cm/sec, and solving for ν in cm^{-1}, the values for the two A_1 vibrations for H_2O are obtained.

For the B_2 type of vibrations, the secular equation in expanded form is

$$\lambda_3 - F_{33}G_{33} = 0 \tag{4-67}$$

When this equation is combined with Eq. (7-66), the following equation is obtained:

$$\nu_3 = (F_{33}G_{33})^{1/2}/2\pi c \tag{4-68}$$

and ν_3 can be solved for by substituting the values for f, d, and α in the F and G matrices.

Equations (4-65) and (4-68) can be used for any nonlinear AB_2 molecule. It is only necessary to substitute the appropriate values for the constants.

The units of the elements of the F matrix are summarized below. The units for bond distance are centimeters; the units for f are dynes/cm. Thus

$$F_{11} = (f_d + f_{dd}) \qquad [\text{dynes/cm}]$$

$$F_{12} = F_{21} = \sqrt{2}\, df_{d\alpha} \qquad [\text{cm}\cdot\text{dynes/cm} = \text{dynes}]$$

$$F_{22} = d^2f_\alpha \qquad [\text{cm}^2\cdot\text{dynes/cm} = \text{dynes}\cdot\text{cm}]$$

$$F_{33} = f_d - f_{dd} \qquad [\text{dynes/cm}]$$

The units of the elements of the G matrix may be summarized as follows:

$$G_{11} = \mu_H + \mu_0(1 + \cos\alpha) \qquad [1/\text{g}]$$

$$G_{12} = G_{21} = -\frac{\mu_0\sqrt{2}\sin\alpha}{d} \qquad [1/\text{g}\cdot\text{cm}]$$

$$G_{22} = \frac{2[\mu_H + \mu_0(1 - \cos\alpha)]}{d^2} \qquad [1/\text{g}\cdot\text{cm}^2]$$

$$G_{33} = \mu_H + \mu_0(1 - \cos\alpha) \qquad [1/\text{g}]$$

For the matrix multiplication used in Eqs. (4-64) and (4-67), the following results are obtained:

$$F_{11} \times G_{11} \qquad [\text{dynes/cm} \times 1/\text{g} = \text{dynes/cm}\cdot\text{g}]$$

$$F_{12} \times G_{12} \qquad [\text{dynes} \times 1/\text{g}\cdot\text{cm} = \text{dynes/cm}\cdot\text{g}]$$

$$F_{22} \times G_{22} \qquad [\text{dynes}\cdot\text{cm} \times 1/\text{g}\cdot\text{cm}^2 = \text{dynes/cm}\cdot\text{g}]$$

$$F_{33} \times G_{33} \qquad [\text{dynes/cm} \times 1/\text{g} = \text{dynes/cm}\cdot\text{g}]$$

It can be observed that the units of the products (FG) are in dynes/cm$\cdot$g or

$$\frac{\text{g}\cdot\text{cm}}{\text{sec}^2} \times \frac{1}{\text{cm}\cdot\text{g}} = \frac{1}{\text{sec}^2} = \text{sec}^{-2}$$

Similarly

$$(F_{11}F_{22} - F_{12}^2)(G_{11}G_{22} - G_{12}^2)$$

$$\left[\left(\frac{\text{dynes}}{\text{cm}\cdot\text{g}}\right)\left(\frac{\text{dynes}}{\text{cm}\cdot\text{g}}\right) = \left(\frac{\text{g}\cdot\text{cm}}{\text{sec}^2} \times \frac{1}{\text{g}\cdot\text{cm}}\right)\left(\frac{\text{g}\cdot\text{cm}}{\text{sec}^2} \times \frac{1}{\text{g}\cdot\text{cm}}\right) = \frac{1}{\text{sec}^4} = \text{sec}^{-4}\right]$$

If one solves for ν in Eq. (7-68), one obtains

$$\nu_3 = (F_{33}G_{33})^{1/2}/2\pi c \qquad \left[\frac{\sqrt{1/\text{sec}^2}}{\text{cm/sec}} = \frac{1/\text{sec}}{\text{cm/sec}} = \frac{1}{\text{cm}} = \text{cm}^{-1}\right]$$

and the value of ν_3 is in cm^{-1}.

4-3. NORMAL COORDINATE TREATMENT OF NH_3 (C_{3v} SYMMETRY)

If we assign a C_{3v} symmetry to ammonia, it can be demonstrated that there will be two A_1 and two E type vibrations (see Chapter 2). The modes of vibration are illustrated in Appendix 9. For the two A_1 vibrations, two symmetry coordinates will be needed. For the two E vibrations, four symmetry coordinates will be necessary, since they are doubly degenerate.

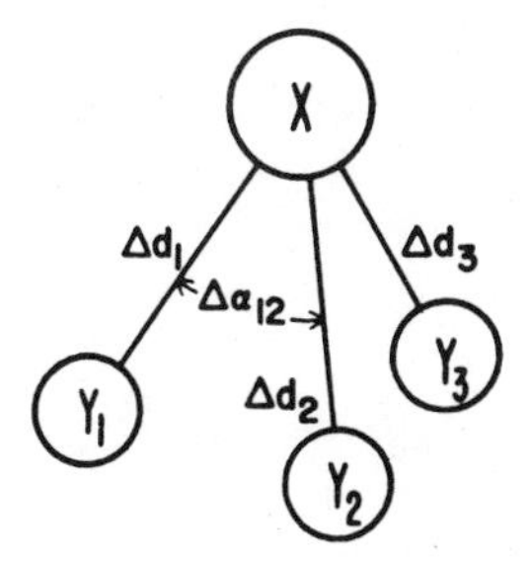

Fig. 4-2. Internal coordinates for NH_3.

The internal coordinates for NH_3 are shown in Fig. 4-2. Only bond stretching and angle deformations need be considered. The six internal coordinates are Δd_1, Δd_2, Δd_3, $\Delta\alpha_{12}$, $\Delta\alpha_{23}$, and $\Delta\alpha_{13}$. Six symmetry coordinates are needed to describe the molecule, and these may be written as follows:

For the A_1 species

$$R_1 = (1/\sqrt{3})\Delta d_1 + (1/\sqrt{3})\Delta d_2 + (1/\sqrt{3})\Delta d_3 \quad (4\text{-}69)$$

$$R_2 = (1/\sqrt{3})\Delta\alpha_{12} + (1/\sqrt{3})\Delta\alpha_{23} + (1/\sqrt{3})\Delta\alpha_{13} \quad (4\text{-}70)$$

For the E species

$$R_{3a} = (2/\sqrt{6})\Delta d_1 - (1/\sqrt{6})\Delta d_2 - (1/\sqrt{6})\Delta d_3 \quad (4\text{-}71)$$

$$R_{3b} = (1/\sqrt{2})\Delta d_2 - (1/\sqrt{2})\Delta d_3 \quad (4\text{-}72)$$

$$R_{4a} = (2/\sqrt{6})\Delta\alpha_{12} - (1/\sqrt{6})\Delta\alpha_{23} - (1/\sqrt{6})\Delta\alpha_{13} \quad (4\text{-}73)$$

$$R_{4b} = (1/\sqrt{2})\Delta\alpha_{23} - (1/\sqrt{2})\Delta\alpha_{13} \quad (4\text{-}74)$$

These coordinates all are normalized and made orthogonal according to Eqs. (4-3) and (4-4). The transformations of the internal coordinates under the covering operations of the C_{3v} group are shown in Table 4-6. Proceeding as for H_2O, we can show that the A_1 symmetry coordinates R_1 and R_2 transform according to the characters for the A_1 species. Table 4-7 shows the character table for the C_{3v} group.

Applying a covering operation to one of a pair of E type vibrations fails to give the same coordinate or its negative. One obtains, instead, a linear combination of the coordinates which form the pair. For example, if the identity operation (E) is applied to the pair R_{3a} and R_{3b}, one obtains

$$E(R_{3a}) = (2/\sqrt{6})\Delta d_1 - (1/\sqrt{6})\Delta d_2 - (1/\sqrt{6})\Delta d_3 = AR_{3a} + BR_{3b} \quad (4\text{-}75)$$

Table 4-6. Transformations of the Internal Coordinates of the NH_3 Molecule Under the Covering Operations of the C_{3v} Group

	E	C_3^+	C_3^-	σ_v^1	σ_v^2	σ_v^3
Δd_1	Δd_1	Δd_3	Δd_2	Δd_1	Δd_3	Δd_2
Δd_2	Δd_2	Δd_1	Δd_3	Δd_3	Δd_2	Δd_1
Δd_3	Δd_3	Δd_2	Δd_1	Δd_2	Δd_1	Δd_3
$\Delta \alpha_{12}$	$\Delta \alpha_{12}$	$\Delta \alpha_{13}$	$\Delta \alpha_{23}$	$\Delta \alpha_{13}$	$\Delta \alpha_{23}$	$\Delta \alpha_{12}$
$\Delta \alpha_{23}$	$\Delta \alpha_{23}$	$\Delta \alpha_{12}$	$\Delta \alpha_{13}$	$\Delta \alpha_{23}$	$\Delta \alpha_{12}$	$\Delta \alpha_{13}$
$\Delta \alpha_{13}$	$\Delta \alpha_{13}$	$\Delta \alpha_{23}$	$\Delta \alpha_{12}$	$\Delta \alpha_{12}$	$\Delta \alpha_{13}$	$\Delta \alpha_{23}$

and

$$E(R_{3b}) = (1/\sqrt{2})\Delta d_2 - (1/\sqrt{2})\Delta d_3 = A'R_{3a} + B'R_{3b} \tag{4-76}$$

Substituting for R_{3a} and R_{3b}, one gets

$$\begin{aligned}(2/\sqrt{6})\Delta d_1 - (1/\sqrt{6})\Delta d_2 - (1/\sqrt{6})\Delta d_3 &= (2A/\sqrt{6})\Delta d_1 \\ &- (A/\sqrt{6})\Delta d_2 - (A/\sqrt{6})\Delta d_3 + (B/\sqrt{2})\Delta d_2 - (B/\sqrt{2})\Delta d_3\end{aligned} \tag{4-77}$$

and

$$\begin{aligned}(1/\sqrt{2})\Delta d_2 - (1/\sqrt{2})\Delta d_3 &= (2A'/\sqrt{6})\Delta d_1 - (A'/\sqrt{6})\Delta d_2 \\ &- (A'/\sqrt{6})\Delta d_3 + (B'/\sqrt{2})\Delta d_2 - (B'/\sqrt{2})\Delta d_3\end{aligned} \tag{4-78}$$

The constants A, B, A', B' can be obtained by equating the coefficients of Eqs. (4-77) and (4-78). For example, using Δd_1 from Eq. (4-77),

$$(2/\sqrt{6})\Delta d_1 = (2A/\sqrt{6})\Delta d_1$$
$$A = 1$$

Similarly, we obtain $B = 0$, $A' = 0$, and $B' = 1$. The matrix formed will be

Table 4-7. Character Table for the C_{3v} Group

C_{3v}	E	$2C_3$	$3\sigma_v$
A_1	1	1	1
A_2	1	1	−1
E	2	−1	0

$$E\begin{bmatrix} R_{3a} \\ R_{3b} \end{bmatrix} = \begin{bmatrix} A & B \\ A' & B' \end{bmatrix} = \begin{bmatrix} 1 & 0 \\ 0 & 1 \end{bmatrix}$$

The sum of the principal diagonal is $+2$ and is the character under E, the identity, for the E type vibration. Likewise it can be demonstrated that for the (C_3^+ or C_3^-) covering operations applied to R_{3a} and R_{3b}, the follow ing results ensue:

$$C_3^+\begin{bmatrix} R_{3a} \\ R_{3b} \end{bmatrix} = \begin{bmatrix} A & B \\ A' & B' \end{bmatrix} = \begin{bmatrix} -1/2 & \sqrt{3}/2 \\ \sqrt{3}/2 & -1/2 \end{bmatrix}$$

and the sum of the terms on the diagonal is -1, which corresponds to the character for C_3 for the E vibration. Similarly, the results for three σ_v's give a character of 0, since

$$\sigma_v\begin{bmatrix} R_{3a} \\ R_{3b} \end{bmatrix} = \begin{bmatrix} A & B \\ A' & B' \end{bmatrix} = \begin{bmatrix} 1 & 0 \\ 0 & -1 \end{bmatrix}$$

Thus, R_{3a} and R_{3b} belong to the E irreducible representation (see Table 4-7).

Thus, all the necessary conditions are satisfied insofar as the symmetry coordinates [Eqs. (4-69) to (4-74)] are concerned, and one can derive the F matrix for NH_3. The potential energy of the molecule takes the form of Eq. (4-19), and for NH_3 can be written

$$\begin{aligned} 2V = {} & f_d[(\Delta d_1)^2 + (\Delta d_2)^2 + (\Delta d_3)^2] + f_\alpha[(d\Delta\alpha_{12})^2 + (d\Delta\alpha_{23})^2 \\ & + (d\Delta\alpha_{13})^2] + 2f_{d\alpha}[\Delta d_1(d\Delta\alpha_{12} + d\Delta\alpha_{13}) + d_2(d\Delta\alpha_{12} + d\Delta\alpha_{23}) \\ & + d_3(d\Delta\alpha_{23} + d\Delta\alpha_{13})] + 2f_{d\alpha'}[\Delta d_1(d\Delta\alpha_{23}) + \Delta d_2(d\Delta\alpha_{13}) \\ & + \Delta d_3(d\Delta\alpha_{12})] + 2f_{\alpha\alpha}[(d\Delta\alpha_{12})(d\Delta\alpha_{23} + d\Delta\alpha_{13}) + (d\Delta\alpha_{23})(d\Delta\alpha_{13})] \\ & + 2f_{dd}[(\Delta d_1)(\Delta d_2) + (\Delta d_1)(\Delta d_3) + (\Delta d_2)(\Delta d_3)] \end{aligned} \quad (4\text{-}79)$$

The U matrix can be derived and is shown in Table 4-8 and the U' matrix is

$$U' = \begin{bmatrix} 1/\sqrt{3} & 0 & 2/\sqrt{6} & 0 & 0 & 0 \\ 1/\sqrt{3} & 0 & -1/\sqrt{6} & 1/\sqrt{2} & 0 & 0 \\ 1/\sqrt{3} & 0 & -1/\sqrt{6} & -1/\sqrt{2} & 0 & 0 \\ 0 & 1/\sqrt{3} & 0 & 0 & 2/\sqrt{6} & 0 \\ 0 & 1/\sqrt{3} & 0 & 0 & -1/\sqrt{6} & 1/\sqrt{2} \\ 0 & 1/\sqrt{3} & 0 & 0 & -1/\sqrt{6} & -1/\sqrt{2} \end{bmatrix}$$

Table 4-8. The U Matrix for NH_3

U		Δd_1	Δd_2	Δd_3	$\Delta\alpha_{12}$	$\Delta\alpha_{23}$	$\Delta\alpha_{13}$
A_1	R_1	$1/\sqrt{3}$	$1/\sqrt{3}$	$1/\sqrt{3}$	0	0	0
	R_2	0	0	0	$1/\sqrt{3}$	$1/\sqrt{3}$	$1/\sqrt{3}$
E	R_{3a}	$2/\sqrt{6}$	$-1/\sqrt{6}$	$-1/\sqrt{6}$	0	0	0
	R_{3b}	0	$1/\sqrt{2}$	$-1/\sqrt{2}$	0	0	0
	R_{4a}	0	0	0	$2/\sqrt{6}$	$-1/\sqrt{6}$	$-1/\sqrt{6}$
	R_{4b}	0	0	0	0	$1/\sqrt{2}$	$-1/\sqrt{2}$

The f matrix is tabulated in Table 4-9 and corresponds to the potential function written for NH_3 [Eq. (4-79)].

By using Eqs. (4-29) and (4-30), the F matrix can be derived. For the two A_1 species of vibration, the F matrix is found to be

$$\begin{bmatrix} F_{11} & F_{12} \\ F_{21} & F_{22} \end{bmatrix} = \begin{bmatrix} f_d + 2f_{dd} & d(f_{d\alpha'} + 2f_{d\alpha}) \\ d(f_{d\alpha'} + 2f_{d\alpha}) & d^2(f_\alpha + 2f_{\alpha\alpha}) \end{bmatrix}$$

The F matrix for the E vibrations may be determined following the same procedure:

$$\begin{bmatrix} F_{11} & F_{12} \\ F_{21} & F_{22} \end{bmatrix} = \begin{bmatrix} f_d - f_{dd} & d(f_{d\alpha'} - f_{d\alpha}) \\ d(f_{d\alpha'} - f_{d\alpha}) & d^2(f_\alpha - f_{\alpha\alpha}) \end{bmatrix}$$

In obtaining the F matrix for the E vibrations, one may use either the pair R_{3a}, R_{4a} or R_{3b}, R_{4b}.

It is necessary to write a g matrix for NH_3 before the G matrix can be obtained. The g matrix for NH_3 is tabulated in Table 4-10.

Table 4-9. The f Matrix for NH_3

f	Δd_1	Δd_2	Δd_3	$\Delta\alpha_{12}$	$\Delta\alpha_{23}$	$\Delta\alpha_{13}$
Δd_1	f_d	f_{dd}	f_{dd}	$df_{d\alpha}$	$df_{d\alpha'}$	$df_{d\alpha}$
Δd_2	f_{dd}	f_d	f_{dd}	$df_{d\alpha}$	$df_{d\alpha}$	$df_{d\alpha'}$
Δd_3	f_{dd}	f_{dd}	f_d	$df_{d\alpha'}$	$df_{d\alpha}$	$df_{d\alpha}$
$\Delta\alpha_{12}$	$df_{d\alpha}$	$df_{d\alpha}$	$df_{d\alpha'}$	$d^2f_{\alpha\alpha}$	$d^2f_{\alpha\alpha}$	$d^2f_{\alpha\alpha}$
$\Delta\alpha_{23}$	$df_{d\alpha'}$	$df_{d\alpha}$	$df_{d\alpha}$	$d^2f_{\alpha\alpha}$	d^2f_α	$d^2f_{\alpha\alpha}$
$\Delta\alpha_{13}$	$df_{d\alpha}$	$df_{d\alpha'}$	$df_{d\alpha}$	$d^2f_{\alpha\alpha}$	$d^2f_{\alpha\alpha}$	d^2f_α

Note: $f_{d\alpha'}$ indicates interaction between Δd and $\Delta\alpha$ having no common bond (e.g., Δd_1 and $\Delta\alpha_{23}$).

Table 4-10. The g Matrix for NH_3

g	Δd_1	Δd_2	Δd_3	$\Delta\alpha_{12}$	$\Delta\alpha_{23}$	$\Delta\alpha_{13}$
Δd_1	g_d	g_{dd}	g_{dd}	$g_{d\alpha}$	$g_{d\alpha'}$	$g_{d\alpha}$
Δd_2	g_{dd}	g_d	g_{dd}	$g_{d\alpha}$	$g_{d\alpha}$	$g_{d\alpha'}$
Δd_3	g_{dd}	g_{dd}	g_d	$g_{d\alpha'}$	$g_{d\alpha}$	$g_{d\alpha}$
$\Delta\alpha_{12}$	$g_{d\alpha}$	$g_{d\alpha}$	$g_{d\alpha'}$	g_α	$g_{\alpha\alpha}$	$g_{\alpha\alpha}$
$\Delta\alpha_{23}$	$g_{d\alpha'}$	$g_{d\alpha}$	$g_{d\alpha}$	$g_{\alpha\alpha}$	g_α	$g_{\alpha\alpha}$
$\Delta\alpha_{13}$	$g_{d\alpha}$	$g_{d\alpha'}$	$g_{d\alpha}$	$g_{\alpha\alpha}$	$g_{\alpha\alpha}$	g_α

The G matrix for the two A_1 vibrations can be determined by matrix multiplication using Eq. (4-49).

The U and U' values for the A_1 vibrations must be used. When this is done, the G matrix for the A_1 vibrations in terms of the g elements results:

$$\begin{bmatrix} G_{11} & G_{12} \\ G_{21} & G_{22} \end{bmatrix} = \begin{bmatrix} g_d + 2g_{dd} & 2g_{d\alpha} + g_{d\alpha'} \\ 2g_{d\alpha} + g_{d\alpha'} & g_\alpha + 2g_{\alpha\alpha} \end{bmatrix}$$

When we substitute values from Appendix 6 for the g elements, they become

$$g_d = G_{rr}^2 = \mu_N + \mu_H$$

$$g_{dd} = G_{rr}^1 = \mu_N \cos\alpha$$

$$g_{d\alpha'} = G_{r\varphi}^1\begin{pmatrix}1\\1\end{pmatrix} = -\frac{2}{d}\,\frac{\cos\alpha(1-\cos\alpha)\mu_N}{\sin\alpha}$$

$$g_{d\alpha} = G_{r\varphi}^2 = -\frac{\mu_N}{d}\sin\alpha$$

$$g_\alpha = G_{\varphi\varphi}^3 = \frac{2}{d^2}\,[\mu_H + \mu_N(1-\cos\alpha)]$$

$$g_{\alpha\alpha} = G_{\varphi\varphi}^2\begin{pmatrix}1\\1\end{pmatrix} = \frac{\mu_H}{d^2}\,\frac{\cos\alpha}{1+\cos\alpha} + \frac{\mu_N}{d^2}\,\frac{(1+3\cos\alpha)(1-\cos\alpha)}{(1+\cos\alpha)}$$

and the following results are obtained:

$$\begin{bmatrix} G_{11} & G_{12} \\ G_{21} & G_{22} \end{bmatrix}$$

$$= \begin{bmatrix} \mu_H + (1+2\cos\alpha)\mu_N & -\dfrac{2}{d}\,\dfrac{\mu_N(1+2\cos\alpha)(1-\cos\alpha)}{\sin\alpha} \\ -\dfrac{2}{d}\,\dfrac{\mu_N(1+2\cos\alpha)(1-\cos\alpha)}{\sin\alpha} & \dfrac{2}{d^2}\left(\dfrac{1+2\cos\alpha}{1+\cos\alpha}\right)[\mu_H + 2\mu_N(1-\cos\alpha)] \end{bmatrix}$$

By similar procedures, it can be shown that the G matrix for the two E type vibrations in terms of g elements is

$$\begin{bmatrix} G_{11} & G_{12} \\ G_{21} & G_{22} \end{bmatrix} = \begin{bmatrix} g_d - g_{dd} & g_{d\alpha'} - g_{d\alpha} \\ g_{d\alpha'} - g_{d\alpha} & g_\alpha - g_{\alpha\alpha} \end{bmatrix}$$

and with substitution for the g elements using the notation of Appendix 6,

$$\begin{bmatrix} G_{11} & G_{12} \\ G_{21} & G_{22} \end{bmatrix}$$

$$= \begin{bmatrix} \mu_H + \mu_N(1 - \cos\alpha) & \dfrac{-\mu_N}{d\sin\alpha}(1 - \cos\alpha)^2 \\ \dfrac{-\mu_N}{d\sin\alpha}(1 - \cos\alpha)^2 & \dfrac{1}{d^2(1+\cos\alpha)}[(2+\cos\alpha)\mu_H + (1-\cos\alpha)^2\mu_N] \end{bmatrix}$$

The F and G matrices for NH_3 have now been obtained. Providing the force constants and other parameters are known, and following the procedure outlined previously, one can obtain a secular equation, which can be solved for λ. Alternatively, one can determine which of the force constants can be set equal to zero if they are unobtainable. In the case of NH_3, we can use ND_3 to obtain additional frequencies and, following the procedure described for H_2O, the force constants can be obtained. For NH_3 and ammonia-like molecules the NCT has been made.[35–37] For the NCT of other C_{3v} molecules see Ziomek and Piotrowski.[38]

4-4. NORMAL COORDINATE TREATMENT OF UF_6 (O_h SYMMETRY)

The hexafluorides of Mo, W, Re, Os, Ir, Pt, U, Np, and Pu have been found to possess O_h symmetry. Molecules of formula AB_6 and possessing O_h symmetry have three Raman-active vibrations ($A_{1g} + E_g + F_{2g}$), two infrared-active vibrations ($2F_{1\mu}$), and one inactive ($F_{2\mu}$) vibration. The modes of vibrations are shown in Appendix 9.

The internal coordinates for an AB_6 molecule (e.g., UF_6) are illustrated in Fig. 4-3. Six bond length changes and 12 interbond angle changes are involved. These are: Δd_1, Δd_2, Δd_3, Δd_4, Δd_5, Δd_6, $\Delta\alpha_{12}$, $\Delta\alpha_{23}$, $\Delta\alpha_{34}$, $\Delta\alpha_{45}$, $\Delta\alpha_{56}$, $\Delta\alpha_{61}$, $\Delta\alpha_{13}$, $\Delta\alpha_{24}$, $\Delta\alpha_{35}$, $\Delta\alpha_{46}$, $\Delta\alpha_{51}$, and $\Delta\alpha_{62}$. Eighteen symmetry coordinates can be considered, three of which will be redundant.[9,10]

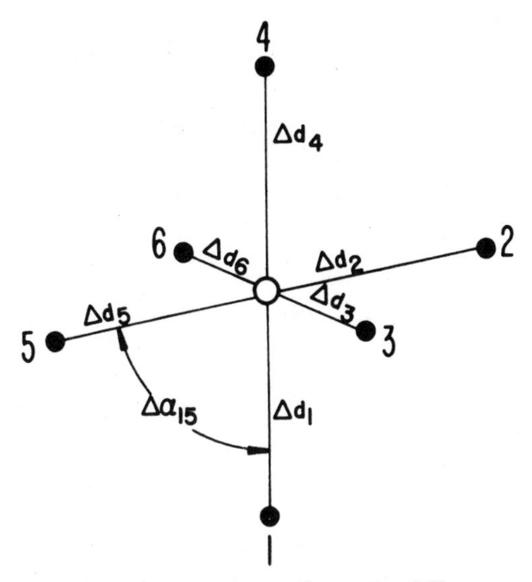

Fig. 4-3. Internal coordinates for UF_6.

A suitable set of orthonormal symmetry coordinates for each type of vibration can be given as follows:

For the A_{1g} vibration

$$R_1 = (1/\sqrt{6})(\Delta d_1 + \Delta d_2 + \Delta d_3 + \Delta d_4 + \Delta d_5 + \Delta d_6) \tag{4-80}$$

For the E_g vibration

$$R_{2a} = (1/\sqrt{12})(\Delta d_1 - 2\Delta d_2 + \Delta d_3 + \Delta d_4 - 2\Delta d_5 + \Delta d_6) \tag{4-81}$$

$$R_{2b} = \tfrac{1}{2}(\Delta d_1 + \Delta d_4 - \Delta d_6 - \Delta d_3) \tag{4-82}$$

For the F_{1u} vibrations

$$R_{3a} = (1/\sqrt{2})(\Delta d_4 - \Delta d_1) \tag{4-83}$$

$$R_{3b} = (1/\sqrt{2})(\Delta d_5 - \Delta d_2) \tag{4-84}$$

$$R_{3c} = (1/\sqrt{2})(\Delta d_6 - \Delta d_3) \tag{4-85}$$

$$R_{4a} = (1/\sqrt{8})(\Delta\alpha_{24}+\Delta\alpha_{34}+\Delta\alpha_{45}+\Delta\alpha_{46} - \Delta\alpha_{12} - \Delta\alpha_{13} - \Delta\alpha_{15} - \Delta\alpha_{16}) \tag{4-86}$$

$$R_{4b} = (1/\sqrt{8})(\Delta\alpha_{23}+\Delta\alpha_{26}+\Delta\alpha_{12}+\Delta\alpha_{24} - \Delta\alpha_{56} - \Delta\alpha_{35} - \Delta\alpha_{15} - \Delta\alpha_{45}) \tag{4-87}$$

$$R_{4c} = (1/\sqrt{8})(\Delta\alpha_{34}+\Delta\alpha_{35}+\Delta\alpha_{13}+\Delta\alpha_{23} - \Delta\alpha_{46} - \Delta\alpha_{56} - \Delta\alpha_{16} - \Delta\alpha_{26}) \tag{4-88}$$

For the F_{2g} vibration

$$R_{5a} = \tfrac{1}{2}(\Delta\alpha_{23} + \Delta\alpha_{56} - \Delta\alpha_{26} - \Delta\alpha_{35}) \tag{4-89}$$

$$R_{5b} = \tfrac{1}{2}(\Delta\alpha_{34} + \Delta\alpha_{16} - \Delta\alpha_{46} - \Delta\alpha_{13}) \tag{4-90}$$

$$R_{5c} = \tfrac{1}{2}(\Delta\alpha_{15} + \Delta\alpha_{24} - \Delta\alpha_{45} - \Delta\alpha_{12}) \tag{4-91}$$

For the F_{2u} vibration

$$R_{6a} = (1/\sqrt{8})(-\Delta\alpha_{24}+\Delta\alpha_{34} - \Delta\alpha_{45}+\Delta\alpha_{46}+\Delta\alpha_{12} - \Delta\alpha_{13}+\Delta\alpha_{15} - \Delta\alpha_{16}) \tag{4-92}$$

$$R_{6b} = (1/\sqrt{8})(\Delta\alpha_{46}+\Delta\alpha_{16}+\Delta\alpha_{35}+\Delta\alpha_{23} - \Delta\alpha_{34} - \Delta\alpha_{13} - \Delta\alpha_{56} - \Delta\alpha_{26}) \tag{4-93}$$

$$R_{6c} = (1/\sqrt{8})(\Delta\alpha_{35}+\Delta\alpha_{56}+\Delta\alpha_{24}+\Delta\alpha_{12} - \Delta\alpha_{23} - \Delta\alpha_{26} - \Delta\alpha_{45} - \Delta\alpha_{15}) \tag{4-94}$$

The three redundant coordinates belong to the A_{1g} type and are listed as follows:

$$R_{a} = \tfrac{1}{2}(\Delta\alpha_{46} + \Delta\alpha_{16} + \Delta\alpha_{13} + \Delta\alpha_{34}) = 0 \tag{4-95}$$

$$R_{b} = \tfrac{1}{2}(\Delta\alpha_{15} + \Delta\alpha_{45} + \Delta\alpha_{24} + \Delta\alpha_{12}) = 0 \tag{4-96}$$

$$R_{c} = \tfrac{1}{2}(\Delta\alpha_{23} + \Delta\alpha_{26} + \Delta\alpha_{56} + \Delta\alpha_{15}) = 0 \tag{4-97}$$

The f matrix for an O_h molecule is tabulated in Table 4-11. The potential function can take the form of Eq. (4-23) and has been expressed by Pistorius.[10] The definition of the f interaction constants are as follows:

d equilibrium AB bond length

f_{dd} bond with a bond at right angles to it

$f_{dd'}$ bond with an opposite bond

$f_{d\alpha}$ angle with one of the bonds forming its sides

$f_{d\alpha'}$ angle with a bond perpendicular to the plane of the angle

$f_{d\alpha''}$ angle with a bond in its plane but not forming one of its sides

$f_{\alpha\alpha}$ angle with an angle adjacent and in the same plane

$f_{\alpha\alpha'}$ angle with an angle in a perpendicular plane, but with a bond in common

$f_{\alpha\alpha''}$ angle with an angle in the same plane but not adjacent

$f_{\alpha\alpha'''}$ angle with an angle in a perpendicular plane, but with no bond in common

Table 4-11. The f Matrix for UF_6

f	Δd_1	Δd_2	Δd_3	Δd_4	Δd_5	Δd_6	$\Delta\alpha_{12}$	$\Delta\alpha_{23}$	$\Delta\alpha_{34}$	$\Delta\alpha_{45}$	$\Delta\alpha_{56}$	$\Delta\alpha_{61}$	$\Delta\alpha_{13}$	$\Delta\alpha_{24}$	$\Delta\alpha_{35}$	$\Delta\alpha_{46}$	$\Delta\alpha_{51}$	$\Delta\alpha_{62}$
Δd_1	f_d	f_{dd}	f_{dd}	$f_{dd'}$	f_{dd}	f_{dd}	$df_{d\alpha}$	$df_{d\alpha'}$	$df_{d\alpha''}$	$df_{d\alpha''}$	$df_{d\alpha'}$	$df_{d\alpha}$	$df_{d\alpha}$	$df_{d\alpha''}$	$df_{d\alpha'}$	$df_{d\alpha''}$	$df_{d\alpha}$	$df_{d\alpha'}$
Δd_2		f_d	f_{dd}	f_{dd}	$f_{dd'}$	f_{dd}	$df_{d\alpha}$	$df_{d\alpha}$	$df_{d\alpha'}$	$df_{d\alpha''}$	$df_{d\alpha''}$	$df_{d\alpha'}$	$df_{d\alpha'}$	$df_{d\alpha}$	$df_{d\alpha''}$	$df_{d\alpha'}$	$df_{d\alpha''}$	$df_{d\alpha}$
Δd_3			f_d	f_{dd}	f_{dd}	$f_{dd'}$	$df_{d\alpha'}$	$df_{d\alpha}$	$df_{d\alpha}$	$df_{d\alpha'}$	$df_{d\alpha''}$	$df_{d\alpha''}$	$df_{d\alpha}$	$df_{d\alpha'}$	$df_{d\alpha}$	$df_{d\alpha''}$	$df_{d\alpha'}$	$df_{d\alpha''}$
Δd_4				f_d	f_{dd}	f_{dd}	$df_{d\alpha''}$	$df_{d\alpha'}$	$df_{d\alpha}$	$df_{d\alpha}$	$df_{d\alpha'}$	$df_{d\alpha''}$	$df_{d\alpha''}$	$df_{d\alpha}$	$df_{d\alpha'}$	$df_{d\alpha}$	$df_{d\alpha''}$	$df_{d\alpha'}$
Δd_5					f_d	f_{dd}	$df_{d\alpha''}$	$df_{d\alpha''}$	$df_{d\alpha'}$	$df_{d\alpha}$	$df_{d\alpha}$	$df_{d\alpha'}$	$df_{d\alpha'}$	$df_{d\alpha''}$	$df_{d\alpha}$	$df_{d\alpha'}$	$df_{d\alpha}$	$df_{d\alpha''}$
Δd_6						f_d	$df_{d\alpha'}$	$df_{d\alpha''}$	$df_{d\alpha''}$	$df_{d\alpha'}$	$df_{d\alpha}$	$df_{d\alpha}$	$df_{d\alpha''}$	$df_{d\alpha'}$	$df_{d\alpha''}$	$df_{d\alpha}$	$df_{d\alpha'}$	$df_{d\alpha}$
$\Delta\alpha_{12}$							d^2f_α	$d^2f_{\alpha\alpha'}$	$d^2f_{\alpha\alpha'''}$	$d^2f_{\alpha\alpha''}$	$d^2f_{\alpha\alpha'''}$	$d^2f_{\alpha\alpha'}$	$d^2f_{\alpha\alpha'}$	$d^2f_{\alpha\alpha}$	$d^2f_{\alpha\alpha'''}$	$d^2f_{\alpha\alpha'''}$	$d^2f_{\alpha\alpha}$	$d^2f_{\alpha\alpha'}$
$\Delta\alpha_{23}$								d^2f_α	$d^2f_{\alpha\alpha'}$	$d^2f_{\alpha\alpha'''}$	$d^2f_{\alpha\alpha''}$	$d^2f_{\alpha\alpha'''}$	$d^2f_{\alpha\alpha'}$	$d^2f_{\alpha\alpha'}$	$d^2f_{\alpha\alpha}$	$d^2f_{\alpha\alpha'''}$	$d^2f_{\alpha\alpha'''}$	$d^2f_{\alpha\alpha}$
$\Delta\alpha_{34}$									d^2f_α	$d^2f_{\alpha\alpha'}$	$d^2f_{\alpha\alpha'''}$	$d^2f_{\alpha\alpha''}$	$d^2f_{\alpha\alpha}$	$d^2f_{\alpha\alpha'}$	$d^2f_{\alpha\alpha'}$	$d^2f_{\alpha\alpha}$	$d^2f_{\alpha\alpha'''}$	$d^2f_{\alpha\alpha'''}$
$\Delta\alpha_{45}$										d^2f_α	$d^2f_{\alpha\alpha'}$	$d^2f_{\alpha\alpha'''}$	$d^2f_{\alpha\alpha'''}$	$d^2f_{\alpha\alpha}$	$d^2f_{\alpha\alpha'}$	$d^2f_{\alpha\alpha'}$	$d^2f_{\alpha\alpha}$	$d^2f_{\alpha\alpha'''}$
$\Delta\alpha_{56}$											d^2f_α	$d^2f_{\alpha\alpha'}$	$d^2f_{\alpha\alpha'''}$	$d^2f_{\alpha\alpha'''}$	$d^2f_{\alpha\alpha}$	$d^2f_{\alpha\alpha'}$	$d^2f_{\alpha\alpha'}$	$d^2f_{\alpha\alpha}$
$\Delta\alpha_{61}$												d^2f_α	$d^2f_{\alpha\alpha}$	$d^2f_{\alpha\alpha'''}$	$d^2f_{\alpha\alpha'''}$	$d^2f_{\alpha\alpha}$	$d^2f_{\alpha\alpha'}$	$d^2f_{\alpha\alpha'}$
$\Delta\alpha_{13}$													d^2f_α	$d^2f_{\alpha\alpha'''}$	$d^2f_{\alpha\alpha'}$	$d^2f_{\alpha\alpha''}$	$d^2f_{\alpha\alpha'}$	$d^2f_{\alpha\alpha'''}$
$\Delta\alpha_{24}$														d^2f_α	$d^2f_{\alpha\alpha'''}$	$d^2f_{\alpha\alpha'}$	$d^2f_{\alpha\alpha''}$	$d^2f_{\alpha\alpha'}$
$\Delta\alpha_{35}$															d^2f_α	$d^2f_{\alpha\alpha'''}$	$d^2f_{\alpha\alpha'}$	$d^2f_{\alpha\alpha''}$
$\Delta\alpha_{46}$																d^2f_α	$d^2f_{\alpha\alpha'''}$	$d^2f_{\alpha\alpha'}$
$\Delta\alpha_{51}$																	d^2f_α	$d^2f_{\alpha\alpha'''}$
$\Delta\alpha_{62}$																		d^2f_α

$f_{ij} = f_{ji}$

Table 4-12. The U Matrix for UF_6

U	Δd_1	Δd_2	Δd_3	Δd_4	Δd_5	Δd_6	$\Delta\alpha_{12}$	$\Delta\alpha_{23}$	$\Delta\alpha_{34}$
A_{1g}	$1/\sqrt{6}$	$1/\sqrt{6}$	$1/\sqrt{6}$	$1/\sqrt{6}$	$1/\sqrt{6}$	$1/\sqrt{6}$	0	0	0
E_g	$1/\sqrt{12}$	$-2/\sqrt{12}$	$1/\sqrt{12}$	$1/\sqrt{12}$	$-2/\sqrt{12}$	$1/\sqrt{12}$	0	0	0
	$\frac{1}{2}$	0	$-\frac{1}{2}$	$\frac{1}{2}$	0	$-\frac{1}{2}$	0	0	0
F_{1g}	0	0	0	0	0	0	$-1/\sqrt{8}$	0	$1/\sqrt{8}$
	0	0	0	0	0	0	$1/\sqrt{8}$	$1/\sqrt{8}$	0
	0	0	0	0	0	0	0	$1/\sqrt{8}$	$1/\sqrt{8}$
F_{1u}	$-1/\sqrt{2}$	0	0	$1/\sqrt{2}$	0	0	0	0	0
	0	$-1/\sqrt{2}$	0	0	$1/\sqrt{2}$	0	0	0	0
	0	0	$-1/\sqrt{2}$	0	0	$1/\sqrt{2}$	0	0	0
F_{2g}	0	0	0	0	0	0	0	$\frac{1}{2}$	0
	0	0	0	0	0	0	0	0	$\frac{1}{2}$
	0	0	0	0	0	0	$-\frac{1}{2}$	0	0
F_{2u}	0	0	0	0	0	0	$1/\sqrt{8}$	0	$1/\sqrt{8}$
	0	0	0	0	0	0	0	$1/\sqrt{8}$	$-1/\sqrt{8}$
	0	0	0	0	0	0	$1/\sqrt{8}$	$-1/\sqrt{8}$	0

Table 4-12 *(Continued)*

U	$\Delta\alpha_{45}$	$\Delta\alpha_{56}$	$\Delta\alpha_{16}$	$\Delta\alpha_{13}$	$\Delta\alpha_{24}$	$\Delta\alpha_{35}$	$\Delta\alpha_{46}$	$\Delta\alpha_{15}$	$\Delta\alpha_{26}$
A_{1g}	0	0	0	0	0	0	0	0	0
E_g	0	0	0	0	0	0	0	0	0
	0	0	0	0	0	0	0	0	0
F_{1g}	$1/\sqrt{8}$	0	$-1/\sqrt{8}$	$-1/\sqrt{8}$	$1/\sqrt{8}$	0	$1/\sqrt{8}$	$-1/\sqrt{8}$	0
	$-1/\sqrt{8}$	$-1/\sqrt{8}$	0	0	$1/\sqrt{8}$	$-1/\sqrt{8}$	0	$-1/\sqrt{8}$	$1/\sqrt{8}$
	0	$-1/\sqrt{8}$	$-1/\sqrt{8}$	$1/\sqrt{8}$	0	$1/\sqrt{8}$	$-1/\sqrt{8}$	0	$-1/\sqrt{8}$
F_{1u}	0	0	0	0	0	0	0	0	0
	0	0	0	0	0	0	0	0	0
	0	0	0	0	0	0	0	0	0
F_{2g}	0	$\frac{1}{2}$	0	0	0	$-\frac{1}{2}$	0	0	$-\frac{1}{2}$
	0	0	$\frac{1}{2}$	$-\frac{1}{2}$	0	0	$-\frac{1}{2}$	0	0
	$-\frac{1}{2}$	0	0	0	$\frac{1}{2}$	0	0	$\frac{1}{2}$	0
F_{2u}	$-1/\sqrt{8}$	0	$-1/\sqrt{8}$	$-1/\sqrt{8}$	$-1/\sqrt{8}$	0	$1/\sqrt{8}$	$1/\sqrt{8}$	0
	0	$-1/\sqrt{8}$	$1/\sqrt{8}$	$-1/\sqrt{8}$	0	$1/\sqrt{8}$	$1/\sqrt{8}$	0	$-1/\sqrt{8}$
	$-1/\sqrt{8}$	$1/\sqrt{8}$	0	0	$1/\sqrt{8}$	$1/\sqrt{8}$	0	$-1/\sqrt{8}$	$-1/\sqrt{8}$

The same nomenclature that was used for the C_{2v} and C_{3v} cases is used for the O_h case. The U matrix for the UF_6 molecule is tabulated in Table 4-12 and the U' matrix in Table 4-13.

By using Eqs. (4-29) and (4-30), the F matrix can be derived. For the A_{1g} vibration, the F matrix is

$$F = f_d + 4f_{dd} + f_{dd'}$$

For the E_g vibration, using either R_{2a} or R_{2b}, the F matrix is

$$F = f_d - 2f_{dd} + f_{dd'}$$

For the two F_{1u} vibrations, and using any one of the pairs R_{3a}, R_{4a}; R_{3b}, R_{4b}; or R_{3c}, R_{4c}, the F matrix is found to be

$$\begin{bmatrix} F_{11} & F_{12} \\ F_{21} & F_{22} \end{bmatrix} = \begin{bmatrix} d^2(f_\alpha - f_{\alpha\alpha''} + 2f_{\alpha\alpha'} - 2f_{\alpha\alpha'''}) & 2d(f_{d\alpha} - f_{d\alpha''}) \\ 2d(f_{d\alpha} - f_{d\alpha''}) & f_d - f_{dd'} \end{bmatrix}$$

For the F_{2g} vibration, the F matrix is

$$F = d^2(f_\alpha + f_{\alpha\alpha''} - 2f_{\alpha\alpha})$$

For the F_{2u} vibration, the F matrix is

$$F = d^2(f_\alpha - f_{\alpha\alpha''} - 2f_{\alpha\alpha'} + 2f_{\alpha\alpha'''})$$

A g matrix can be written for UF_6, similar to the f matrix in Table 4-11. The G matrices for the A_{1g}, E_g, F_{1u}, F_{2g}, and F_{2u} vibrations can then be determined, as has been previously illustrated, using Eq. (4-49), again using the proper U and U' values. The resulting G matrices in terms of the g elements for the UF_6 molecule are:

For the A_{1g} vibration

$$G = g_d + 4g_{dd} + g_{dd'}$$

For the E_g vibration

$$G = g_d - 2g_{dd} + g_{dd'}$$

For the two F_{1u} vibrations

$$\begin{bmatrix} G_{11} & G_{12} \\ G_{21} & G_{22} \end{bmatrix} = \begin{bmatrix} g_\alpha - g_{\alpha\alpha''} + 2g_{\alpha\alpha'} - 2g_{\alpha\alpha'''} & 2(g_{d\alpha} - g_{d\alpha''}) \\ 2(g_{d\alpha} - g_{d\alpha''}) & g_d - g_{dd'} \end{bmatrix}$$

Table 4-13. The U' Matrix for UF_6

$$
U' = \begin{bmatrix}
1/\sqrt{6} & 1/\sqrt{12} & \frac{1}{2} & 0 & 0 & 0 & -1/\sqrt{2} & 0 & 0 & 0 & 0 & 0 & 0 & 0 & 0 \\
1/\sqrt{6} & -2/\sqrt{12} & 0 & 0 & 0 & 0 & 0 & -1/\sqrt{2} & 0 & 0 & 0 & 0 & 0 & 0 & 0 \\
1/\sqrt{6} & 1/\sqrt{12} & -\frac{1}{2} & 0 & 0 & 0 & 0 & 0 & -1/\sqrt{2} & 0 & 0 & 0 & 0 & 0 & 0 \\
1/\sqrt{6} & 1/\sqrt{12} & \frac{1}{2} & 0 & 0 & 0 & 1/\sqrt{2} & 0 & 0 & 0 & 0 & 0 & 0 & 0 & 0 \\
1/\sqrt{6} & -2/\sqrt{12} & 0 & 0 & 0 & 0 & 0 & 1/\sqrt{2} & 0 & 0 & 0 & 0 & 0 & 0 & 0 \\
1/\sqrt{6} & 1/\sqrt{12} & -\frac{1}{2} & 0 & 0 & 0 & 0 & 0 & 1/\sqrt{2} & 0 & 0 & 0 & 0 & 0 & 0 \\
0 & 0 & 0 & -1/\sqrt{8} & 1/\sqrt{8} & 0 & 0 & 0 & 0 & 0 & 0 & -\frac{1}{2} & 1/\sqrt{8} & 0 & 1/\sqrt{8} \\
0 & 0 & 0 & 0 & 1/\sqrt{8} & 1/\sqrt{8} & 0 & 0 & 0 & \frac{1}{2} & 0 & 0 & 0 & 1/\sqrt{8} & -1/\sqrt{8} \\
0 & 0 & 0 & 1/\sqrt{8} & 0 & 1/\sqrt{8} & 0 & 0 & 0 & 0 & \frac{1}{2} & 0 & 1/\sqrt{8} & -1/\sqrt{8} & 0 \\
0 & 0 & 0 & 1/\sqrt{8} & -1/\sqrt{8} & 0 & 0 & 0 & 0 & 0 & 0 & -\frac{1}{2} & -1/\sqrt{8} & 0 & -1/\sqrt{8} \\
0 & 0 & 0 & 0 & -1/\sqrt{8} & -1/\sqrt{8} & 0 & 0 & 0 & \frac{1}{2} & 0 & 0 & 0 & -1/\sqrt{8} & 1/\sqrt{8} \\
0 & 0 & 0 & -1/\sqrt{8} & 0 & -1/\sqrt{8} & 0 & 0 & 0 & 0 & \frac{1}{2} & 0 & -1/\sqrt{8} & 1/\sqrt{8} & 0 \\
0 & 0 & 0 & -1/\sqrt{8} & 0 & 1/\sqrt{8} & 0 & 0 & 0 & 0 & -\frac{1}{2} & 0 & -1/\sqrt{8} & -1/\sqrt{8} & 0 \\
0 & 0 & 0 & 1/\sqrt{8} & 1/\sqrt{8} & 0 & 0 & 0 & 0 & 0 & 0 & \frac{1}{2} & -1/\sqrt{8} & 0 & 1/\sqrt{8} \\
0 & 0 & 0 & 0 & -1/\sqrt{8} & 1/\sqrt{8} & 0 & 0 & 0 & -\frac{1}{2} & 0 & 0 & 0 & 1/\sqrt{8} & 1/\sqrt{8} \\
0 & 0 & 0 & 1/\sqrt{8} & 0 & -1/\sqrt{8} & 0 & 0 & 0 & 0 & -\frac{1}{2} & 0 & 1/\sqrt{8} & 1/\sqrt{8} & 0 \\
0 & 0 & 0 & -1/\sqrt{8} & -1/\sqrt{8} & 0 & 0 & 0 & 0 & 0 & 0 & \frac{1}{2} & 1/\sqrt{8} & 0 & -1/\sqrt{8} \\
0 & 0 & 0 & 0 & 1/\sqrt{8} & -1/\sqrt{8} & 0 & 0 & 0 & -\frac{1}{2} & 0 & 0 & 0 & -1/\sqrt{8} & -1/\sqrt{8}
\end{bmatrix}
$$

For the F_{2g} vibration

$$G = (g_\alpha + g_{\alpha\alpha''} - 2g_{\alpha\alpha})$$

For the F_{2u} vibration

$$G = (g_\alpha - g_{\alpha\alpha''} - 2g_{\alpha\alpha'} + 2g_{\alpha\alpha'''})$$

When the g values of Appendix 6, are substituted into the g elements for an O_h molecule, the G matrices are:

For the A_{1g} vibration

$$G = \mu_B$$

For the E_g vibration

$$G = \mu_B$$

For the F_{1u} vibrations

$$\begin{bmatrix} G_{11} & G_{12} \\ G_{21} & G_{22} \end{bmatrix} = \begin{bmatrix} 2\mu_B[1 + 4(\mu_A/\mu_B)]/d^2 & -4\mu_A/d \\ -4\mu_A/d & \mu_B[1 + 2(\mu_A/\mu_B)] \end{bmatrix}$$

For the F_{2g} vibration

$$G = \frac{4\mu_B}{d^2}$$

For the F_{2u} vibration

$$G = \frac{2\mu_B}{d^2}$$

where, if the molecule under consideration is UF_6, B is the fluorine atom and A the uranium atom.

With the use of Eq. (4-10) and defining m as the mass of the B atom and M as the mass of the A atom, the following secular equations can be obtained:[9]

For the A_{1g} vibration

$$m\lambda_1 = f_d + 4f_{dd} + f_{dd'} \tag{4-98}$$

For the E_g vibration

$$m\lambda_2 = f_d - 2f_{dd} + f_{dd'} \tag{4-99}$$

For the F_{1u} vibration

$$\begin{aligned} m\lambda_3 + m\lambda_4 = {} & 2[1 + 4(m/M)](f_\alpha - f_{\alpha\alpha''} + 2f_{\alpha\alpha'} - 2f_{\alpha\alpha'''}) \\ & + [1 + 2(m/M)](f_d - f_{dd'}) - 16(m/M)(f_{d\alpha} - f_{d\alpha''}) \end{aligned} \tag{4-100}$$

For the F_{1u} vibration

$$m\lambda_3 m\lambda_4 = 2[1 + 6(m/M)][(f_d - f_{dd'})(f_\alpha - f_{\alpha\alpha''} + 2f_{\alpha\alpha'} - 2f_{\alpha\alpha'''}) - 4(f_{d\alpha} - f_{d\alpha''})^2] \quad (4\text{-}101)$$

For the F_{2g} vibration

$$m\lambda_5 = 4(f_\alpha + f_{\alpha\alpha''} - 2f_{\alpha\alpha}) \quad (4\text{-}102)$$

For the F_{2u} vibration

$$m\lambda_6 = 2(f_\alpha - f_{\alpha\alpha''} - 2f_{\alpha\alpha'} + 2f_{\alpha\alpha'''}) \quad (4\text{-}103)$$

Since there are seven independent potential constants in Eqs. (4-98) to (4-103), it is impossible to calculate unique values from the six frequencies for a molecule of O_h symmetry. Several physical assumptions would have to be made. In the case of the general valence force field (GVFF) the seven force constants must be reduced to five or less than the number of observed frequencies. The usual procedure is to assume that force constants associated with a stretch–bend and a bend–bend interaction are zero if the two internal coordinates do not share a common bond. This will reduce the force constants to five. For other force fields, similar assumptions can be made to reduce the number of force constants. The Urey–Bradley force field (UBFF) and the orbital valence force field (OVFF) require four force constants for an octahedral molecule. Even with any possible modification of these latter force fields with the introduction of an additional force constant (five force constants, six frequencies) the calculation is still possible.

Aside from physical assumptions, it is possible to use frequencies of isotopic species to reduce the number of potential constants to be determined Additionally, attempts to transfer force constants for a series of molecules can be made to make the problem solvable.

For further discussion, see Claassen,[9] Heath and Linnett,[11] and Venkateswarlu and Sundaram.[12]

4-5. SOME RESULTS OF NCT OF MOLECULES*

Tables 4-14 to 4-19 present a comparison of experimentally observed values with those calculated by NCT.

* For a detailed NCT of $CHCl_3$, see Colthup, Daly, and Wiberley.[2] For a treatment of CH_3Cl, see Meister and Cleveland.[8] For a NCT of a large molecule, see Nakamoto,[13] and for a NCT of a C_{3v} molecule, see Schatz.[35–38]

Table 4-14. Comparison of Calculated NCT Data and Experimental Data for H_2O[8]

Fundamental	Frequency, cm^{-1}	
	Calculated	Experimental
ν_3	3937	3936
ν_1	3825	3825
ν_2	1653	1654

Table 4-15. Comparison of Calculated NCT Data and Experimental Data for $CHCl_3$

Fundamental and species	Frequency, cm^{-1}	
	Calculated	Experimental
$\nu_4(E)$	1203	1216
$\nu_5(E)$	773	757
$\nu_6(E)$	261	261
$\nu_1(A_1)$	3025	3019
$\nu_2(A_1)$	668	668
$\nu_3(A_1)$	373	368

Table 4-16. Comparison of Calculated NCT Data and Experimental Data for CH_4

Fundamental and species	Frequency, cm^{-1}	
	Calculated	Experimental
$\nu_4(F)$	3157	3020
$\nu_1(A_1)$	3030	2914
$\nu_2(E)$	1390	(1526)
$\nu_3(F)$	1358	1306

Table 4-17. Comparison of Calculated NCT Data and Experimental Data for CH_3Cl

	Frequency, cm^{-1}	
Species	Calculated	Experimental
A_1	2883	2920
A_1	1371	1355
A_1	758	732
E	3001	3047
E	1468	1460
E	1012	1020

A recent study compares five force fields as applied to several octahedral hexahalogen molecules.[24] Table 4-18 summarizes some results for several hexafluorides. The modified orbital valence force field (MOVFF) gave better all-around agreement for the observed frequencies. Table 4-19 summarizes a similar study made with several tetrafluorides[25] comparing three force fields. Only small differences were observed between the OVFF and UBFF for T_d molecules. For other results of molecules in O_h symmetry, see Refs. 9, 10, 12, 26–28, and 34. For results on AB_4 molecules of T_d symmetry see Refs. 15–19, and for AB_4 molecules of D_{4h} symmetry see Refs. 20–23.

4-6. THE PRODUCT RULE

The replacement of an atom in a molecule with an isotopic atom of the same element will change the potential energy function and configuration of the molecule by negligible amounts.[30–31] The frequency of vibration, however, can be altered considerably, as in the case in the substitution of deuterium for hydrogen, where the change in mass is quite large. These shifts in frequencies on isotopic substitution can be used in making the assignment of certain frequencies, by observing which bands change, and to what extent they are altered. The vibrational frequencies involving the isotope effect are quantitatively related by the Teller–Redlich product rule.[32] The rule can be summarized as follows: If one considers that the roots of the secular equation (4-9), where $| GF - E\lambda | = 0$, are $\lambda_1, \lambda_2, \ldots, \lambda_n$, then it is possible that for a second molecule

$$\lambda_1', \lambda_2', \ldots, \lambda_n' = | G' | | F | \tag{4-104}$$

Table 4-18. Comparison of Calculated NCT Data and Experimental Data for Several Hexafluorides of O_h Symmetry[24],*

Hexafluoride		Frequencies, cm^{-1}						Force constants, mdyn/Å					Average % deviation	
		$\nu_1(A_{1g})$	$\nu_2(E_g)$	$\nu_3(F_{1u})$	$\nu_4(F_{1u})$	$\nu_5(F_{2g})$	$\nu_6(F_{2u})$	K f_r	F f_{rr}	$H(D)$ f_α	$k(F')$ $f_{\alpha\alpha}$	h $f_{r\alpha}$	(ν_1, ν_2, ν_3)	(ν_4, ν_5, ν_6)
MoF_6	Obs.	741	643	741	262	312	(122)							
	MUBFF	797	604	727	270	286	124	3.29	0.92	−0.28	0.14	0.03	5.2	4.2
	UBFF	795	594	773	270	280	124	3.43	0.91	−0.30	−0.13		5.1	5.0
	OVFF	756	619	756	262	308	122	3.97	0.61	−0.34	−0.10		2.6	0.5
	MOVFF	763	628	741	263	308	122	3.95	0.60	−0.34	−0.10	0.15	1.8	0.6
	GVFF	741	643	741	262	312	221	4.79	0.25	0.26	−0.01	0.06	0.0	
TcF_6	Obs.	712.9	639	748	275	297	(145)							
	MUBFF	739	620	745	274	274	148	3.89	0.55	−0.09	0.01	0.02	2.4	4.3
	UBFF	739	618	748	274	272	148	3.99	0.53	−0.10	−0.08		2.3	4.7
	OVFF	734	621	750	266	293	145	4.06	0.49	−0.18	−0.08		2.0	0.7
	MOVFF	735	623	748	266	293	145	4.05	0.49	−0.18	−0.08	0.02	1.9	0.7
	GVFF	713	639	748	265	297	210	4.69	0.19	0.25	0.00	0.01	0.0	
RuF_6	Obs.	675	624	735	275	283	(186)							
	MUBFF	674	625	735	273	287	185	4.17	0.22	0.11	0.04	0.01	0.1	1.0
	UBFF	674	621	740	273	287	185	4.20	0.22	0.10	−0.04		0.4	0.9
	OVFF	693	612	731	276	279	187	3.97	0.35	0.08	−0.04		1.7	0.7
	MOVFF	692	610	735	276	279	189	3.98	0.35	0.08	−0.04	−0.03	1.6	0.6
	GVFF	675	624	735	275	283	200	4.44	0.12	0.25	0.01	−0.01	0.0	
RhF_6	Obs.	634	592	724	283	269	(189)							
	MUBFF	633	593	724	271	285	183	3.94	0.17	0.13	−0.13	0.01	0.1	4.4

	UBFF	637	598	713	270	285	183	3.98	0.14	0.14	−0.04		1.0	4.5
	OVFF	652	590	706	285	265	190	3.80	0.24	0.21	−0.05		1.9	0.9
	MOVFF	647	582	724	284	266	190	3.84	0.25	0.20	−0.05	−0.17	1.3	0.6
	GVFF	634	592	724	283	269	190	3.98	0.10	0.27	0.03	−0.11	0.0	
$[PdF_6]$	Estd.	590	525	711	280	258	191							
	MUBFF	585	528	711	261	279	182	3.44	0.21	0.10	−0.47	0.01	0.5	6.6
	UBFF	592	552	656	262	279	181	3.38	0.14	0.13	−0.03		4.4	6.6
	OVFF	601	547	653	283	255	191	3.30	0.18	0.27	−0.05		4.7	0.8
	MOVFF	588	526	711	280	258	191	3.47	0.21	0.23	−0.04	−0.46	0.2	0.1
	GVFF	590	525	711	280	258	182	3.17	0.14	0.34	0.08	−0.27	0.0	
WF_6	Obs.	771	677	711	258	320	(127)							
	MUBFF	820	642	700	265	300	128	3.59	0.89	−0.24	0.40	0.04	4.4	3.3
	UBFF	809	616	735	266	292	129	3.79	0.88	−0.28	−0.14		5.8	4.5
	OVFF	777	641	741	258	317	127	4.32	0.61	−0.32	−0.11		3.4	0.4
	MOVFF	793	662	711	259	316	127	4.28	0.60	−0.31	−0.11	0.36	1.7	0.6
	GVFF	771	676	713	258	320	226	5.29	0.26	0.32	0.02	0.34	0.1	
ReF_6	Obs.	753.7	671	715	257	295	(147)							
	MUBFF	773	657	713	262	284	148	4.18	0.56	−0.09	0.26	0.02	1.7	2.3
	UBFF	767	639	735	262	281	149	4.28	0.57	−0.11	−0.10		3.1	2.7
	OVFF	753	648	738	257	294	147	4.49	0.47	−0.14	−0.09		2.3	0.2
	MOVFF	765	663	715	257	293	147	4.45	0.46	−0.13	−0.09	0.27	0.9	0.3
	GVFF	754	671	715	257	295	209	5.18	0.22	0.26	0.01	0.14	0.0	
OsF_6	Obs.	730.7	668	720	268	276	(205)							
	MUBFF	721	676	720	254	296	197	4.75	0.21	0.13	0.20	0.00	0.8	5.5
	UBFF	716	662	738	254	296	196	4.75	0.25	0.11	−0.03		1.8	5.6

Table 4-18 (*continued*)

Hexafluoride		Frequencies, cm^{-1}						Force constants, mdyn/Å					Average % deviation	
		$\nu_1(A_{1g})$	$\nu_2(E_g)$	$\nu_3(F_{1u})$	$\nu_4(F_{1u})$	$\nu_5(F_{2g})$	$\nu_6(F_{2u})$	K f_r	F f_{rr}	$H(D)$ f_α	$k(F')$ $f_{\alpha\alpha}$	h $f_{r\alpha}$	(ν_1, ν_2, ν_3)	(ν_4, ν_5, ν_6)
	OVFF	721	659	737	268	277	205	4.68	0.28	0.22	−0.04		1.7	0.2
	MOVFF	729	669	720	268	276	205	4.65	0.28	0.23	−0.04	0.19	0.1	0.1
	GVFF	731	668	720	268	276	195	5.10	0.16	0.26	0.02	0.06	0.0	
IrF_6	Obs.	701	645	719	276	258	(206)							
	MUBFF	690	653	719	247	287	190	4.62	0.17	0.14	0.03	0.01	0.9	9.8
	UBFF	690	651	722	248	287	190	4.65	0.17	0.13	−0.03		1.0	9.8
	OVFF	690	650	723	275	260	206	4.69	0.16	0.36	−0.04		1.0	0.4
	MOVFF	692	652	719	275	259	206	4.68	0.16	0.36	−0.04	0.04	0.8	0.3
	GVFF	701	645	719	276	258	182	4.75	0.14	0.27	0.04	−0.12	0.0	
PtF_6	Obs.	656.4	601	705	273	242	(211)							
	MUBFF	645	610	705	232	275	186	4.25	0.15	0.13	−0.20	0.00	1.1	13.4
	UBFF	650	620	685	233	275	186	4.22	0.13	0.14	−0.02		2.3	13.4
	OVFF	644	622	689	272	244	210	4.35	0.07	0.46	−0.03		2.5	0.5
	MOVFF	638	615	705	271	245	210	4.40	0.08	0.45	−0.03	−0.16	1.7	0.7
	GVFF	656	601	705	273	242	171	4.13	0.13	0.31	0.07	−0.28	0.0	
PtF_6^{2-} ‡; solid K_2PtF_6	Obs.	600	576	571	281	210	149							
	MUBFF	598	578	573	222	237	138	3.32	0.08	0.11	0.37	0.02	0.3	13.7
	UBFF	587	542	606	222	237	138	3.30	0.14	0.06	−0.05		4.7	13.7

	OVFF	566	547	618	269	216	148	3.67	−0.02	0.37	−0.10		6.4	2.5
	MOVFF	587	588	573	274	213	148	3.82	−0.14	0.52	−0.13	0.58	1.6	1.3
	GVFF	603	556	589	281	210	149	3.53	0.10	0.25	0.06	0.22	2.3	0.0
UF_6	Obs.	667	535	624	186	201	(140)							
	MUBFF	630	553	627	178	214	135	3.28	0.31	−0.04	−0.07	0.00	3.2	4.8
	UBFF	630	558	620	179	214	135	3.27	0.29	−0.03	−0.03		3.5	4.8
	OVFF	612	565	620	185	205	139	3.45	0.19	0.05	−0.02		4.8	1.1
	MOVFF	610	562	624	185	205	139	3.46	0.19	0.05	−0.02	−0.04	4.6	1.1
	GVFF	667	535	624	186	201	142	3.40	0.30	0.22	0.05	−0.32	0.0	
NpF_6	Obs.	648	528	624	199	208	(165)							
	MUBFF	611	547	627	184	228	154	3.30	0.25	0.01	−0.13	−0.01	3.3	7.9
	UBFF	612	556	612	186	228	153	3.27	0.23	0.02	−0.02		4.2	7.9
	OVFF	601	560	613	198	213	164	3.37	0.17	0.14	−0.01		5.0	1.2
	MOVFF	596	554	624	198	213	164	3.41	0.17	0.14	−0.01	−0.10	4.3	1.2
	GVFF	648	528	624	199	208	147	3.29	0.26	0.24	0.06	−0.03	0.0	
PuF_6	Obs.	628	523	616	206	211	(173)							
	MUBFF	597	540	618	188	234	159	3.25	0.22	0.03	−0.13	−0.01	2.9	9.0
	UBFF	598	550	603	190	233	158	3.22	0.20	0.05	−0.01		4.0	9.0
	OVFF	591	552	603	205	215	172	3.29	0.15	0.19	−0.01		4.5	1.1
	MOVFF	586	546	616	205	215	172	3.33	0.16	0.18	−0.01	−0.01	3.7	1.2
	GVFF	628	523	616	206	211	149	3.21	0.23	0.25	0.06	−0.31	0.0	

* All spectroscopic data measured on gases except where indicated. ( ) Values obtained from overtones, combination bands, or site splitting. [ ] Theoretical molecule.

† Oscillated slightly.

‡ Calculated from the equation $\nu_6 = \nu_5/\sqrt{2}$.

Table 4-19. Observed and Calculated Fundamental Frequencies for Main-Family Tetrafluorides (cm^{-1})*

Molecule		$A_1(\nu_1)$	$E(\nu_2)$	$F_2(\nu_3)$	$F_2(\nu_4)$	Average % deviation	
						ν_1, ν_3	ν_2, ν_4
BeF_4^{2-}	Obs.	547(R)	255(R)	800(R)	385(R)		
	OVFF	548	275	801	364	0.15	6.7
	UBFF	553	267	798	369	1.38	4.4
BF_4^- (solid)	Obs.	769(R)	353(R)	984(R)†	524(R)		
	OVFF	770	359	984	519	0.06	1.4
	UBFF	774	365	983	510	0.39	3.0
BF_4^- (melt)	Obs.	777(R)	360(R)	1070(R)	533(R)		
	OVFF	777	385	1071	508	0.05	5.8
	UBFF	784	375	1068	514	0.93	2.5
CF_4	Obs.	908(R)	435(R)	1283(IR/R)	632(IR/R)		
	OVFF	909	442	1283	625	0.08	1.3
	UBFF	918	457	1280	605	0.64	4.3
SiF_4	Obs.	801(R)	264(R)	1032(IR/R)	389(IR/R)		
	OVFF	801	278	1032	377	0.03	4.1
	UBFF	811	291	1025	356	0.92	9.4
GeF_4	Obs.	738(R)	205‡	800(IR)	260(IR)		
	OVFF	737	212	801	254	0.11	2.8
	UBFF	743	219	796	242	0.62	6.9
TiF_4 (gas)	Obs.	712(R)	185(R)	793(R)	209(R)		
	OVFF	709	190	795	206	1.5	1.6
	UBFF	709	177	795	216	1.5	3.4

* (R) Frequency obtained from Raman data. (IR) Frequency obtained from infrared data.
† Average of observed doublet.
‡ Estimated to give best fit.

Further,

$$\frac{\lambda_1'\lambda_2' \ldots \lambda_n'}{\lambda_1\lambda_2 \ldots \lambda_n} = \frac{|G'|}{|G|} \tag{4-105}$$

since from Eq. (4-66)

$$\nu = \frac{1}{2\pi c}\sqrt{\lambda}$$

then

$$\frac{\nu_1'\nu_2' \ldots \nu_n'}{\nu_1\nu_2 \ldots \nu_n} = \frac{\sqrt{|G'|}}{\sqrt{|G|}} \tag{4-106}$$

Since the determinants of G are related to μ_k (reciprocal of mass of an atom k), Eq. (4-105) becomes

$$\frac{\nu_1'\nu_2' \ldots \nu_n'}{\nu_1\nu_2 \ldots \nu_n} = \left(\frac{\mu_1'\mu_2' \ldots \mu_n'}{\mu_1\mu_2 \ldots \mu_n}\right)^{1/2} \tag{4-107}$$

and

$$\frac{\nu_1'\nu_2' \ldots \nu_n'}{\nu_1\nu_2 \ldots \nu_n} = \left(\frac{m_1m_2 \ldots m_n}{m_1'm_2' \ldots m_n'}\right)^{1/2} \tag{4-108}$$

where the m_n values are the masses of the representative atoms; the primes represent the isotopic molecules. The result holds only if the symmetry species contain no translation or rotation. The product rule expressed in Eq. (4-108) can be verified by using pairs of molecules such as H_2O and D_2O or CH_4 and CD_4. For further discussion see Wilson, Decius, and Cross.[1]

4-7. THE SUM RULE

Another rule which can be applied to isotopically substituted molecules is the sum rule. This rule relates the sums of the squares of the frequencies of isotopic molecules. The basis for the rule is the fact that the sum of squares of the frequencies is a linear function of the reciprocal masses of the atoms. For example for the molecules HOH, DOD, and HOD the rule states that the sum of the squares of the frequencies are related by

$$\Sigma\, \nu^2(\text{HOD}) + \Sigma\, \nu^2(\text{DOH}) = 2\,\Sigma\, \nu^2(\text{HOD}) = \Sigma\, \nu^2(\text{HOH}) + \Sigma\, \nu^2(\text{DOD}) \tag{4-109}$$

For further discussion see Decius and Wilson.[32] Both the product rule and the sum rule may be helpful for certain molecules to provide supplementary data for the calculations of force constants.

Table 4-20. Summary of F and G Matrices for Typical Molecules Using GVFF and UBFF*

1. Octahedral AB_6 Molecules—O_h

A_{1g} Species

$F = f_d + 4f_{dd} + f_{dd'}$ $\quad G = \mu_B$

E_g Species

$F = f_d - 2f_{dd} + f_{dd'}$ $\quad G = \mu_B$

$2F_{1u}$ Species

$F_{11} = d^2(f_\alpha - f_{\alpha\alpha''} + 2f_{\alpha\alpha'} - 2f_{\alpha\alpha'''})$ $\quad G_{11} = (2\mu_B/d^2)(1 + 4\mu_A/\mu_B)$

$F_{12} = F_{21} = 2d(f_{d\alpha} - f_{d\alpha''})$ $\quad G_{12} = G_{21} = -4\mu_A/d$

$F_{22} = f_d - f_{dd'}$ $\quad G_{22} = \mu_B(1 + 2\mu_A/\mu_B)$

F_{2g} Species

$F = d^2(f_\alpha + f_{\alpha\alpha''} - 2f_{\alpha\alpha})$ $\quad G = 4\mu_B/d^2$

F_{2u} Species

$F = d^2(f_\alpha - f_{\alpha\alpha''} - 2f_{\alpha\alpha'} + 2f_{\alpha\alpha'''})$ $\quad G = 2\mu_B/d^2$

For definitions of the primed force constants (f), see page 158

A_{1g} Species

$F^* = K + 4F$

E_g Species

$F^* = K + 0.7F$

$2F_{1u}$ Species

$F^*_{11} = d^2(H + 0.65F)$

$F^*_{12} = 0.9dF$

$F^*_{22} = K + 1.8F$

F_{2g} Species

$F^* = d^2(H + 0.55F)$

F_{2u} Species

$F^* = d^2(H + 0.45F)$

2. Square-Planar AB_4 Molecules—D_{4h}

A_{1g} Species

$F = f_d + 2f_{dd} + f_{dd'}$ $\quad G = \mu_B$

B_{1g} Species

$F = f_d - 2f_{dd} + f_{dd'}$ $\quad G = \mu_B$

* Asterisks refer to UBFF constants.

Table 4-20 (*continued*)

B_{2g} Species

$F = d^2(f_\alpha - 2f_{\alpha\alpha} + f_{\alpha\alpha'})$ $\qquad G = 4\mu_B/d^2$

$2E_u$ Species

$F_{11} = f_d - f_{dd'}$ $\qquad G_{11} = 2\mu_A + \mu_B$

$F_{12} = F_{21} = \sqrt{2}\,d(f_{d\alpha} - f_{d\alpha'})$ $\qquad G_{12} = G_{21} = -2\sqrt{2}\,\mu_A/d$

$F_{22} = d^2(f_\alpha - f_{\alpha\alpha'})$ $\qquad G_{22} = (2/d^2)(\mu_B + 2\mu_A)$

The primed force constants indicate interaction having no common bond.

A_{1g} Species

$F^* = K + 2F$

B_{1g} Species

$F^* = K - 0.2F$

B_{2g} Species

$F^* = d^2(H + 0.55F)$

$2E_u$ Species

$F_{11}^* = K + 0.9F$

$F_{12}^* = -(\sqrt{2}/d)(0.45F)$

$F_{22}^* = d^2(H + 0.55F)$

3. Tetrahedral AB_4 Molecules—T_d

A_1 Species

$F = f_d + 3f_{dd}$ $\qquad G = \mu_B$

E Species

$F = d^2(f_\alpha - 2f_{\alpha\alpha} + f_{\alpha\alpha'})$ $\qquad G = 3\mu_B/d^2$

$2F_2$ Species

$F_{11} = d^2(f_\alpha - f_{\alpha\alpha'})$ $\qquad G_{11} = (1/d^2)[(16/3)\mu_A + 2\mu_B]$

$F_{12} = F_{21} = d\sqrt{2}(f_{d\alpha} - f_{d\alpha'})$ $\qquad G_{12} = G_{21} = -(8/3d)\mu_A$

$F_{22} = f_d - f_{dd}$ $\qquad G_{22} = (4/3)\mu_A + \mu_B$

The primed force constants indicate interaction having no common bond.

A_1 Species

$F^* = K + 4F$

E Species

$F^* = d^2(H + 0.37F)$

Table 4-20 (*continued*)

$2F_2$ Species

$F_{11}^* = d^2(H + \frac{1}{2}F)$

$F_{12}^* = \frac{3}{5}\,dF$

$F_{22}^* = K + \frac{6}{5}F$

4. Planar AB_3 Molecules—D_{3h}

A_1' Species

$F = f_d + f_{dd}$ $\qquad G = \mu_B$

A_2'' Species

$F = d^2 f_\beta$ $\qquad G = (4/d^2)(\mu_B + 3\mu_A)$

$2E'$ Species

$F_{11} = f_d - f_{dd}$ $\qquad G_{11} = \frac{1}{2}(\mu_A + \mu_B)$

$F_{12} = E_{21} = (df_{d\alpha} - df_{d\alpha'})$ $\qquad G_{12} = G_{21} = -(3\sqrt{3}/2d)\mu_A$

$F_{22} = (d^2 f_\alpha - d^2 f_{\alpha\alpha})$ $\qquad G_{22} = (3/d^2)(\mu_B + \frac{3}{2}\mu_A)$

The primed force constants indicate interaction having no common bond.

A_1' Species

$F^* = K + 3F$

A_2'' Species

$F^* = d^2 H$

$2E'$ Species

$F_{11}^* = K + 0.675F$

$F_{12}^* = -0.9\,(\sqrt{3}/4)\,dF$

$F_{22}^* = d^2(H + 0.325F)$

5. Pyramidal AB_3 Molecules—C_{3v}

$2A_1$ Species

$F_{11} = f_d + 2f_{dd}$ $\qquad G_{11} = \mu_B + (1 + 2\cos\alpha)\mu_A$

$F_{12} = F_{21} = d(f_{d\alpha'} + 2f_{d\alpha})$ $\qquad G_{12} = G_{21} = \dfrac{-2}{d}\mu_A \dfrac{(1 + 2\cos\alpha)(1 - \cos\alpha)}{\sin\alpha}$

$F_{22} = d^2(f_\alpha + 2f_{\alpha\alpha})$ $\qquad G_{22} = \dfrac{2}{d^2}\left(\dfrac{1 + 2\cos\alpha}{1 + \cos\alpha}\right)[\mu_B + 2\mu_A(1 - \cos\alpha)]$

Table 4-20 *(continued)*

2*E* Species

$F_{11} = f_d - f_{dd}$ | $G_{11} = \mu_B + \mu_A(1 - \cos\alpha)$

$F_{12} = F_{21} = d(f_{d\alpha'} - f_{d\alpha})$ | $G_{12} = G_{21} = -(\mu_A/d \sin\alpha)(1 - \cos\alpha)^2$

$F_{22} = d^2(f_\alpha - f_{\alpha\alpha})$ | $G_{22} = [(1/d^2)(1 + \cos\alpha)] \times [(2 + \cos\alpha)\mu_B + (1 - \cos\alpha)^2\mu_A]$

The primed force constants indicate interaction having no common bond.

2A_1 Species

$F_{11}^* = K + 4F\sin^2\tfrac{1}{2}\alpha$

$F_{12}^* = 1.8\ dF\sin\tfrac{1}{2}\alpha\cos\tfrac{1}{2}\alpha$

$F_{22}^* = d^2[H + F(\cos^2\tfrac{1}{2}\alpha + 0.1\sin^2\tfrac{1}{2}\alpha)]$

2*E* Species

$F_{11}^* = K + (\sin^2\tfrac{1}{2}\alpha - 0.3\cos^2\tfrac{1}{2}\alpha)F$

$F_{12}^* = -0.9\ dF\sin\tfrac{1}{2}\alpha\cos\tfrac{1}{2}\alpha$

$F_{22}^* = d^2[H + F(\cos^2\tfrac{1}{2}\alpha + 0.1\sin^2\tfrac{1}{2}\alpha)]$

6. Bent AB_2 Molecules—C_{2v}

2A_1 Species

$F_{11} = f_d + f_{dd}$ | $G_{11} = \mu_A(1 + \cos\alpha) + \mu_B$

$F_{12} = F_{21} = \sqrt{2}\,df_{d\alpha}$ | $G_{12} = G_{21} = -(\sqrt{2}/d)\mu_A\sin\alpha$

$F_{22} = d^2f_\alpha$ | $G_{22} = (2/d^2)[\mu_B + \mu_A(1 - \cos\alpha)]$

B_2 Species

$F = f_d - f_{dd}$ | $G = \mu_A(1 - \cos\alpha) + \mu_B$

2A_1 Species

$F_{11}^* = K + 2F\sin^2\tfrac{1}{2}\alpha$

$F_{12}^* = 0.9\sqrt{2}\,dF\sin\tfrac{1}{2}\alpha\cos\tfrac{1}{2}\alpha$

$F_{22}^* = d^2[H + F(\cos^2\tfrac{1}{2}\alpha + 0.1\sin^2\tfrac{1}{2}\alpha)]$

B_2 Species

$F^* = K - 0.2F\cos^2\tfrac{1}{2}\alpha$

* F^* denotes F matrix element in UBFF. F' is taken as $-(1/10)F$ in all cases and molecular tension is ignored.

4-8. SUMMARY

The calculation of the vibrational frequencies of molecules is theoretically possible if the following conditions are met:

1. A suitable force field such as the GVFF or UBFF or modifications thereof can be selected.
2. Physical parameters such as bond distances and bond angles are available for the molecule or can be borrowed from a related molecule.
3. Force constants are known for the molecule or can be borrowed from a related molecule.

Generally, a complete set of force constants is not available. The borrowing process is, at best, only qualitative. Many of the interaction constants are unobtainable, for there are more constants than frequencies. A good deal of these must be therefore left out, which further increases the qualitative aspect of the results. It has been suggested[(33)] that for widely separated frequencies, the interaction constants can be ignored and set equal to zero. To circumvent some of these problems, frequency assignment of experimental data is made and force constants which fit the secular equation are calculated. In addition, isotopic compounds can be prepared, which give additional frequencies, and additional equations, from which the unknown constants can then be calculated.

The advent of computers has certainly aided the NCT of molecules, since computers can compensate for some of the inherent difficulties and problems in the method. The multiplication of matrices and the solving of the secular equation becomes simplified and less time-consuming, and this is a distinct improvement from the days of the hand calculator. Programs for the NCT of molecules are available; see, e.g., Yeranos[(39)] and Schachtschneider and Snyder.[(40)]

A summary of the F and G matrices for typical molecules is presented in Table 4-20.

PROBLEMS

1. Derive the F and G matrices for a planar AB_4 molecule of D_{4h} symmetry in GVFF.

References: K. Nakamoto, *Infrared Spectra of Inorganic and Coordination Compounds*, J. Wiley & Sons, New York (1970), pp. 266–267; A. Maccoll, *Proc. Roy. Soc. N. S. Wales*, **77**:130 (1943).

2. Derive the F and G matrices for a planar AB_3 molecule of D_{3h} symmetry in the GVFF.
3. Derive the F and G matrices for a metal bidentate–nitrate complex in C_{2v} symmetry in the GVFF field.

References: R. E. Hester and W. E. L. Grossman, *Inorg. Chem.*, **5**:1308 (1966). See also G. Topping, *Spec. Acta*, **21**:1743 (1965), for a treatment of a monodentate complex of nitrate; H. Brintzinger and R. E. Hester, *Inorg. Chem.*, **5**:980 (1966); S. C. Wait, Jr., A. T. Ward, and G. J. Janz, *J. Chem. Phys.*, **45**:133 (1966).

REFERENCES

1. E. B. Wilson, J. C. Decius, and P. C. Cross, *Molecular Vibrations*, McGraw-Hill, New York (1955).
2. N. B. Colthup, L. H. Daly, and S. E. Wiberley, *Introduction to Infrared and Raman Spectroscopy*, Academic Press, New York (1975).
3. A. G. Meister and F. F. Cleveland, *Am. J. Phys.*, **14**:13 (1946).
4. S. M. Ferigle and A. G. Meister, *J. Chem. Phys.*, **19**:982 (1951).
5. J. C. Decius, *J. Chem. Phys.*, **17**:1315 (1949).
6. E. B. Wilson, *J. Chem. Phys.*, **7**:1047 (1939).
7. E. B. Wilson, *J. Chem. Phys.*, **9**:76 (1941).
8. A. G. Meister and F. F. Cleveland, *Molecular Spectra, II*, Publs. Ill. Inst. Technol. (1948).
9. H. H. Claassen, *J. Chem. Phys.*, **30**:968 (1959).
10. C. W. F. T. Pistorius, *J. Chem. Phys.*, **29**:1328 (1958).
11. D. F. Heath and J. W. Linnett, *Trans. Faraday Soc.*, **45**:264 (1949).
12. K. Venkateswarlu and S. Sundaram, *Z. Physik. Chem.*, **9**:174 (1956).
13. K. Nakamoto, *Infrared Spectra of Inorganic and Coordination Compounds*, J. Wiley & Sons, New York (1970).
14. P. N. Schatz, *J. Chem. Phys.*, **29**:481 (1958).
15. J. W. Linnett, *Quart. Revs.* (*London*), **1**:73 (1947).
16. L. A. Woodward, *Trans. Faraday Soc.*, **54**:1271 (1958).
17. C. W. F. T. Pistorius, *J. Chem. Phys.*, **28**:514 (1958).
18. K. Venkateswarlu and S. Sundaram, *J. Chem. Phys.*, **23**:2365 (1955).
19. S. Sundaram, *J. Chem. Phys.*, **33**:708 (1960).
20. H. Stammreich and R. Forneris, *Spec. Acta*, **16**:363 (1960).
21. J. D. S. Goulden, A. Maccoll, and D. J. Millen, *J. Chem. Soc.*, **1950**:1635.
22. A. Maccoll, *Proc. Roy. Soc. N. S. Wales*, **77**:130 (1943).
23. A. Sabatini, L. Sacconi, and V. Schettino, *Inorg. Chem.*, **3**:1775 (1964).
24. P. LaBonville, J. R. Ferraro, M. C. Wall, and L. J. Basile, *Coord. Chem. Rev.* **7**:257 (1972).
25. L. J. Basile, J. R. Ferraro, P. LaBonville, and M. C. Wall, *Coord. Chem. Rev.*, **11**:21 (1973).
26. J. Gaunt, *Trans. Faraday Soc.*, **50**:546 (1954).
27. J. Hiraishi, I. Nakagawa, and T. Shimanouchi, *Spec. Acta*, **20**:819 (1964).
28. H. F. Shurvell, *Canadian Spec.*, **12**:1156 (1967).
29. G. Herzberg, *Spectra of Diatomic Molecules*, Van Nostrand, New York (1950).

30. E. B. Wilson, *Ann. Rev. Phys. Chem.*, **2**:151 (1951).
31. O. Redlich, *Z. Phys. Chem. (B)*, **28**:371 (1935).
32. J. B. Decius and E. B. Wilson, *J. Chem. Phys.*, **19**:1409 (1951).
33. B. E. Crawford and S. R. Brinkley, *J. Chem. Phys.*, **9**:69 (1941).
34. H. Kim, P. A. Souder, and H. H. Claassen, *J. Mol. Spec.*, **26**:46 (1968).
35. S. Sundaram, F. Suszek, and F. F. Cleveland, *J. Chem. Phys.*, **32**:251 (1960).
36. G. DeAlti, G. Costa, and V. Galasso, *Spec. Acta*, **20**:965 (1964).
37. M. Pariseau, E. Wu, and J. Overend, *J. Chem. Phys.*, **39**:217 (1963).
38. J. S. Ziomek and E. A. Piotrowski, *J. Chem. Phys.* **34**:1087 (1961).
39. W. A. Yeranos, *Bull. Soc. Chim. Belg.*, **74**:414 (1965).
40. J. H. Schachtschneider and R. G. Snyder, *Spec. Acta*, **19**:117 (1963).

Chapter 5

APPLICATIONS OF GROUP THEORY FOR THE DETERMINATION OF MOLECULAR STRUCTURE

5-1. INTRODUCTION

The purpose of this chapter is to illustrate with examples taken from the literature the usefulness of group theory in solving problems of molecular structure. In many cases the spectroscopic method of determining molecular structure is far from unequivocal. However, at the least, very important inferences can be drawn by means of this method. As always, it remains for the experimenter to substantiate his conclusions by using a more conclusive method—perhaps X-ray, electron, or neutron diffraction techniques. The importance of the spectroscopic method is readily apparent, however, where the situation is such that diffraction techniques are not possible.

However, we cannot overlook the difficulties involved in the spectroscopic method. Until the advent of the recent high-resolution Raman instruments, many Raman spectra were very poorly resolved. Poor resolution also was a problem in infrared experiments accomplished with prism instruments, and conclusions drawn from poorly resolved spectra are of course subject to criticism. The conclusions originally derived for metallic carbonyls are illustrative of this point. The original infrared research in the 5 μm region was performed with NaCl prism instruments. Certain conclusions regarding structure inferred from these data were later shown to be incorrect when the spectra were obtained with LiF or CaF_2 prisms or with a grating instrument. Additional bands were observed, and new structural assessments had to be made. Further, certain conclusions drawn from infrared data alone or from Raman data alone are again subject to criticism. In addition, all regions of the infrared or Raman spectra must be investigated. For example, it is necessary that the far-infrared spectrum of a compound be obtained if a complete vibrational analysis of the molecule is desired.

5-2. PROCEDURE USED IN DETERMINING THE STRUCTURE OF A MOLECULE

The following procedure is recommended in making a determination of the structure of a molecule.

1. Obtain the best possible Raman and infrared spectra of the compound from 4000 to 50 cm^{-1}.
2. Attempt to make the Raman and infrared spectra on the same state of matter. In addition, make certain the measurements are complete. Avoid leaving portions of the spectra blank.
3. On the basis of the empirical formula of the compound, select all possible structural configurations of the molecule.
4. For each configuration or model derive the total number of fundamentals allowed and the Raman and infrared selection rules. Use the procedures outlined in Chapter 2.
5. Compare the theoretical results with the experimental data for each model, and choose the model which fits best.
6. Make frequency assignments on the basis of the chosen model.
7. Make a NCT of molecule if possible, following the procedures outlined in Chapter 4.

5-3. EXAMPLES ILLUSTRATING THE USE OF GROUP THEORY IN DETERMINING MOLECULAR STRUCTURE

Several examples illustrating the use of group theory in solving problems of molecular structure are presented.

Iodine Heptafluoride (IF_7)

Eight possible structures can be cited for IF_7. Table 5-1 lists the selection rules for these structures and compares them with the observed results.[1] The structures with low symmetry can be ruled out for obvious reasons. Of the more symmetrical structures, only the D_{5h} (pentagonal bipyramid) and the D_{7h} (plane heptagon) structures are reasonably compatible with the experimental data. Structure D_{7h} was eliminated on the basis of the polarized Raman spectrum, since two polarized Raman bands were found, and since on the basis of a D_{7h} model only one is predicted from the selection rules. In the D_{5h} model, the following species and number of vibrations are expected:

Raman—$A_1'(2)$, $E_1''(1)$, $E_2'(2)$; Infrared—$A_2''(2)$, $E_1'(3)$

Table 5-1. Selection Rules for the Possible Structures of IF_7

	Total frequencies	Raman		Infrared	Coincidences	
		total	polarized*		total	polarized
D_{7h}	10	3	1	3	0	0
C_{7v}	10	7	2	4	4	2
C_{6v}	12	9	3	6	6	3
D_{5h}	11	5	2	5	0	0
C_{5v}	11	11	4	8	8	4
C_{3v}	12	11	5	11	11	5
C_3	12	12	6	12	12	6
C_{2v}	18	18	7	16	16	7
Observed	11	7	2	4	0	0

* The state of polarization of a Raman line is determined by the depolarization ratio ϱ. The depolarization ratio is defined as the ratio of the intensity of the Raman line recorded with the incident light polarized perpendicular to the scattered light to the intensity of the line recorded with the incident light polarized parallel to the direction of the scattered light, $\varrho = I_{\perp}/I_{\parallel}$. A line is said to be depolarized if ϱ has a value of 6/7 or greater, and polarized if ϱ has a value of less than 6/7.

The E_2'' type is spectroscopically inactive. The D_{5h} model fits better, as two of the observed Raman lines can be ascribed to difference tones. The missing infrared band probably lies below 250 cm^{-1}, in a region which was not investigated.

The frequency assignments made on the basis of a D_{5h} model are listed in Table 5-2. Two polarized Raman frequencies are expected from I–F stretching vibrations, and these were assigned at 635 and 678 cm^{-1} (A_1' type). The three remaining Raman lines were assigned to E_2' type (2) and the E_1'' type (1). The lowest frequency of the three was assigned to the E_2' type, and the two bands at 360 and 511 cm^{-1} to the E_1'' and E_2' types, respectively. The I–F infrared-active stretching vibration of type A_2'' was assigned at 670 cm^{-1}. Two of the E_1' type were assigned at 426 and 547 cm^{-1}, and the A_2'' type at 368 cm^{-1}. The third E_1' type lies probably below 250 cm^{-1}, in a region which was not investigated. The remaining infrared bands above 900 cm^{-1} can be suitably assigned to summation tones or overtones. The spectroscopically inactive vibration E_2'', on the basis of a normal coordinate treatment, has been estimated to be at 350 cm^{-1}. Confirmation for the D_{5h} structure comes from nuclear magnetic resonance data[(2)] and electron diffraction studies.[(3)] Figure 5-1 shows the pentagonal bipyramid structure for the IF_7 molecule.

Table 5-2. Frequency Assignments[1] for IF_7 on Basis of a D_{5h} Structure

Raman spectrum, $\Delta\nu$, cm^{-1}	Assignment	Infrared spectrum, ν, cm^{-1}
174 (W)	$(426 - 250) = A_1'$	
198 (W)	$(511 - 313) = A_1'$	
	E_1'	(250) (?)*
313 (d) (M)	E_2'	
(350) (?)*	E_2''	
360 (d) (M)	E_1''	
	A_2''	368 (S)
	E_1'	426 (S)
511 (d) (S)	E_2'	
	E_1'	547 (M)
595 (VW) (?)	IF_5-impurity	
635 (p) (VS)	A_1'	
	A_2''	670 (VS)
678 (p) (VS)	A_1'	
710 (VW) (?)	IF_5-impurity	
	$(250 + 678) = E_1'$	918 (W)
	$(3 \times 313) = E_1'$	927 (M)
	$(426 + 511) = E_1'$	937 (M)
	$(360 + 670) = E_1'$	1030 (W)
	$(426 + 635) = E_1'$	1060 (W)
	$(426 + 678) = E_1'$	1101 (W)
	$(547 + 635) = E_1'$	1180 (W)
	$(3 \times 426) = (E_1')$	1260 (W)
	$(635 + 670) = A_2''$	1300 (M)
	$((2 \times 511) + 426) = E_1'$	1420 (W)
	$(3 \times 511) = E_1'$	1502 (W)

* Predicted.

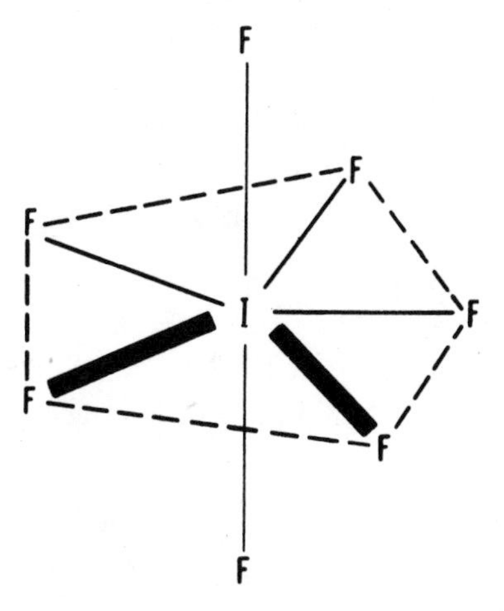

Fig. 5-1. Pentagonal bipyramidal IF_7 molecule.

Recent normal coordinate analysis[4] of IF_7 has resulted in some reassignments of the frequencies from those made by Lord *et al.*[1] This does not alter the discussion presented nor the conclusions reached.

Iodine Pentafluoride (IF_5)

Seven possible models can be formulated for iodine pentafluoride.[1] Table 5-3 lists the selection rules for these models and the observed experimental results for IF_5. The structures with low symmetry are not compatible with the experimental results. The high-symmetry structures D_{5h} and D_{3h}

Table 5-3. Selection Rules for the Possible Structures of IF_5

	Total frequencies	Raman		Infrared	Coincidences	
		total	polarized		total	polarized
D_{5h}	7	3	1	3	0	0
D_{3h}	8	6	2	5	3	0
C_{5v}	7	7	2	4	4	2
C_{4v}	9	9	3	6	6	3
C_{3v}	8	8	4	8	8	4
C_{2v}	12	12	5	11	11	5
C	12	12	8	12	12	8
Observed	9	8	3	4 (5)*	3 (5)*	2

* Numbers in parentheses indicate more recent data.[5,6]

can be ruled out on the basis of the coincidences and their Raman polarizations. Three polarized Raman lines are expected for a C_{4v} symmetry and two for a C_{5v} symmetry. Three are observed. Although the C_{4v} model was selected, the choice made is not without difficulty. For the C_{4v} model, the following species and number of vibrations are expected:

Raman—$A_1(3)$, $B_1(2)$, $B_2(1)$, $E(3)$; Infrared—$A_1(3)$, $E(3)$

The frequency assignments made on the basis of a C_{4v} model were originally made by Lord *et al.*,[1] followed by more recent assignments made by

Table 5-4. Frequency Assignments for IF_5 on Basis of C_{4v} Structure

Raman spectra $\Delta\nu$, cm^{-1}			Assignments*
Lord *et al.*[1]	Begun *et al.*[5]	Gillespie and Clase[6]	
710 (p) (S) } 693 (p) (S) }	698 (p) (6)	705 (p) (9) } 694 (p) (9) }	$2E(\nu_9) + A_1(\nu_3)$ $A_1(\nu_1)$
(645)?†		635 (0)	$E(\nu_7)$
605 (d) (VS)			
	593 (p) (10)	598 (p) (11)	$A_1(\nu_2)$
572 (d) (VS)	575 (8)	574 (12)	$B_1(\nu_4)$
375 (p) (S)	374 (2)	376 (4)	$E(\nu_8)$
317 (p) (M)	315 (p) (1)	316 (p) (2)	$A_1(\nu_3)$
275 (d) (S)	273 (2)	275 (4)	$B_2(\nu_6)$
192 (W)	189 (0)	191 (1)	$E(\nu_9)$
Infrared spectra ν, cm^{-1}			
p 721, q 712, *r* 703 } (M)	710 (S)	685 (S) (Sh)	$A_1(\nu_1)$
(693)?†			
645 (S)	640 (VS)		$E(\nu_7)$
	[595]‡	590 (VS) (V.Br)	$A_1(\nu_2)$
372 (M)	372 (M)	364 (M)	$E(\nu_8)$
		350 (Sh)	
318 (M)	318 (M)	305 (M)	$A_1(\nu_3)$
[192]‡			

* $B_1(\nu_5)$ not observed.
† Missing.
‡ [ ] = predicted.

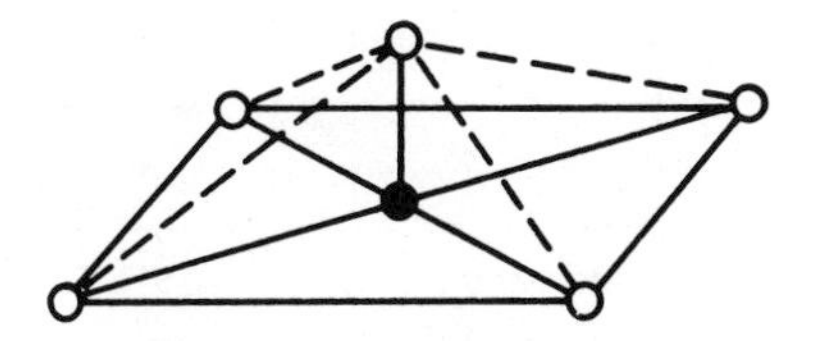

Fig. 5-2. The structure of IF_5.

Begun *et al.*[5] and Gillespie and Clase.[6] Table 5-4 tabulates the Raman data as well as the infrared frequencies, and records the assignments. Although some discrepancies remain between observations, results are consistent for a C_{4v} model. Figure 5-2 shows the structure of IF_5. The tetragonal pyramidal structure has been supported by nuclear magnetic resonance studies.[3]

AB_4 Molecules

For an AB_4 molecule, several structures are possible. These are illustrated in Fig. 5-3. Table 5-5 shows the expected infrared and Raman fun-

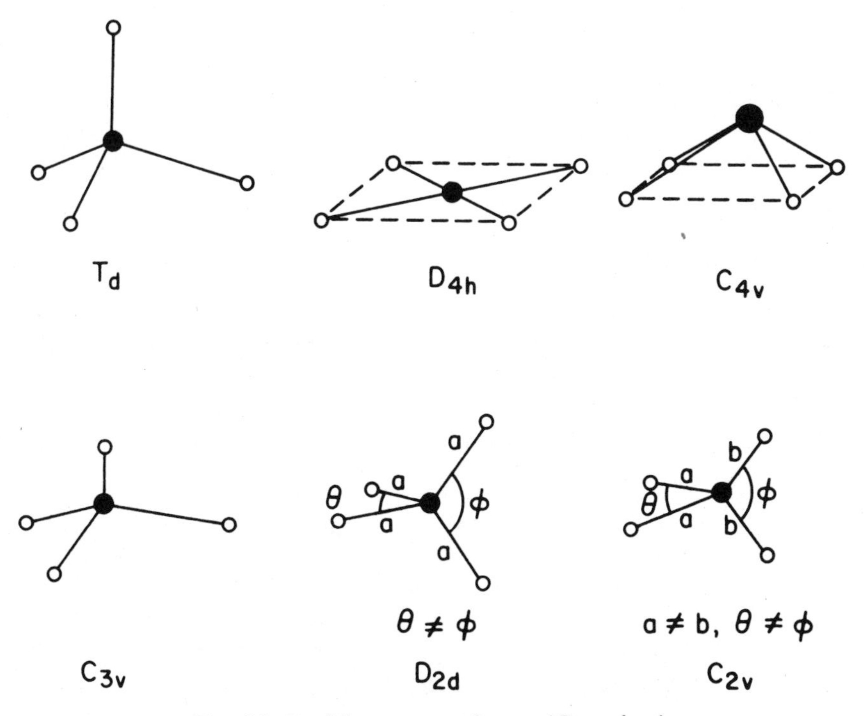

Fig. 5-3. Possible structures for an AB_4 molecule.

Table 5-5. Selection Rules for the Possible Structures of AB_4 Type Molecules

	Total frequencies	Raman fundamentals	Raman polarized	Infrared fundamentals	Coincidences, total
T_d	4	4	1	2	2
D_{4h}	7	3	1	3	0
D_{2d}	7	7	2	4	4
C_{4v}	7	7	2	4	4
C_{3v}	6	6	3	6	6
C_{2v}	9	9	4	8	8
C_1	9	9	9	9	9
Observed for					
SF_4	9	5 certain, 8 possible	1	5 certain,* 8 possible	5 certain, 8 possible
CCl_4	4	4	1	2	2
XeF_4	6	3	0	3	0

* The infrared spectrum was obtained only down to 400 cm^{-1}, and a complete count of coincidences was therefore impossible.

damentals for the different models obtained from the selection rules, and the observed results for SF_4, CCl_4, and XeF_4.

For SF_4,[7] the C_1 model was excluded on the basis of the polarization experiments. The C_{4v} and D_{2d} models are excluded since more infrared bands are observed experimentally than required. The D_{4h} model has a center of symmetry, and therefore has no coincidences, and at least five coincidences were observed. The T_d model can be eliminated as lacking the number of infrared and Raman bands which are observed. The choice between the C_{3v} and C_{2v} models was made on the basis of vapor phase band contours. The band contours were more like those of an asymmetric top molecule, and it was concluded that the molecule was of C_{2v} symmetry.

For the C_{2v} model the following species and number of vibrations are expected:

Raman—$A_1(4)$, $A_2(1)$, $B_1(2)$, $B_2(2)$; Infrared—$A_1(4)$, $B_1(2)$, $B_2(2)$

Frequency assignments made on the basis of a C_{2v} model are listed in Table 5-6. The SF_4 case is interesting, for it is an example in which one is not intellectually happy about the interpretations made from the spectroscopic results. However, nuclear magnetic resonance studies do support a C_{2v} structure.[8,9]

Table 5-6. Frequency Assignments for SF_4 on Basis of a C_{2v} Structure

Raman spectrum, $\Delta\nu$, cm^{-1}	Assignment	Infrared spectrum, ν, cm^{-1}
239 (W)	A_1	(235) (?)*
401 (VW)	A_2	
463 (d) (M)	B_2	463 (VW)
536 (d) (VS)	B_1	532 (S)
	A_1	557 (M)
620, 760 (W, VB)	$(239 + 401)A_2$?	(645) (?)*
	A_1	715 (M)
	B_2	728 (VS)
	$(532 + 235)B_2$	768
858 (d) (M)	B_1	867 (VS)
898 (p) (VS)	A_1	889 (VS)
	$(728 - 235)B_2$	961 (VW)
	$(2 \times 532)A_1$	1070 (VW)
	$(557 + 532)B_1$	1091 (W)
	$(867 + 235)B_1$	1098 (W)
	$(2 \times 557)A_1$ or $(645 + 463)B_1$	1114 (W)
	$(889 + 235)A_1$	1125 (VW)
	$(532 + 645)B_2$ or $(463 + 718)B_2$	1177 (VW)
	$(728 + 557)B_2$	1281 (M)
	$(728 + 645)B_1$	1369 (VW)
	$(867 + 557)B_1$, $(889 + 532)B_1$	1421 (W)
	$(889 + 728)B_2$	1617 (W)
	$(235 + 715 + 728)B_2$, $(889 + 557 + 235)A_1$	1678 (VW)
	$(3 \times 557)A_1$	1710 (VW)
	$(2 \times 867)A_1$	1727 (VW)
	$(889 + 867)B_1$	1744 (M)
	$(2 \times 889)A_1$	1778 (VW)

* Deduced from combinations, not observed.

For CCl_4, the observed spectrum shows four Raman fundamentals (one polarized band) and two infrared fundamentals, with two coincidences. Structure D_{4h} can immediately be ruled out on the basis of the coincidences found in the spectrum. The D_{2d}, C_{4v}, and C_{3v} structures are eliminated on the basis that more fundamentals are necessary for these models than are experimentally found. The C_{2v} and C_1 models can be eliminated because a high number of coincidences are required and only two are found. The final choice is the T_d structure. For a T_d model the following species and number of vibrations are expected:

$$\text{Raman}-A_1(1),\ E(1),\ F_2(2); \qquad \text{Infrared}-F_2(2)$$

Figure 5-4 illustrates the four normal modes of vibration for a tetrahedral AB_4 molecule. Table 5-7 contains the frequency assignments made for CCl_4 on the basis of a T_d structure. Reasonable assignments can be made if one assigns the triplet at about 450 cm^{-1} to the C–Cl stretching vibration to the various chloride isotopes, and the doublet at about 775 cm^{-1} to Fermi resonance between F_2 and the combination $A_1 + F_2$.

Xenon tetrafluoride is another example of an AB_4 type of molecule. The vibrational spectrum of the molecule was recently investigated.[10] The infrared spectrum of XeF_4 vapor had strong bands at 123, 291, and 586 cm^{-1}. The Raman spectrum of the solid showed very intense peaks at 502 and 543 cm^{-1} and weaker bands at 235 and 442 cm^{-1}. The fact that no coincidences are found indicates a center of symmetry. This allows one to reject all of the other models indicated in Table 5-5, except D_{4h}. For a D_{4h} model the following species and number of vibrations are expected:

$$\text{Raman}-A_{1g}(1),\ B_{1g}(1),\ B_{2g}(1); \qquad \text{Infrared}-A_{2u}(1),\ E_u(2)$$

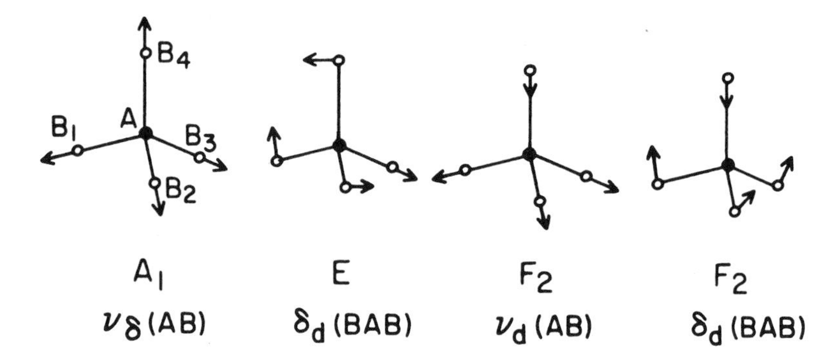

Fig. 5-4. Normal modes of vibration of a tetrahedral AB_4 molecule.

Table 5-7. Frequency Assignments for CCl_4 on Basis of a T_d Structure*

Raman spectrum, $\Delta\nu$, cm^{-1}	Assignment	Infrared spectrum, ν, cm^{-1}
145 (VW)	$(458 - 145) = F_2$	145 (VW)
218 (d) (S)	E	
314 (d) (S)	F_2	305
434 (VW)	$(2 \times 218) = (A_1 + E)$	
455 (p) (VS)	A_1 $CCl_2^{35}Cl_2^{37}$	
458 (p) (VS)	A_1 $CCl_3^{35}Cl^{37}$	
462 (p) (VS)	A_1 CCl_4^{35}	
	$(2 \times 314) = (A_1 + E + E_2)$	635 (W)
762 (d) (M)	F_2 and	763 (VS)
791 (d) (M)	$(458 + 314 \text{ or } 305) = F_2$	797 (VS)
	$(762 + 218) = F_2$	982 (M)
	$(458 + 218 + 305) = F_2$	1006 (W)
	$(762 + 305) = (A_1 + E + F_2)$	1068 (W)
		1107 (W)
	$((2 \times 458) + 305) = F_2$	1218 (M)
	$(458 + 768) = F_2$	1253 (M)
	$(2 \times 768) = (A_1+E+F_2)$	1529 (M)
1539 (p) (VW)	$(2 \times 458) + (2 \times 305) = (A_1+E+F_2)$	1546 (S)
	$458 + 768 + 305 = (A_1+E+F_2)$	1575 (M)

* G. Herzberg, *Molecular Spectra and Molecular Structure, II, Infrared and Raman Spectra of Polyatomic Molecules*, Van Nostrand, New York (1945).

Figure 5-5 illustrates the normal modes of vibrations for a square planar AB_4 molecule. Table 5-8 contains the frequency assignments made for XeF_4 on the basis of a D_{4h} structure. The two intense bands in the Raman spectrum at 502 and 543 cm^{-1} were assigned to the two stretching frequencies B_{2g} and A_{1g}, with the higher frequency being inferred to be the symmetrical vibration, since no polarization data were obtained. The 235 cm^{-1} band was assigned to the B_{1g} type. The infrared data were assigned as

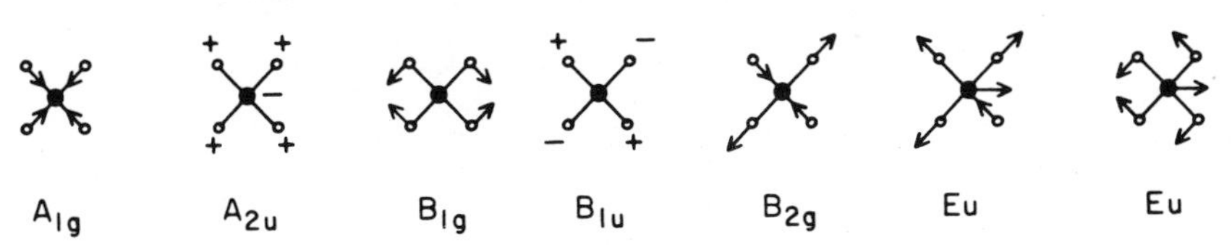

Fig. 5-5. Normal modes of vibrations of a square planar AB_4 molecule.

Table 5-8. Frequency Assignments for XeF_4 on Basis of a D_{4h} Structure

Raman spectrum, $\Delta\nu$, cm^{-1}	Assignment	Infrared spectrum, ν, cm^{-1}
	E_u	123 (S)
235 (W)	B_{1g}	
	A_{2u}	291 (S)
442 (W)	$(2 \times 221) = A_{1g}$	
502 (S)	B_{2g}	
543 (S)	A_{1g}	
	E_u	586 (S)

follows: the 586 cm^{-1} band to the E_u type, the 291 cm^{-1} band to the A_{2u} type, and the 123 cm^{-1} to the other E_u type. The bands found in the 1100 cm^{-1} region were assigned as combinations. The extra Raman band at 442 cm^{-1} cannot be a fundamental, and was assigned as an overtone of the B_{1u} type, which is an allowed overtone (i.e., 2×221). The B_{1u} fundamental species is not allowed by the selection rules, but the even overtones of B_{1u} become active in the Raman spectrum. This is an excellent illustration of the usefulness of the selection rules for overtones for a molecule. Ibers and Hamilton[11] have obtained the X-ray diffraction spectrum of solid XeF_4 and have found it to be square planar within experimental error.

N_4S_4 Molecule

Lippincott and Tobin[12] determined the structure of N_4S_4 using spectroscopic data. The possible structures are the ring structure or the caged structure (Figs. 5-6 and 5-7, respectively). Both would have a D_{2d} symmetry. The caged structure is the preferred one, because of the similarity of its spectrum to $N_4S_4H_4$, which would have single-bonded N–S links, as con-

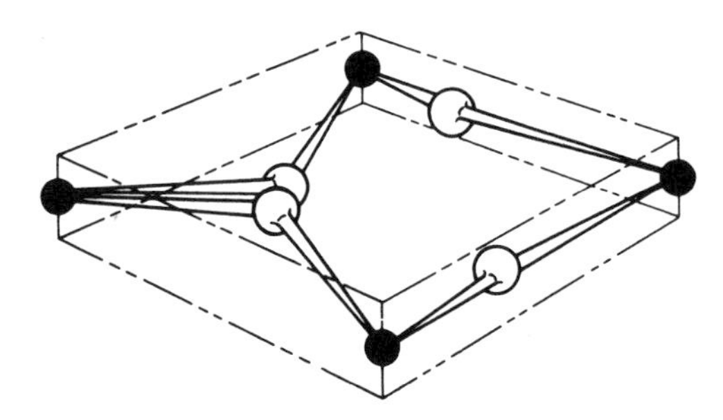

Fig. 5-6. Ring structure for N_4S_4 (N = black, S = white).

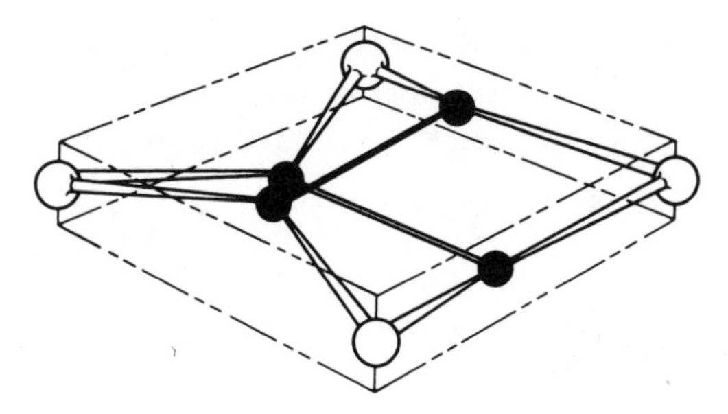

Fig. 5-7. Caged structure for N_4S_4.

Table 5-9. Frequency Assignments for N_4S_4 on Basis of a D_{2d} Structure

Raman spectrum (dioxane), $\Delta\nu$, cm^{-1}	Assignment	Infrared spectrum (Nujol),* ν, cm^{-1}
	A_2	173 (calc.)‡
177 (S)	A_1, B_2†	
213 (S)	B_1, E†	
347 (W)	E	347 (S)
	(177 + 213)	397 (W)
	$(2 \times 213) = A_2$	412 (W)
	A_2	460 (calc.)‡
519 (M)	B_2	519 (M)
	(347 + 177)	531
561 (M)	E	552 (S)
		557 (S)
615 (W)	A_1	
655 (?)		
692 (?)	E	696 (S)
720 (W)		719 (S)
	(934 − 177)	762 (W)
785 (VW)	B_1	792 (VW)
882 (VW)	A_1	
934 (VW)	B_2	925 (S)
	(785 + 213)	1000 (W)
	(2 × 519)§	{1025 (W), 1040 (W)}

* The far-infrared spectrum beyond 300 cm^{-1} was not measured.
† Two vibrations coincident with each other.
‡ Not observed in infrared, forbidden by selection rules.
§ Forbidden by selection rules in infrared.

trasted to the ring structure, which would have resonating N–S bonds with partial double-bond character. For the caged structure of D_{2d} symmetry the following species and number of vibrations are expected:

Raman$-A_1(3)$, $B_1(2)$, $B_2(3)$, $E(4)$; Infrared$-B_2(3)$, $E(4)$

with the $2A_2$ type being unallowed in the Raman and infrared spectra. The data have been assigned in terms of a D_{2d} model in Table 5-9.

5-4. PRACTICE PROBLEM

Determine the structure of N_2O_4.

Reference: G. H. Begun, *J. Mol. Spec.*, **4**: 388 (1960).

Infrared data (gas): ~381 (M), ~429 (S), ~695 (M), 750 (VS), 1262 (S), 1495 (W), 1748 (S), 1937 (W), 2124 (W), 2579 (W), 2630 (M), 2747 (W), 2971 (M), 3111 (M), 3440 (W), 4253 (VW), 4645 (VW), 5076 (VW).

Raman data (solid): 266 (VS), 482 (M), 672 (W), 808 (S), 1326 (M), 1380 (S), 1712 (M).

At least six models can be formulated for dinitrogen tetroxide. These are illustrated in Fig. 5-8. The selection rules for each model are tabulated in Table 5-10. There are five infrared absorptions and five Raman lines of sufficient intensity to be classified as fundamentals. The striking thing about the experimental data is the lack of coincidences. The C_s, D_{2d}, and D_2

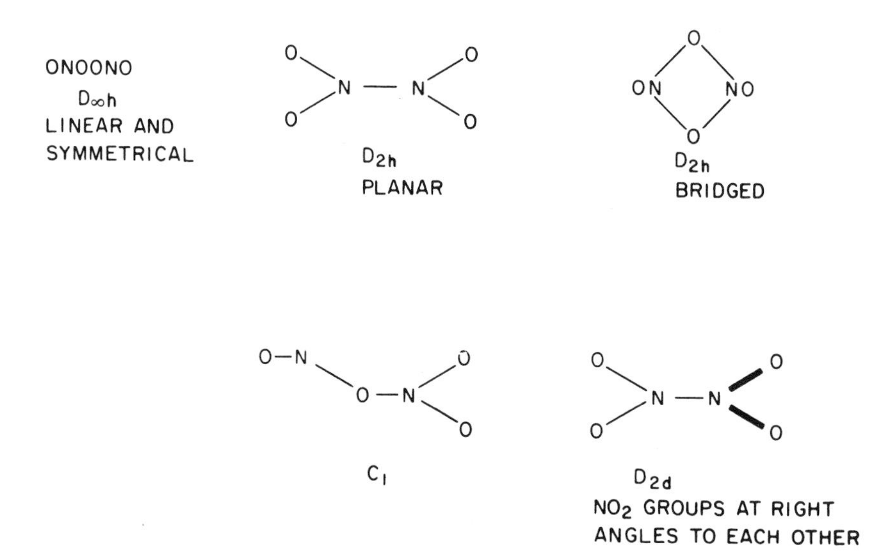

Fig. 5-8. Possible models for N_2O_4 (D_2 model is not illustrated).

Table 5-10. Selection Rules for Possible Models of N_2O_4

Model	Number of fundamentals		
	Infrared	Raman	Coincidences
$D_{\infty h}$	4	5	0
D_{2h} (planar)	5	6	0
C_s	12	12	12
D_{2h} (bridged)	6	6	0
D_{2d}	5	9	5
D_2 (rotated form)	8	12	8
Observed	5	5	0

structures can easily be eliminated, since the observed Raman and infrared spectra show fewer absorptions than are predicted by these structures. The $D_{\infty h}$ can be eliminated on the basis that the chemical behavior of N_2O_4 does not fit a peroxide structure. Thus, the choice is between the

Table 5-11. Vibration Type for Each Species—D_{2h} Model

Species	Fundamental	Vibration
A_g	ν_1	NO stretching
	ν_2	deformation
	ν_3	N–N stretching
A_u	ν_4	torsion
B_{1g}	ν_5	NO stretching
	ν_6	NO_2 rock
B_{1u}	ν_7	NO_2 wag
B_{2g}	ν_8	NO_2 wag
B_{2u}	ν_9	NO stretching
	ν_{10}	NO_2 rock
B_{3u}	ν_{11}	NO stretching
	ν_{12}	deformation

Table 5-12. Frequency Assignments for N_2O_4 on Basis of a D_{2h} Planar Structure

Raman spectrum, $\Delta\nu$, cm^{-1}	Assignment	Infrared spectrum, ν, cm^{-1}
266 (VS)	A_g	
	B_{2u}	~381 (M)
	B_{1u}	429 (S)
482 (M)	B_{1g}	
672 (W)	B_{2g}	
	$(266 + 429) = B_{1u}$	~695 (M)
	B_{3u}	750 (VS)
808 (S)	A_g	
	B_{3u}	1262 (S)
1326 (M)	$(2 \times 672) = A_g$	
1380 (S)	A_g	
	$(1748 - 266) = B_{2u}$	1495 (W)
1712 (M)	B_{1g}	
	B_{2u}	1748 (S)
	$(1262 + 672) = B_{1u}$	1937 (W)
	$(1380 + 750) = B_{3u}$	2124 (W)
	$[(2 \times 672) + 1262] = B_{3u}$	2579 (W)
	$(1380 + 1262) = B_{3u}$	2630 (M)
	$[1262 + (2 \times 750)] = B_{3u}$	2747 (W)
	$(1712 + 1262) = B_{2u}$	2971 (M)
	$(1380 + 1748) = B_{2u}$	3111 (M)
	$(1712 + 1748) = B_{3u}$	3440 (W)
	$[1748 + (2 \times 1262)] = B_{2u}$	4253 (VW)
	$[(2 \times 1712) + 1262] = B_{2u}$	4645 (VW)
	$(3 \times 1748) = B_{2u}$	5076 (VW)

D_{2h} planar or bridged structures. The bridged structure requires fitting in a sixth infrared fundamental vibration, which is difficult to do. The twelve fundamental vibrations of N_2O_4 for a D_{2h} model are as follows: $3A_g + A_u + 2B_{1g} + B_{1u} + B_{2g} + 2B_{2u} + 2B_{3u}$, with the A_u type being inactive. All of the even types (g) are Raman active, and all of the uneven types (u) infrared active. Table 5-11 indicates the type of vibration for each species. Table 5-12 shows the frequency assignments made on the basis of a planar D_{2h} symmetry. It can be observed that reasonable assignments are possible. Electron[12,13] and X-ray[4,14] diffraction studies confirm this structure.

PROBLEMS

1. Determine the structure of CH_3SnCl_3 from spectroscopic data.

Reference: W. F. Edgell and C. H. Ward, *J. Mol. Spec.*, **8**: 343 (1962).
Infrared data: 525 (M), 548 (S), 562 (M), 588 (W), 609 (W), 652 (W), 800 (VVS), 1090 (W), 1200 (S), 1346 (W), 1402 (S, B), 1541 (VVW), 1561 (VW), 1601 (M, Br), 1748 (M), 2000 (W), 2195 (W), 2382 (M), 2783 (W), 2940 (S), 3033 (S), 3559 (W), 3646 (W), 3741 (W), 3830 (M), 4221 (W), 4419 (M).
Raman data (numerical values in parentheses are the values for the depolarization factor, ϱ): 112 (0.76) (M), 142 (0.86) (M), 363 (0.26) (S), 550 (0.43) (M), 1200 (p) (W), 2932 (p) (W), 3024 (d) (W).

2. Determine the structure of the TeF_5^- ion in several complexes from spectroscopic data.

Reference: N. N. Greenwood, A. C. Sarma, and B. P. Straughan, *J. Chem. Soc.*, **1966:** 1446 (1966).
Infrared data ($CsTeF_5$): 164 (MW), 283 (M), 336 (M), 466 (VS, B), 618 (Ms).
Raman data ($CsTeF_5$): 231 (MW), 282 (MW), 338 (MW), 472 (S), 504 (S), 572 (W), 611 (VS).

3. Determine the structure of MoF_6 from spectroscopic data.

References: (1) J. Gaunt, *Trans. Faraday Soc.*, **49**: 1122 (1953). (2) K. N. Tanner and A. B. F. Duncan, *J. Am. Chem. Soc.*, **73**: 1164 (1951).
Infrared data[1]: 434, 456, 742 (VVS), 763, 833, 882, 914, 983, 1004, 1052, 1156, 1262, 1384, 1481, 1590, 1743, 2032, 2128, 2218.
Raman data[2]: 763 (p)* (S), 641 (d) (W), 319 (d) (W).

Hint: Infer an infrared fundamental at 269 cm^{-1} from combination bands, since no far-infrared studies are available.

4. Determine the structure of BrF_5 from spectroscopic data.

References: (1) C. V. Stephenson and E. A. Jones, *J. Chem. Phys.*, **20**: 1830 (1952). (2) T. G. Burke and E. A. Jones, *J. Chem. Phys.*, **19**: 1611 (1951).
Infrared data[2]: 690 (VS), 645 (VW), 583 (VS), 418 (M $\rightarrow$ S) (no data <400 cm^{-1}).
Raman data[1]: 683 (p) (S), 626 (d) (W), 572 (p) (S), 536 (d) (S), 481 (d) (W), 415 (d) (M), 365 (p) (M), 315 (d) (M), 244 (d) (M).

* Totally polarized ($\varrho = 0$).

5. Determine the structure of B_2Cl_4 from spectroscopic data.

Reference: D. E. Mann and L. Fano, *J. Chem. Phys.*, **26**: 1665 (1957).

Infrared data: 291 (VS), 454 (W, Sh), 478 (S), 513 (S), 617 (VS), 641 (W, Sh), 688 (S), 730 (VVS), 745 (M, Sh), 771 (VW, Sh), 797 (VVW, Sh), 818 (M), 845 (VW, B), 917 (VVS), 1026 (S), 1053 (W, Sh), 1089 (W), 1122 (W), 1149 (VW), 1202 (W, Sh), 1227 (M), 1253 (W), 1319 (M), 1647 (M), 1695 (W), 1825 (W), 1901 (W, Br), 2041 (M), 2119 (W).

Raman data: 107 (d?), 180 (M), 225 (VVW), 291 (M), 347 (p) (VW), 401 (p), 520 (VVW), 640 (VVW), 729 (VVW), 800 (VW), 875 (VVW), 917 (M), 946 (W), 1131 (p) (M), 1162 (p) (W).

6. Determine the structure of ClF_3 from spectroscopic data.

Reference: H. H. Claassen, B. Weinstock, and J. G. Malm, *J. Chem. Phys.*, **28**: 285 (1958).

Infrared data: (gas) 319 (M), 332 (M), 364 (W), 434 (W), 518 (S), 535 (S), 694 (VS), 703 (VS), 713 (VS), 741 (S), 761 (S), 845 (M), 957 (M), 1022 (M), 1223 (S), 1273 (W), 1338 (VW), 1466 (S), 1451 (S), 1488 (S), 1505 (S).

Raman data: (gas) 529 (VS), 746 (W), 753 (S).

7. Determine the structure of PF_5 from spectroscopic data.

Reference: H. S. Gutowsky and A. D. Liehr, *J. Chem. Phys.*, **20**: 1652 (1952).

Infrared data: (gas) 487, 532, 848, 892.

Raman data: (gas) 487 (S), 851 (M), 893 (M); (liquid) 486 (W), 531 (W), 840 (VS), 890 (VS).

8. Determine the structure of POF_3 from spectroscopic data.

Reference: H. S. Gutowsky and A. D. Liehr, *J. Chem. Phys.*, **20:** 1652 (1952).

Infrared data: 473 (S), 485 (M → S), 690 (VW), 830 (W), 873 (M → S), 946 (W), 957 (W), 990 (VS), 1275 (W), 1330 (M), 1345 (M), 1360 (W), 1415 (M → S), 1462 (W), 1472 (W), 1740 (VW), 1855 (M), 2295 (W), 2815 (VW).

Raman data: 337 (d) 476 (d), 875 (p), 982 (d), 1395 (p).

Answers

1. C_{3v} symmetry.
2. C_{4v} symmetry.
3. O_h symmetry.
4. C_{4v} symmetry.
5. D_{2d} symmetry.
6. C_{2v} symmetry.
7. C_{3v} symmetry.
8. C_{3v} symmetry.

REFERENCES

1. R. C. Lord, M. A. Lynch, W. C. Schumb, and E. J. Slowinski, *J. Am. Chem. Soc.*, **72**:522 (1950).
2. H. S. Gutowsky and C. J. Hoffman, *J. Chem. Phys.*, **19**:1259 (1951).
3. R. E. La Villa and S. H. Bauer, *J. Chem. Phys.*, **33**:182 (1960).
4. R. K. Khanna, *J. Mol. Spec.*, **8**:134 (1962).
5. G. M. Begun, W. H. Fletcher, and D. F. Smith, *J. Chem. Phys.*, **42**:2236 (1965).
6. R. J. Gillespie and H. J. Clase, *J. Chem. Phys.*, **47**:1071 (1967).
7. R. E. Dodd, L. A. Woodward, and H. L. Roberts, *Trans. Faraday Soc.*, **52**:1052 (1956).
8. F. A. Cotton, J. W. George, and J. S. Waugh, *J. Chem. Phys.*, **28**:994 (1958).
9. E. L. Muetterties and W. D. Phillips, *J. Am. Chem. Soc.*, **81**:1084 (1959).
10. H. H. Claassen, C. L. Chernick, and J. G. Malm, *J. Am. Chem. Soc.*, **88**:1927 (1963).
11. J. A. Ibers and W. C. Hamilton, *Science*, **139**:106 (1963).
12. E. K. Lippincott and M. C. Tobin, *J. Chem. Phys.*, **21**:1559 (1953).
13. D. W. Smith and K. Hedberg, *J. Chem. Phys.*, **25**:1282 (1956).
14. J. J. Broadley and J. M. Robertson, *Nature*, **164**:914 (1949).

Appendix 1

CHARACTER TABLES

The character tables for the more common point groups are presented. The last two columns of each character table list the infrared and Raman activity of the particular species. If one or more of the components of the polarizability α_{xx}, α_{xy}, etc., is listed in the row for a certain species in the last column, that species is Raman active. Similarly, if one or more of the translation components T_x, T_y, and T_z is listed in the next-to-last column, the species is infrared active. The components of the change in the dipole moment (μ_x, μ_y, μ_z) should be listed in this column as well as the components of translation (T_x, T_y, T_z). However, to save space, since they always occur together (both are vectors and transform in the same way if the symmetry operation is carried out similarly), the former are omitted in the character tables in this appendix. Also listed in the next-to-last column are the components for the rotational coordinates R_x, R_y, R_z. The subscripts x, y, z indicate the direction in which the translation or dipole moment change occurs for a particular vibrational species. When the components of translation, change in dipole moment, or polarizability are degenerate, they are enclosed in parentheses.

1. The C_s, C_i, and C_n Groups

C_s	E	σ_{xy}	Activity	
			IR	Raman
A'	1	1	T_x, T_y, R_z	$\alpha_{xx}, \alpha_{yy}, \alpha_{zz}, \alpha_{xy}$
A''	1	-1	T_z, R_x, R_y	α_{yz}, α_{xz}

C_i	E	i	Activity	
			IR	Raman
A_g	1	1	R_x, R_y, R_z	$\alpha_{xx}, \alpha_{yy}, \alpha_{zz}$
A_u	1	−1	T_x, T_y, T_z	

C_2	E	C_2^z	Activity	
			IR	Raman
A	1	1	T_z, R_z	$\alpha_{xx}, \alpha_{yy}, \alpha_{zz}, \alpha_{xy}$
B	1	−1	T_x, T_y, R_x, R_y	α_{yz}, α_{xz}

C_3	E	C_3	C_3^2	Activity	
				IR	Raman
A	1	1	1	T_z, R_z	$\alpha_{xx} + \alpha_{yy}, \alpha_{zz}$
E	$\begin{cases}1\\1\end{cases}$	$\begin{matrix}\varepsilon\\\varepsilon^*\end{matrix}\Big\}$	$\begin{matrix}\varepsilon^*\\\varepsilon\end{matrix}\Big\}$	(T_x, T_y) (R_x, R_y)	$(\alpha_{xx} - \alpha_{yy}, \alpha_{xy}), (\alpha_{xz}, \alpha_{yz})$

$\varepsilon = e^{2\pi i/3}$.

C_4	E	C_4	C_2	C_4^3	Activity	
					IR	Raman
A	1	1	1	1	T_z, R_z	$\alpha_{xx} + \alpha_{yy}, \alpha_{zz}$
B	1	−1	1	−1		$\alpha_{xx} - \alpha_{yy}, \alpha_{xy}$
E	$\begin{cases}1\\1\end{cases}$	$\begin{matrix}i\\-i\end{matrix}$	$\begin{matrix}-1\\-1\end{matrix}$	$\begin{matrix}-i\\i\end{matrix}\Big\}$	$(T_x, T_y)(R_x, R_y)$	$(\alpha_{xz}, \alpha_{yz})$

C_5	E	C_5	C_5^2	C_5^3	C_5^4	Activity	
						IR	Raman
A	1	1	1	1	1	T_z, R_z	$\alpha_{xx}+\alpha_{yy}, \alpha_{zz}$
E_1	$\begin{Bmatrix}1\\1\end{Bmatrix}$	ε ε^*	ε^2 ε^{2*}	ε^{2*} ε^2	ε^* ε	$(T_x, T_y)(R_x, R_y)$	$(\alpha_{xz}, \alpha_{yz})$
E_2	$\begin{Bmatrix}1\\1\end{Bmatrix}$	ε^2 ε^{2*}	ε^* ε	ε ε^*	ε^{2*} ε^2		$(\alpha_{xx}-\alpha_{yy}, \alpha_{xy})$

$\varepsilon = e^{2\pi i/5}$.

C_6	E	C_6	C_3	C_2	C_3^2	C_5^5	Activity	
							IR	Raman
A	1	1	1	1	1	1	T_z, R_z	$\alpha_{xx}+\alpha_{yy}, \alpha_{zz}$
B	1	-1	1	-1	1	-1		
E_1	$\begin{Bmatrix}1\\1\end{Bmatrix}$	ε ε^*	$-\varepsilon^*$ $-\varepsilon$	-1 -1	$-\varepsilon$ $-\varepsilon^*$	ε^* ε	$(T_x, T_y)(R_x, R_y)$	$(\alpha_{xz}, \alpha_{yz})$
E_2	$\begin{Bmatrix}1\\1\end{Bmatrix}$	$-\varepsilon^*$ $-\varepsilon$	$-\varepsilon$ $-\varepsilon^*$	1 1	$-\varepsilon^*$ $-\varepsilon$	$-\varepsilon$ $-\varepsilon^*$		$(\alpha_{xx}-\alpha_{yy}, \alpha_{xy})$

$\varepsilon = e^{2\pi i/6}$.

2. The C_{nv} Groups

C_{2v}	E	C_2	$\sigma_v(xz)$	$\sigma_v(yz)$	Activity	
					IR	Raman
A_1	1	1	1	1	T_z	$\alpha_{xx}, \alpha_{yy}, \alpha_{zz}$
A_2	1	1	-1	-1	R_z	α_{xy}
B_1	1	-1	1	-1	T_x, R_y	α_{xz}
B_2	1	-1	-1	1	T_y, R_x	α_{yz}

C_{3v}	E	$2C_3$	$3\sigma_v$	Activity: IR	Activity: Raman
A_1	1	1	1	T_z	$\alpha_{xx}+\alpha_{yy}, \alpha_{zz}$
A_2	1	1	-1	R_z	
E	2	-1	0	$(T_x, T_y), (R_x, R_y)$	$(\alpha_{xx}-\alpha_{yy}, \alpha_{xy}), (\alpha_{yz}, \alpha_{xz})$

C_{4v}	E	$2C_4^z$	C_2^z	$2\sigma_v$	$2\sigma_d$	Activity: IR	Activity: Raman
A_1	1	1	1	1	1	T_z	$\alpha_{xx}+\alpha_{yy}, \alpha_{zz}$
A_2	1	1	1	-1	-1	R_z	
B_1	1	-1	1	1	-1		$\alpha_{xx}-\alpha_{yy}$
B_2	1	-1	1	-1	1		α_{xy}
E	2	0	-2	0	0	$(T_x, T_y), (R_x, R_y)$	$(\alpha_{yz}, \alpha_{xz})$

C_{5v}	E	$2C_5$	$2C_5^2$	$5\sigma_v$	Activity: IR	Activity: Raman
A_1	1	1	1	1	T_z	$\alpha_{xx}+\alpha_{yy}, \alpha_{zz}$
A_2	1	1	1	-1	R_z	
E_1	2	2 cos 72°	2 cos 144°	0	$(T_x, T_y), (R_x, R_y)$	$(\alpha_{xz}, \alpha_{yz})$
E_2	2	2 cos 144°	2 cos 72°	0		$(\alpha_{xx}-\alpha_{yy}, \alpha_{xy})$

C_{6v}	E	$2C_6$	$2C_3$	C_2	$3\sigma_v$	$3\sigma_d$	Activity: IR	Activity: Raman
A_1	1	1	1	1	1	1	T_z	$\alpha_{xx}+\alpha_{yy}, \alpha_{zz}$
A_2	1	1	1	1	-1	-1	R_z	
B_1	1	-1	1	-1	1	-1		
B_2	1	-1	1	-1	-1	1		
E_1	2	1	-1	-2	0	0	$(T_x, T_y), (R_x, R_y)$	$(\alpha_{xz}, \alpha_{yz})$
E_2	2	-1	-1	2	0	0		$(\alpha_{xx}-\alpha_{yy}), (\alpha_{xy})$

3. The C_{nh} Groups

C_{2h}	E	C_2	i	σ_h	Activity	
					IR	Raman
A_g	1	1	1	1	R_z	$\alpha_{xx}, \alpha_{yy}, \alpha_{zz}, \alpha_{xy}$
A_u	1	1	−1	−1	T_z	
B_g	1	−1	1	−1	R_x, R_y	α_{yz}, α_{xz}
B_u	1	−1	−1	1	T_x, T_y	

C_{3h}	E	$2C_3$	σ_h	$2S_3$	Activity	
					IR	Raman
A'	1	1	1	1	R_z	$\alpha_{xx} + \alpha_{yy}, \alpha_{zz}$
A''	1	1	−1	−1	T_z	
E'	2	−1	2	−1	(T_x, T_y)	$(\alpha_{xx} - \alpha_{yy}, \alpha_{xy})$
E''	2	−1	−2	1	(R_x, R_y)	$(\alpha_{xz}, \alpha_{yz})$

C_{4h}	E	C_4	C_2	C_4^3	i	S_4^3	σ_h	S_4	Activity	
									IR	Raman
A_g	1	1	1	1	1	1	1	1	R_z	$\alpha_{xx}+\alpha_{yy}, \alpha_{zz}$
B_g	1	−1	1	−1	1	−1	1	−1		$\alpha_{xx}-\alpha_{yy}, \alpha_{xy}$
E_g	{1	i	−1	$-i$	1	i	−1	$-i$}	(R_x, R_y)	$(\alpha_{xz}, \alpha_{yz})$
	{1	$-i$	−1	i	1	$-i$	−1	i}		
A_u	1	1	1	1	−1	−1	−1	−1	T_z	
B_u	1	−1	1	−1	−1	1	−1	1		
E_u	{1	i	−1	$-i$	−1	$-i$	1	i}	(T_x, T_y)	
	{1	$-i$	−1	i	−1	i	1	$-i$}		

C_{5h}	E	$2C_5$	$2C_5^2$	σ_h	$2S_5^3$	$2S_5^2$	Activity	
							IR	Raman
A'	1	1	1	1	1	1	R_z	$\alpha_{xx}+\alpha_{yy}$, α_{zz}
A''	1	1	1	-1	-1	-1	T_z	
E_1'	2	2 cos 72°	2 cos 144°	2	2 cos 72°	2 cos 144°	(T_x, T_y)	
E_1''	2	2 cos 72°	2 cos 144°	-2	-2 cos 72°	-2 cos 144°	(R_x, R_y)	$(\alpha_{xz}, \alpha_{yz})$
E_2'	2	2 cos 144°	2 cos 72°	2	2 cos 144°	2 cos 72°		$(\alpha_{xx}-\alpha_{yy}, \alpha_{xy})$
E_2''	2	2 cos 144°	2 cos 72°	-2	-2 cos 144°	-2 cos 72°		

C_{6h}	E	$2C_6$	$2C_6^2 \equiv C_3$	$C_6^3 \equiv C_2''$	σ_h	$2S_6$	$2S_3$	$S_2 \equiv i$	Activity	
									IR	Raman
A_g	1	1	1	1	1	1	1	1	R_z	$\alpha_{xx}+\alpha_{yy}$, α_{zz}
A_u	1	1	1	1	-1	-1	-1	-1	T_z	
B_g	1	-1	1	-1	-1	1	-1	1		
B_u	1	-1	1	-1	1	-1	1	-1		
E_{1g}	2	1	-1	-2	-2	-1	1	2	R_x, R_y	α_{xz}, α_{yz}
E_{1u}	2	1	-1	-2	2	1	-1	-2	T_x, T_y	
E_{2g}	2	-1	-1	2	2	-1	-1	2		$\alpha_{xx}-\alpha_{yy}$, α_{xy}
E_{2u}	2	-1	-1	2	-2	1	1	-2		

4. The D_n Groups

D_2	E	$C_2(z)$	$C_2(y)$	$C_2(x)$	Activity	
					IR	Raman
A	1	1	1	1		α_{xx}, α_{yy}, α_{zz}
B_1	1	1	-1	-1	T_z, R_z	α_{xy}
B_2	1	-1	1	-1	T_y, R_y	α_{xz}
B_3	1	-1	-1	1	T_x, R_x	α_{yz}

D_3	E	$2C_3$	$3C_2$	Activity	
				IR	Raman
A_1	1	1	1		$\alpha_{xx} + \alpha_{yy}, \alpha_{zz}$
A_2	1	1	−1	T_z, R_z	
E	2	−1	0	$(T_x, T_y), (R_x, R_y)$	$(\alpha_{xx} - \alpha_{yy}, \alpha_{xy})$
					$(\alpha_{xz}, \alpha_{yz})$

D_4	E	$2C_4$	$C_2(= C_4^2)$	$2C_2'$	$2C_2''$	Activity	
						IR	Raman
A_1	1	1	1	1	1		$\alpha_{xx} + \alpha_{yy}, \alpha_{zz}$
A_2	1	1	1	−1	−1	T_z, R_z	
B_1	1	−1	1	1	−1		$\alpha_{xx} - \alpha_{yy}$
B_2	1	−1	1	−1	1		α_{xy}
E	2	0	−2	0	0	$(T_x, T_y), (R_x, R_y)$	$(\alpha_{xz}, \alpha_{yz})$

D_5	E	$2C_5$	$2C_5^2$	$5C_2$	Activity	
					IR	Raman
A_1	1	1	1	1		$\alpha_{xx} + \alpha_{yy}, \alpha_{zz}$
A_2	1	1	1	−1	T_z, R_z	
E_1	2	2 cos 72°	2 cos 144°	0	$(T_x, T_y), (R_x, R_y)$	$(\alpha_{xz}, \alpha_{yz})$
E_2	2	2 cos 144°	2 cos 72°	0		$(\alpha_{xx} - \alpha_{yy}, \alpha_{xy})$

D_6	E	$2C_6$	$2C_3$	C_2	$3C_2'$	$3C_2''$	Activity IR	Activity Raman
A_1	1	1	1	1	1	1		$\alpha_{xx} + \alpha_{yy}, \alpha_{zz}$
A_2	1	1	1	1	−1	−1	T_z, R_z	
B_1	1	−1	1	−1	1	−1		
B_2	1	−1	1	−1	−1	1		
E_1	2	1	−1	−2	0	0	$(T_x, T_y), (R_x, R_y)$	$(\alpha_{xz}, \alpha_{yz})$
E_2	2	−1	−1	2	0	0		$(\alpha_{xx} - \alpha_{yy}, \alpha_{xy})$

5. The D_{nh} Groups

D_{2h}	E	C_2^z	C_2^y	C_2^x	i	σ_{xy}	σ_{xz}	σ_{yz}	Activity IR	Activity Raman
A_g	1	1	1	1	1	1	1	1		$\alpha_{xx}, \alpha_{yy}, \alpha_{zz}$
A_u	1	1	1	1	−1	−1	−1	−1		
B_{1g}	1	1	−1	−1	1	1	−1	−1	R_z	α_{xy}
B_{1u}	1	1	−1	−1	−1	−1	1	1	T_z	
B_{2g}	1	−1	1	−1	1	−1	1	−1	R_y	α_{xz}
B_{2u}	1	−1	1	−1	−1	1	−1	1	T_y	
B_{3g}	1	−1	−1	1	1	−1	−1	1	R_x	α_{yz}
B_{3u}	1	−1	−1	1	−1	1	1	−1	T_x	

D_{3h}	E	$2C_3$	$3C_2$	σ_h	$2S_3$	$3\sigma_v$	Activity	
							IR	Raman
A_1'	1	1	1	1	1	1		$\alpha_{xx} + \alpha_{yy}, \alpha_{zz}$
A_1''	1	1	1	−1	−1	−1		
A_2'	1	1	−1	1	1	−1	R_z	
A_2''	1	1	−1	−1	−1	1	T_z	
E'	2	−1	0	2	−1	0	(T_x, T_y)	$(\alpha_{xx} - \alpha_{yy}, \alpha_{xy})$
E''	2	−1	0	−2	1	0	(R_x, R_y)	$(\alpha_{yz}, \alpha_{xz})$

D_{4h}	E	$2C_4$	C_2^z	$2C_2$	$2C_2'$	i	$2S_4$	σ_h	$2\sigma_v$	$2\sigma_d$	Activity	
											IR	Raman
A_{1g}	1	1	1	1	1	1	1	1	1	1		$\alpha_{xx}+\alpha_{yy}, \alpha_{zz}$
A_{1u}	1	1	1	1	1	−1	−1	−1	−1	−1		
A_{2g}	1	1	1	−1	−1	1	1	1	−1	−1	R_z	
A_{2u}	1	1	1	−1	−1	−1	−1	−1	1	1	T_z	
B_{1g}	1	−1	1	1	−1	1	−1	1	1	−1		$\alpha_{xx} - \alpha_{yy}$
B_{1u}	1	−1	1	1	−1	−1	1	−1	−1	1		
B_{2g}	1	−1	1	−1	1	1	−1	1	−1	1		α_{xy}
B_{2u}	1	−1	1	−1	1	−1	1	−1	1	−1		
E_g	2	0	−2	0	0	2	0	−2	0	0	(R_x, R_y)	$(\alpha_{yz}, \alpha_{xz})$
E_u	2	0	−2	0	0	−2	0	2	0	0	(T_x, T_y)	

D_{5h}	E	$2C_5$	$2C_5^2$	$5C_2$	σ_h	$2S_5$	$2S_5^3$	$5\sigma_v$	Activity	
									IR	Raman
A_1'	1	1	1	1	1	1	1	1		$\alpha_{xx} + \alpha_{yy}, \alpha_{zz}$
A_2'	1	1	1	−1	1	1	1	−1	R_z	
E_1'	2	2 cos 72°	2 cos 144°	0	2	2 cos 72°	2 cos 144°	0	(T_x, T_y)	
E_2'	2	2 cos 144°	2 cos 72°	0	2	2 cos 144°	2 cos 72°	0		$(\alpha_{xx} - \alpha_{yy}, \alpha_{xy})$
A_1''	1	1	1	1	−1	−1	−1	−1		
A_2''	1	1	1	−1	−1	−1	−1	1	T_z	
E_1''	2	2 cos 72°	2 cos 144°	0	−2	−2 cos 72°	−2 cos 144°	0	(R_x, R_y)	$(\alpha_{xz}, \alpha_{yz})$
E_2''	2	2 cos 144°	2 cos 72°	0	−2	−2 cos 144°	−2 cos 72°	0		

D_{6h}	E	$2C_6$	$2C_3$	C_2	$3C_2'$	$3C_2''$	i	$2S_3$	$2S_6$	σ_h	$3\sigma_d$	$3\sigma_v$	Activity	
													IR	Raman
A_{1g}	1	1	1	1	1	1	1	1	1	1	1	1		$\alpha_{xx} + \alpha_{yy}, \alpha_{zz}$
A_{2g}	1	1	1	1	−1	−1	1	1	1	1	−1	−1	R_z	
B_{1g}	1	−1	1	−1	1	−1	1	−1	1	−1	1	−1		
B_{2g}	1	−1	1	−1	−1	1	1	−1	1	−1	−1	1		
E_{1g}	2	1	−1	−2	0	0	2	1	−1	−2	0	0	(R_x, R_y)	$(\alpha_{xz}, \alpha_{yz})$
E_{2g}	2	−1	−1	2	0	0	2	−1	−1	2	0	0		$(\alpha_{xx} - \alpha_{yy}, \alpha_{xy})$
A_{1u}	1	1	1	1	1	1	−1	−1	−1	−1	−1	−1		
A_{2u}	1	1	1	1	−1	−1	−1	−1	−1	−1	1	1	T_z	
B_{1u}	1	−1	1	−1	1	−1	−1	1	−1	1	−1	1		
B_{2u}	1	−1	1	−1	−1	1	−1	1	−1	1	1	−1		
E_{1u}	2	1	−1	−2	0	0	−2	−1	1	2	0	0	(T_x, T_y)	
E_{2u}	2	−1	−1	2	0	0	−2	1	1	−2	0	0		

6. The D_{nd} Groups

D_{2d}	E	$2S_4^z$	C_2^z	$2C_2$	$2\sigma_d$	Activity	
						IR	Raman
A_1	1	1	1	1	1		$\alpha_{xx}+\alpha_{yy}$, α_{zz}
A_2	1	1	1	−1	−1	R_z	
B_1	1	−1	1	1	−1		$\alpha_{xx}-\alpha_{yy}$
B_2	1	−1	1	−1	1	T_z	α_{xy}
E	2	0	−2	0	0	(T_x, T_y), (R_x, R_y)	$(\alpha_{yz}, \alpha_{xz})$

D_{3d}	E	$2C_3$	$3C_2$	i	$2S_6$	$3\sigma_d$	Activity	
							IR	Raman
A_{1g}	1	1	1	1	1	1		$\alpha_{xx}+\alpha_{yy}$, α_{zz}
A_{2g}	1	1	−1	1	1	−1	R_z	
E_g	2	−1	0	2	−1	0	(R_x, R_y)	$(\alpha_{xx}-\alpha_{yy}, \alpha_{xy})$, $(\alpha_{xz}, \alpha_{yz})$
A_{1u}	1	1	1	−1	−1	−1		
A_{2u}	1	1	−1	−1	−1	1	T_z	
E_u	2	−1	0	−2	1	0	(T_x, T_y)	

D_{4d}	E	$2S_8$	$2C_4$	$2S_8^3$	C_2	$4C_2'$	$4\sigma_d$	Activity	
								IR	Raman
A_1	1	1	1	1	1	1	1		$\alpha_{xx}+\alpha_{yy}$, α_{zz}
A_2	1	1	1	1	1	−1	−1	R_z	
B_1	1	−1	1	−1	1	1	−1		
B_2	1	−1	1	−1	1	−1	1	T_z	
E_1	2	$\sqrt{2}$	0	$-\sqrt{2}$	−2	0	0	(T_x, T_y)	
E_2	2	0	−2	0	2	0	0		$(\alpha_{xx}-\alpha_{yy}, \alpha_{xy})$
E_3	2	$-\sqrt{2}$	0	$\sqrt{2}$	−2	0	0	(R_x, R_y)	$(\alpha_{xz}, \alpha_{yz})$

D_{5d}	E	$2C_5$	$2C_5^2$	$5C_2$	i	$2S_{10}^3$	$2S_{10}$	$5\sigma_d$	Activity	
									IR	Raman
A_{1g}	1	1	1	1	1	1	1	1		$\alpha_{xx} + \alpha_{yy}, \alpha_{zz}$
A_{2g}	1	1	1	−1	1	1	1	−1	R_z	
E_{1g}	2	2 cos 72°	2 cos 144°	0	2	2 cos 72°	2 cos 144°	0	(R_x, R_y)	$(\alpha_{yz}, \alpha_{zx})$
E_{2g}	2	2 cos 144°	2 cos 72°	0	2	2 cos 144°	2 cos 72°	0		$(\alpha_{xx} - \alpha_{yy}, \alpha_{xy})$
A_{1u}	1	1	1	1	−1	−1	−1	−1		
A_{2u}	1	1	1	−1	−1	−1	−1	1	T_z	
E_{1u}	2	2 cos 72°	2 cos 144°	0	−2	−2 cos 72°	−2 cos 144°	0	(T_x, T_y)	
E_{2u}	2	2 cos 144°	2 cos 72°	0	−2	−2 cos 144°	−2 cos 72°	0		

D_{6d}	E	$2S_{12}$	$2C_6$	$2S_4$	$2C_3$	$2S_{12}^5$	C_2	$6C_2'$	$6\sigma_d$	Activity	
										IR	Raman
A_1	1	1	1	1	1	1	1	1	1		$\alpha_{xx} + \alpha_{yy}, \alpha_{zz}$
A_2	1	1	1	1	1	1	1	−1	−1	R_z	
B_1	1	−1	1	−1	1	−1	1	1	−1		
B_2	1	−1	1	−1	1	−1	1	−1	1	T_z	
E_1	2	$\sqrt{3}$	1	0	−1	$-\sqrt{3}$	−2	0	0	(T_x, T_y)	
E_2	2	1	−1	−2	−1	1	2	0	0		$(\alpha_{xx} - \alpha_{yy}, \alpha_{xy})$
E_3	2	0	−2	0	2	0	−2	0	0		
E_4	2	−1	−1	2	−1	−1	2	0	0		
E_5	2	$-\sqrt{3}$	1	0	−1	$\sqrt{3}$	−2	0	0	(R_x, R_y)	$(\alpha_{yz}, \alpha_{zx})$

7. The S_n Groups

S_4	E	S_4	C_2	S_4^3	Activity	
					IR	Raman
A	1	1	1	1	R_z	$\alpha_{xx}, + \alpha_{yy}, \alpha_{zz}$
B	1	-1	1	-1	T_z	$\alpha_{xx} - \alpha_{yy}, \alpha_{xy}$
E	$\begin{cases}1\\1\end{cases}$	$\begin{matrix}i\\-i\end{matrix}$	$\begin{matrix}-1\\-1\end{matrix}$	$\begin{matrix}-i\\i\end{matrix}\Big\}$	(T_x, T_y) (R_x, R_y)	$(\alpha_{yz}, \alpha_{zx})$

S_6	E	C_3	C_3^2	i	S_6^5	S_6	Activity	
							IR	Raman
A_g	1	1	1	1	1	1	R_z	$\alpha_{xx} + \alpha_{yy}, \alpha_{zz}$
E_g	$\begin{cases}1\\1\end{cases}$	$\begin{matrix}\varepsilon\\\varepsilon^*\end{matrix}$	$\begin{matrix}\varepsilon^*\\\varepsilon\end{matrix}$	$\begin{matrix}1\\1\end{matrix}$	$\begin{matrix}\varepsilon\\\varepsilon^*\end{matrix}$	$\begin{matrix}\varepsilon^*\\\varepsilon\end{matrix}\Big\}$	(R_x, R_y)	$(\alpha_{xx} - \alpha_{yy}, \alpha_{xy})$, $(\alpha_{yz}, \alpha_{zx})$
A_u	1	1	1	-1	-1	-1	T_z	
E_u	$\begin{cases}1\\1\end{cases}$	$\begin{matrix}\varepsilon\\\varepsilon^*\end{matrix}$	$\begin{matrix}\varepsilon^*\\\varepsilon\end{matrix}$	$\begin{matrix}-1\\-1\end{matrix}$	$\begin{matrix}-\varepsilon\\-\varepsilon^*\end{matrix}$	$\begin{matrix}-\varepsilon^*\\-\varepsilon\end{matrix}\Big\}$	(T_x, T_y)	

$S_6 = C_3 \times i; \quad \varepsilon = e^{2\pi i/3}.$

S_8	E	S_8	C_4	S_8^3	C_2	S_8^5	C_4^3	S_8^7	Activity	
									IR	Raman
A	1	1	1	1	1	1	1	1	R_z	$(\alpha_{xx} + \alpha_{yy}, \alpha_{zz})$
B	1	−1	1	−1	1	−1	1	−1	T_z	
E_1	$\begin{cases}1\\1\end{cases}$	ε ε^*	i $-i$	$-\varepsilon^*$ $-\varepsilon$	−1 −1	$-\varepsilon$ $-\varepsilon^*$	$-i$ i	ε^* ε	(T_x, T_y) (R_x, R_y)	
E_2	$\begin{cases}1\\1\end{cases}$	i $-i$	−1 −1	$-i$ i	1 1	i $-i$	−1 −1	$-i$ i		$(\alpha_{xx} - \alpha_{yy}, \alpha_{xy})$
E_3	$\begin{cases}1\\1\end{cases}$	$-\varepsilon^*$ $-\varepsilon$	$-i$ i	ε ε^*	−1 −1	ε^* ε	i $-i$	$-\varepsilon$ $-\varepsilon^*$		$(\alpha_{yz}, \alpha_{zx})$

$\varepsilon = e^{2\pi i/8}$.

8. The Cubic Groups

T	E	$4C_3$	$4C_3^2$	$3C_2$	Activity	
					IR	Raman
A	1	1	1	1		$\alpha_{xx} + \alpha_{yy} + \alpha_{zz}$
E	$\begin{Bmatrix} 1 \\ 1 \end{Bmatrix}$	$\begin{matrix} \varepsilon \\ \varepsilon^* \end{matrix}$	$\begin{matrix} \varepsilon^* \\ \varepsilon \end{matrix}$	$\begin{matrix} 1 \\ 1 \end{matrix}$		$(\alpha_{xx} + \alpha_{yy} - 2\alpha_{zz}, \alpha_{xx} - \alpha_{yy})$
F	3	0	0	-1	(T_x, T_y, T_z); (R_x, R_y, R_z)	$\alpha_{xy}, \alpha_{xz}, \alpha_{yz}$

$\varepsilon = e^{2\pi i/3}$.

T_h	E	$4C_3$	$4C_3^2$	$3C_2$	i	$4S_6$	$4S_6^5$	$3\sigma_h$	Activity	
									IR	Raman
A_g	1	1	1	1	1	1	1	1		$\alpha_{xx} + \alpha_{yy} + \alpha_{zz}$
A_u	1	1	1	1	−1	−1	−1	−1		
E_g	1	ε	ε^*	1	1	ε	ε^*	1		$(\alpha_{xx} + \alpha_{yy} - 2\alpha_{zz}, \alpha_{xx} - \alpha_{yy})$
	1	ε^*	ε	1	1	ε^*	ε	1		
E_u	1	ε	ε^*	1	−1	$-\varepsilon^*$	$-\varepsilon^*$	−1		
	1	ε^*	ε	1	−1	ε^*	$-\varepsilon$	−1		
F_g	3	0	0	−1	3	0	0	−1	(R_x, R_y, R_z)	$(\alpha_{xy}, \alpha_{xz}, \alpha_{yz})$
F_u	3	0	0	−1	−3	0	0	1	(T_x, T_y, T_z)	

$\varepsilon = e^{2\pi i/3}$.

T_d	E	$8C_3$	$3C_2$	$6S_4$	$6\sigma_d$	Activity	
						IR	Raman
A_1	1	1	1	1	1		$\alpha_{xx} + \alpha_{yy} + \alpha_{zz}$
A_2	1	1	1	−1	−1		
E	2	−1	2	0	0		$(\alpha_{xx} + \alpha_{yy} - 2\alpha_{zz}, \alpha_{xx} - \alpha_{yy})$
F_1	3	0	−1	1	−1	(R_x, R_y, R_z)	
F_2	3	0	−1	−1	1	(T_x, T_y, T_z)	$(\alpha_{xy}, \alpha_{yz}, \alpha_{xz})$

O	E	$8C_3$	$3C_2$	$6C_4$	$6C_2'$	Activity	
						IR	Raman
A_1	1	1	1	1	1		$\alpha_{xx} + \alpha_{yy} + \alpha_{zz}$
A_2	1	1	1	−1	−1		
E	2	−1	2	0	0		$(\alpha_{xx} + \alpha_{yy} - 2\alpha_{zz}$, $\alpha_{xx} - \alpha_{yy})$
F_1	3	0	−1	1	−1	(T_x, T_y, T_z) (R_x, R_y, R_z)	
F_2	3	0	−1	−1	1		$(\alpha_{xy}, \alpha_{yz}, \alpha_{zx})$

O_h	E	$8C_3$	$3C_2$	$6C_4$	$6C_2'$	i	$8S_6$	$3\sigma_h$	$6S_4$	$6\sigma_d$	Activity	
											IR	Raman
A_{1g}	1	1	1	1	1	1	1	1	1	1		$\alpha_{xx} + \alpha_{yy} + \alpha_{zz}$
A_{2g}	1	1	1	−1	−1	1	1	1	−1	−1		
E_g	2	−1	2	0	0	2	−1	2	0	0		$(2\alpha_{zz} - \alpha_{xx} - \alpha_{yy}, \alpha_{xx} - \alpha_{yy})$
F_{1g}	3	0	−1	1	−1	3	0	−1	1	−1	(R_x, R_y, R_z)	
F_{2g}	3	0	−1	−1	1	3	0	−1	−1	1		$(\alpha_{xy}, \alpha_{xz}, \alpha_{yz})$
A_{1u}	1	1	1	1	1	−1	−1	−1	−1	−1		
A_{2u}	1	1	1	−1	−1	−1	−1	−1	1	1		
E_u	2	−1	2	0	0	−2	1	−2	0	0		
F_{1u}	3	0	−1	1	−1	−3	0	1	−1	1	(T_x, T_y, T_z)	
F_{2u}	3	0	−1	−1	1	−3	0	1	1	−1		

9. The $C_{\infty v}$ and $D_{\infty h}$ Groups

$C_{\infty v}$	E	$\dots 2C^{(z)}_{2\pi/\varphi}$	$\dots C^{(z)}_{2}$	$\infty\sigma_v$	Activity	
					IR	Raman
Σ^+	1	1	1	1	T_z	$\alpha_{xx} + \alpha_{yy}, \alpha_{zz}$
Σ^-	1	1	1	-1	R_z	
Π	2	$2 \cos \varphi$	-2	0	(T_x, T_y) (R_x, R_y)	$(\alpha_{xz}, \alpha_{yz})$
Δ	2	$2 \cos 2\varphi$	2	0		$(\alpha_{xx} - \alpha_{yy}, \alpha_{xy})$
φ	2	$2 \cos 3\varphi$	-2	0		

$D_{\infty h}$	E	$\ldots 2C_{2\pi/\varphi}$	$\ldots C_2^{(z)}$	$\infty\sigma_v$	i	$\ldots 2S_{2\pi/\varphi}$	$\ldots \sigma_h$	∞C_2	Activity	
									IR	Raman
Σ_g^+	1	1	1	1	1	1	1	1		$\alpha_{xx} + \alpha_{yy}, \alpha_{zz}$
Σ_u^+	1	1	1	1	−1	−1	−1	−1	T_z	
Σ_g^-	1	1	1	−1	1	1	1	−1	R_z	
Σ_u^-	1	1	1	−1	−1	−1	−1	1		
Π_g	2	$2\cos\varphi$	−2	0	2	$-2\cos\varphi$	−2	0	(R_x, R_y)	$(\alpha_{xz}, \alpha_{yz})$
Π_u	2	$2\cos\varphi$	−2	0	−2	$2\cos\varphi$	2	0	(T_x, T_y)	
Δ_g	2	$2\cos 2\varphi$	2	0	2	$2\cos 2\varphi$	2	0		$(\alpha_{xx} - \alpha_{yy}, \alpha_{xy})$
Δ_u	2	$2\cos 2\varphi$	2	0	−2	$-2\cos 2\varphi$	−2	0		
φ_g	2	$2\cos 3\varphi$	−2	0	2	$-2\cos 3\varphi$	−2	0		
φ_u	2	$2\cos 3\varphi$	−2	0	−2	$2\cos 3\varphi$	2	0		

Appendix 2

DESCRIPTION OF SYMBOLISM USED IN THE INTERNATIONAL TABLES FOR X-RAY CRYSTALLOGRAPHY

The space group *Pbca*/D_{2h}^{15} taken from the *International Tables for X-Ray Crystallography*[1] is used to describe further characteristics of the space group (Fig. A2-1).

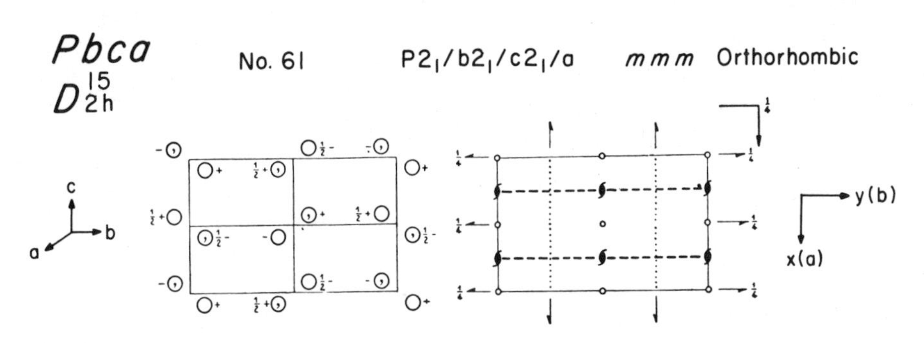

Origin at $\bar{1}$

Number of positions, Wyckoff notation, and point symmetry			Co-ordinates of equivalent positions	Conditions limiting possible reflections
				General:
8	*c*	1	x,y,z; $\frac{1}{2}+x,\frac{1}{2}-y,\bar{z}$; $\bar{x},\frac{1}{2}+y,\frac{1}{2}-z$; $\frac{1}{2}-x,\bar{y},\frac{1}{2}+z$; $\bar{x},\bar{y},\bar{z}$; $\frac{1}{2}-x,\frac{1}{2}+y,z$; $x,\frac{1}{2}-y,\frac{1}{2}+z$; $\frac{1}{2}+x,y,\frac{1}{2}-z$.	*hkl*: No conditions 0*kl*: $k=2n$ *h*0*l*: $l=2n$ *hk*0: $h=2n$ *h*00: $(h=2n)$ 0*k*0: $(k=2n)$ 00*l*: $(l=2n)$
				Special: as above, plus
4	*b*	$\bar{1}$	$0,0,\frac{1}{2}$; $\frac{1}{2},\frac{1}{2},\frac{1}{2}$; $0,\frac{1}{2},0$; $\frac{1}{2},0,0$.	} *hkl*: $h+k, k+l, (l+h)=2n$
4	*a*	$\bar{1}$	$0,0,0$; $\frac{1}{2},\frac{1}{2},0$; $0,\frac{1}{2},\frac{1}{2}$; $\frac{1}{2},0,\frac{1}{2}$.	

Symmetry of special projections

(001) *pgm*; $a'=a/2$, $b'=b$ (100) *pgm*; $b'=b/2$, $c'=c$ (010) *pgm*; $c'=c/2$, $a'=a$

Fig. A2-1.

The abbreviated symmetry elements are given across the top of Fig. A2-1 at the left corner; *Pbca* is the Hermann–Mauguin designation, and D_{2h}^{15} is the Schoenflies notation. This is followed by the space group number (in this case No. 61 of a total of 230), followed by the "full" space group symbols (these include other symmetry elements if they are present), followed by the crystal class out of 32 (*mmm*) and the crystal system (out of seven)—orthorhombic. The symbol $P2_1/b2_1/c2_1/a$ indicates the unit cell is "primitive," that the symmetry element 2_1 may be associated with the crystallographic axes a, b, and c respectively. The two-fold screw axes are designated as 2_1 and indicate that a one-half unit cell translation follows the normal two-fold (180°) rotation. The glide planes are all perpendicular to the respective crystallographic axes.

The top left-hand diagram symbolizes the way in which the "asymmetric unit" is repeated with the unit cell (circles and plus or minus signs are used; the plus sign indicates lying above plane of the page and the negative sign indicates lying below the plane of the page). The *upper left-hand corner* of the rectangle is the origin, the x axis extends down the page, and the y axis extends across the page (from left to right).

The symbols in the right-hand diagram indicate which symmetry elements are present and their positions. The symbol ↾ represents one two-fold screw axis perpendicular to both a–c and b–c planes located at $x = 0$, $z = \frac{1}{4}$ and $z = 0$, $y = \frac{1}{4}$, respectively. The two-fold axis $x = \frac{1}{4}$, $z = 0$ parallel to c and perpendicular to the a–b plane is denoted by the symbol §. The broken lines represent glide planes seen edge-on and the small circles represent centers of symmetry.

Under the heading "conditions limiting possible reflections" the systematically absent reflections for the space group are listed.

The listing of "coordinates of equivalent positions" gives the allowable atom positional parameters for sites of various symmetries within the space group.

REFERENCE

1. N. F. M. Henry and K. Lonsdale (eds.), *International Tables for X-Ray Crystallography*, Vol. 1, The Kynoch Press, Birmingham, England (1965).

Appendix 3

SITE SYMMETRIES FOR THE 230 SPACE GROUPS

The site symmetries for the 230 space groups are given. The symmetries can be used directly for a Bravais primitive cell, and the space groups are numbered from 1 to 230 to make for easier reference. The sequence of numbering corresponds to that used in the *International Tables for X-Ray Crystallography*.[1] In using these tables, it should be noted that when the site symmetry is C_p, C_{pv}, or C_s and $p = 1, 2, 3$, etc., the number of sites is infinite.

Note: The site symmetry is described by several symbols. For the example

Space group 25 *pmm*2 C_{2v}^1: $4C_{2v}(1)$; $4C_s(2)$; $C_1(4)$

the coefficient represents the nonequivalent sets of sites (in this case $4C_{2v}$ sites) with one equivalent site per set shown in parentheses; an infinite number of C_s sites, each having an occupation number of two equivalent atoms and there are four such nonequivalent sets; an infinite number of C_1 sites with four equivalent sites per set. For further discussion on this subject see Couture[3] and Irish and Brooker.[4]

Space group[1]		Site symmetries[2]
1 *P*1	C_1^1	$C_1(1)$
2 $P\bar{1}$	C_i^1	$8C_i(1)$; $C_1(2)$
3 *P*2	C_2^1	$4C_2(1)$; $C_1(2)$
4 $P2_1$	C_2^2	$C_1(2)$
5 B_2 or C_2	C_2^3	$2C_2(1)$; $C_1(2)$
6 *Pm*	C_s^1	$2C_s(1)$; $C_1(2)$
7 *Pb* or *Pc*	C_s^2	$C_1(2)$
8 *Bm* or *Cm*	C_s^3	$C_s(1)$; $C_1(2)$
9 *Bb* or *Cc*	C_s^4	$C_1(2)$
10 *P*2/*m*	C_{2h}^1	$8C_{2h}(1)$; $4C_2(2)$; $2C_s(2)$; $C_1(4)$

Appendix 3 (*continued*)

Space group[1]		Site symmetries[2]
11 $P2_1/m$	C_{2h}^2	$4C_i(2)$; $C_s(2)$; $C_1(4)$
12 $B2/m$ or $C2/m$	C_{2h}^3	$4C_{2h}(1)$; $2C_i(2)$; $2C_2(2)$; $C_s(2)$; $C_1(4)$
13 $P2/b$ or $P2/c$	C_{2h}^4	$4C_i(2)$; $2C_2(2)$; $C_1(4)$
14 $P2_1/b$ or $P2_1/c$	C_{2h}^5	$4C_i(2)$; $C_1(4)$
15 $B2/b$ or $C2/c$	C_{2h}^6	$4C_i(2)$; $C_2(2)$; $C_1(4)$
16 $P222$	D_2^1	$8D_2(1)$; $12C_2(2)$; $C_1(4)$
17 $P222_1$	D_2^2	$4C_2(2)$; $C_1(4)$
18 $P2_12_12$	D_2^3	$2C_2(2)$; $C_1(4)$
19 $P2_12_12_1$	D_2^4	$C_1(4)$
20 $C222_1$	D_2^5	$2C_2(2)$; $C_1(4)$
21 $C222$	D_2^6	$4D_2(1)$; $7C_2(2)$; $C_1(4)$
22 $F222$	D_2^7	$4D_2(1)$; $6C_2(2)$; $C_1(4)$
23 $I222$	D_2^8	$4D_2(1)$; $6C_2(2)$; $C_1(4)$
24 $I2_12_12_1$	D_2^9	$3C_2(2)$; $C_1(4)$
25 $Pmm2$	C_{2v}^1	$4C_{2v}(1)$; $4C_s(2)$; $C_1(4)$
26 $Pmc2_1$	C_{2v}^2	$2C_s(2)$; $C_1(4)$
27 $Pcc2$	C_{2v}^3	$4C_2(2)$; $C_1(4)$
28 $Pma2$	C_{2v}^4	$2C_2(2)$; $C_s(2)$; $C_1(4)$
29 $Pca2_1$	C_{2v}^5	$C_1(4)$
30 $Pnc2$	C_{2v}^6	$2C_2(2)$; $C_1(4)$
31 $Pmn2_1$	C_{2v}^7	$C_s(2)$; $C_1(4)$
32 $Pba2$	C_{2v}^8	$2C_2(2)$; $C_1(4)$
33 $Pna2_1$	C_{2v}^9	$C_1(4)$
34 $Pnn2$	C_{2v}^{10}	$2C_2(2)$; $C_1(4)$
35 $Cmm2$	C_{2v}^{11}	$2C_{2v}(1)$; $C_2(2)$; $2C_s(2)$; $C_1(4)$
36 $Cmc2_1$	C_{2v}^{12}	$C_s(2)$; $C_1(4)$
37 $Ccc2$	C_{2v}^{13}	$3C_2(2)$; $C_1(4)$
38 $Amm2$	C_{2v}^{14}	$2C_{2v}(1)$; $3C_s(2)$; $C_1(4)$
39 $Abm2$	C_{2v}^{15}	$2C_2(2)$; $C_s(2)$; $C_1(4)$
40 $Ama2$	C_{2v}^{16}	$C_2(2)$; $C_s(2)$; $C_1(4)$
41 $Aba2$	C_{2v}^{17}	$C_2(2)$; $C_1(4)$
42 $Fmm2$	C_{2v}^{18}	$C_{2v}(1)$; $C_2(2)$; $2C_s(2)$; $C_1(4)$
43 $Fdd2$	C_{2v}^{19}	$C_2(2)$; $C_1(4)$
44 $Imm2$	C_{2v}^{20}	$2C_{2v}(1)$; $2C_s(2)$; $C_1(4)$
45 $Iba2$	C_{2v}^{21}	$2C_2(2)$; $C_1(4)$

Appendix 3 (*continued*)

Space group[1]		Site symmetries[2]
46 *Ima*2	C_{2v}^{22}	$C_2(2)$; $C_s(2)$; $C_1(4)$
47 *Pmmm*	D_{2h}^{1}	$8D_{2h}(1)$; $12C_{2v}(2)$; $6C_s(4)$; $C_1(8)$
48 *Pnnn*	D_{2h}^{2}	$4D_2(2)$; $2C_i(4)$; $6C_2(4)$; $C_1(8)$
49 *Pccm*	D_{2h}^{3}	$4C_{2h}(2)$; $4D_2(2)$; $8C_2(4)$; $C_s(4)$; $C_1(8)$
50 *Pban*	D_{2h}^{4}	$4D_2(2)$; $2C_i(4)$; $6C_2(4)$; $C_1(8)$
51 *Pmma*	D_{2h}^{5}	$4C_{2h}(2)$; $2C_{2v}(2)$; $2C_2(4)$; $3C_s(4)$; $C_1(8)$
52 *Pnna*	D_{2h}^{6}	$2C_i(4)$; $2C_2(4)$; $C_1(8)$
53 *Pmna*	D_{2h}^{7}	$4C_{2h}(2)$; $3C_2(4)$; $C_s(4)$; $C_1(8)$
54 *Pcca*	D_{2h}^{8}	$2C_i(4)$; $3C_2(4)$; $C_1(8)$
55 *Pbam*	D_{2h}^{9}	$4C_{2h}(2)$; $2C_2(4)$; $2C_s(4)$; $C_1(8)$
56 *Pccn*	D_{2h}^{10}	$2C_i(4)$; $2C_2(4)$; $C_1(8)$
57 *Pbcm*	D_{2h}^{11}	$2C_i(4)$; $C_2(4)$; $C_s(4)$; $C_1(8)$
58 *Pnnm*	D_{2h}^{12}	$4C_{2h}(2)$; $2C_2(4)$; $C_s(4)$; $C_1(8)$
59 *Pmmn*	D_{2h}^{13}	$2C_{2v}(2)$; $2C_i(4)$; $2C_s(4)$; $C_1(8)$
60 *Pbcn*	D_{2h}^{14}	$2C_i(4)$; $C_2(4)$; $C_1(8)$
61 *Pbca*	D_{2h}^{15}	$2C_i(4)$; $C_1(8)$
62 *Pnma*	D_{2h}^{16}	$2C_i(4)$; $C_s(4)$; $C_1(8)$
63 *Cmcm*	D_{2h}^{17}	$2C_{2h}(2)$; $C_{2v}(2)$; $C_i(4)$; $C_2(4)$; $2C_s(4)$; $C_1(8)$
64 *Cmca*	D_{2h}^{18}	$2C_{2h}(2)$; $C_i(4)$; $2C_2(4)$; $C_s(4)$; $C_i(8)$
65 *Cmmm*	D_{2h}^{19}	$4D_{2h}(1)$; $2C_{2h}(2)$; $6C_{2v}(2)$; $C_2(4)$; $4C_s(4)$; $C_1(8)$
66 *Cccm*	D_{2h}^{20}	$2D_2(2)$; $4C_{2h}(2)$; $5C_2(4)$; $C_s(4)$; $C_1(8)$
67 *Cmma*	D_{2h}^{21}	$2D_2(2)$; $4C_{2h}(2)$; $C_{2v}(2)$; $5C_2(4)$; $2C_s(4)$; $C_1(8)$
68 *Ccca*	D_{2h}^{22}	$2D_2(2)$; $2C_i(4)$; $4C_2(4)$; $C_1(8)$
69 *Fmmm*	D_{2h}^{23}	$2D_{2h}(1)$; $3C_{2h}(2)$; $D_2(2)$; $3C_{2v}(2)$; $3C_2(4)$; $3C_s(4)$; $C_1(8)$
70 *Fddd*	D_{2h}^{24}	$2D_2(2)$; $2C_i(4)$; $3C_2(4)$; $C_1(8)$
71 *Immm*	D_{2h}^{25}	$4D_{2h}(1)$; $6C_{2v}(2)$; $C_i(4)$; $3C_s(4)$; $C_1(8)$
72 *Ibam*	D_{2h}^{26}	$2D_2(2)$; $2C_{2h}(2)$; $C_i(4)$; $4C_2(4)$; $C_s(4)$; $C_1(8)$
73 *Ibca*	D_{2h}^{27}	$2C_i(4)$; $3C_2(4)$; $C_1(8)$
74 *Imma*	D_{2h}^{28}	$4C_{2h}(2)$; $C_{2v}(2)$; $2C_2(4)$; $2C_s(4)$; $C_1(8)$
75 *P*4	C_4^{1}	$2C_4(1)$; $C_2(2)$; $C_1(4)$
76 $P4_1$	C_4^{2}	$C_1(4)$
77 $P4_2$	C_4^{3}	$3C_2(2)$; $C_1(4)$
78 $P4_3$	C_4^{4}	$C_1(4)$
79 14	C_4^{5}	$C_4(1)$; $C_2(2)$; $C_1(4)$
80 $I4_1$	C_4^{6}	$C_2(2)$; $C_1(4)$

Appendix 3 (*continued*)

Space group[1]		Site symmetries[2]
81 $P\bar{4}$	S_4^1	$4S_4(1)$; $3C_2(2)$; $C_1(4)$
82 $I\bar{4}$	S_4^2	$4S_4(1)$; $2C_2(2)$; $C_1(4)$
83 $P4/m$	C_{4h}^1	$4C_{4h}(1)$; $2C_{2h}(2)$; $2C_4(2)$; $C_2(4)$; $2C_s(4)$; $C_1(8)$
84 $P4_2/m$	C_{4h}^2	$4C_{2h}(2)$; $2S_4(2)$; $3C_2(4)$; $C_s(4)$; $C_1(8)$
85 $P4/n$	C_{4h}^3	$2S_4(2)$; $C_4(2)$; $2C_i(4)$; $C_2(4)$; $C_1(8)$
86 $P4_2/n$	C_{4h}^4	$2S_4(2)$; $2C_i(4)$; $2C_2(4)$; $C_1(8)$
87 $I4/m$	C_{4h}^5	$2C_{4h}(1)$; $C_{2h}(2)$; $S_4(2)$; $C_4(2)$; $C_i(4)$; $C_2(4)$; $C_s(4)$; $C_1(8)$
88 $I4_1/a$	C_{4h}^6	$2S_4(2)$; $2C_i(4)$; $C_2(4)$; $C_1(8)$
89 $P422$	D_4^1	$4D_4(1)$; $2D_2(2)$; $2C_4(2)$; $7C_2(4)$; $C_1(8)$
90 $P42_12$	D_4^2	$2D_2(2)$; $C_4(2)$; $3C_2(4)$; $C_1(8)$
91 $P4_122$	D_4^3	$3C_2(4)$; $C_1(8)$
92 $P4_12_12$	D_4^4	$C_2(4)$; $C_1(8)$
93 $P4_222$	D_4^5	$6D_2(2)$; $9C_2(4)$; $C_1(8)$
94 $P4_22_12$	D_4^6	$2D_2(2)$; $4C_2(4)$; $C_1(8)$
95 $P4_322$	D_4^7	$3C_2(4)$; $C_1(8)$
96 $P4_32_12$	D_4^8	$C_2(4)$; $C_1(8)$
97 $I422$	D_4^9	$2D_4(1)$; $2D_2(2)$; $C_4(2)$; $5C_2(4)$; $C_1(8)$
98 $I4_122$	D_4^{10}	$2D_2(2)$; $4C_2(4)$; $C_1(8)$
99 $P4mm$	C_{4v}^1	$2C_{4v}(1)$; $C_{2v}(2)$; $3C_s(4)$; $C_1(8)$
100 $P4bm$	C_{4v}^2	$C_4(2)$; $C_{2v}(2)$; $C_s(4)$; $C_1(8)$
101 $P4_2cm$	C_{4v}^3	$2C_{2v}(2)$; $C_2(4)$; $C_s(4)$; $C_1(8)$
102 $P4_2nm$	C_{4v}^4	$C_{2v}(2)$; $C_2(4)$; $C_s(4)$; $C_1(8)$
103 $P4cc$	C_{4v}^5	$2C_4(2)$; $C_2(4)$; $C_1(8)$
104 $P4nc$	C_{4v}^6	$C_4(2)$; $C_2(4)$; $C_1(8)$
105 $P4_2mc$	C_{4v}^7	$3C_{2v}(2)$; $2C_s(4)$; $C_1(8)$
106 $P4_2bc$	C_{4v}^8	$2C_2(4)$; $C_1(8)$
107 $I4mm$	C_{4v}^9	$C_{4v}(1)$; $C_{2v}(2)$; $2C_s(4)$; $C_1(8)$
108 $I4cm$	C_{4v}^{10}	$C_4(2)$; $C_{2v}(2)$; $C_s(4)$; $C_1(8)$
109 $I4_1md$	C_{4v}^{11}	$C_{2v}(2)$; $C_s(4)$; $C_1(8)$
110 $I4_1cd$	C_{4v}^{12}	$C_2(4)$; $C_1(8)$
111 $P\bar{4}2m$	D_{2d}^1	$4D_{2d}(1)$; $2D_2(2)$; $2C_{2v}(2)$; $5C_2(4)$; $C_s(4)$; $C_1(8)$
112 $P\bar{4}2c$	D_{2d}^2	$4D_2(2)$; $2S_4(2)$; $7C_2(4)$; $C_1(8)$
113 $P\bar{4}2_1m$	D_{2d}^3	$2S_4(2)$; $C_{2v}(2)$; $C_2(4)$; $C_s(4)$; $C_1(8)$
114 $P\bar{4}2_1c$	D_{2d}^4	$2S_4(2)$; $2C_2(4)$; $C_1(8)$
115 $P\bar{4}m2$	D_{2d}^5	$4D_{2d}(1)$; $3C_{2v}(2)$; $2C_2(4)$; $2C_s(4)$; $C_1(8)$

Appendix 3 (*continued*)

Space group[1]		Site symmetries[2]
116 $P\bar{4}c2$	D_{2d}^{6}	$2D_2(2)$; $2S_4(2)$; $5C_2(4)$; $C_1(8)$
117 $P\bar{4}b2$	D_{2d}^{7}	$2S_4(2)$; $2D_2(2)$; $4C_2(4)$; $C_1(8)$
118 $P\bar{4}n2$	D_{2d}^{8}	$2S_4(2)$; $2D_2(2)$; $4C_2(4)$; $C_1(8)$
119 $I\bar{4}m2$	D_{2d}^{9}	$4D_{2d}(1)$; $2C_{2v}(2)$; $2C_2(4)$; $C_s(4)$; $C_1(8)$
120 $I\bar{4}c2$	D_{2d}^{10}	$D_2(2)$; $2S_4(2)$; $D_2(2)$; $4C_2(4)$; $C_1(8)$
121 $I\bar{4}2m$	D_{2d}^{11}	$2D_{2d}(1)$; $D_2(2)$; $S_4(2)$; $C_{2v}(2)$; $3C_2(4)$; $C_s(4)$; $C_1(8)$
122 $I\bar{4}2d$	D_{2d}^{12}	$2S_4(2)$; $2C_2(4)$; $C_1(8)$
123 $P4/mmm$	D_{4h}^{1}	$4D_{4h}(1)$; $2D_{2h}(2)$; $2C_{4v}(2)$; $7C_{2v}(4)$; $5C_s(8)$; $C_1(16)$
124 $P4/mcc$	D_{4h}^{2}	$D_4(2)$; $C_{4h}(2)$; $D_4(2)$; $C_{4h}(2)$; $C_{2h}(4)$; $D_2(4)$; $2C_4(4)$; $4C_2(8)$; $C_s(8)$; $C_1(16)$
125 $P4/nbm$	D_{4h}^{3}	$2D_4(2)$; $2D_{2d}(2)$; $2C_{2h}(4)$; $C_4(4)$; $C_{2v}(4)$; $4C_2(8)$; $C_s(8)$; $C_1(16)$
126 $P4/nnc$	D_{4h}^{4}	$2D_4(2)$; $D_2(4)$; $S_4(4)$; $C_4(4)$; $C_i(8)$; $4C_2(8)$; $C_1(8)$
127 $P4/mbm$	D_{4h}^{5}	$2C_{4h}(2)$; $2D_{2h}(2)$; $C_4(4)$; $3C_{2v}(4)$; $3C_s(8)$; $C_1(16)$
128 $P4/mnc$	D_{4h}^{6}	$2C_{4h}(2)$; $C_{2h}(4)$; $D_2(4)$; $C_4(4)$; $2C_2(8)$; $C_s(8)$; $C_1(16)$
129 $P4/nmm$	D_{4h}^{7}	$2D_{2d}(2)$; $C_{4v}(2)$; $2C_{2h}(4)$; $C_{2v}(4)$; $2C_2(8)$; $2C_s(8)$; $C_1(16)$
130 $P4/ncc$	D_{4h}^{8}	$D_2(4)$; $S_4(4)$; $C_4(4)$; $C_i(8)$; $2C_2(8)$; $C_1(16)$
131 $P4_2/mmc$	D_{4h}^{9}	$4D_{2h}(2)$; $2D_{2d}(2)$; $7C_{2v}(4)$; $C_2(8)$; $3C_s(8)$; $C_1(16)$
132 $P4_2/mcm$	D_{4h}^{10}	$D_{2h}(2)$; $D_{2d}(2)$; $D_{2h}(2)$; $D_{2d}(2)$; $D_2(4)$; $C_{2h}(4)$; $4C_{2v}(4)$; $3C_2(8)$; $2C_s(8)$; $C_1(16)$
133 $P4_2/nbc$	D_{4h}^{11}	$3D_2(4)$; $S_4(4)$; $C_i(8)$; $5C_2(8)$; $C_1(16)$
134 $P4_2/nnm$	D_{4h}^{12}	$2D_{2d}(2)$; $2D_2(4)$; $2C_{2h}(4)$; $C_{2v}(4)$; $5C_2(8)$; $C_s(8)$; $C_1(16)$
135 $P4_2/mbc$	D_{4h}^{13}	$C_{2h}(4)$; $S_4(4)$; $C_{2h}(4)$; $D_2(4)$; $3C_2(8)$; $C_s(8)$; $C_1(16)$
136 $P4_2/mnm$	D_{4h}^{14}	$2D_{2h}(2)$; $C_{2h}(4)$; $S_4(4)$; $3C_{2v}(4)$; $C_2(8)$; $2C_s(8)$; $C_1(16)$
137 $P4_2/nmc$	D_{4h}^{15}	$2D_{2d}(2)$; $2C_{2v}(4)$; $C_i(8)$; $C_2(8)$; $C_s(8)$; $C_1(16)$
138 $P4_2/ncm$	D_{4h}^{16}	$D_2(4)$; $S_4(4)$; $2C_{2h}(4)$; $C_{2v}(4)$; $3C_2(8)$; $C_s(8)$; $C_1(16)$
139 $I4/mmm$	D_{4h}^{17}	$2D_{4h}(1)$; $D_{2h}(2)$; $D_{2d}(2)$; $C_{4v}(2)$; $C_{2h}(4)$; $4C_{2v}(4)$; $C_2(8)$; $3C_s(8)$; $C_1(16)$
140 $I4/mcm$	D_{4h}^{18}	$D_4(2)$; $D_{2d}(2)$; $C_{4h}(2)$; $D_{2h}(2)$; $C_{2h}(4)$; $C_4(4)$; $2C_{2v}(4)$; $2C_2(8)$; $2C_s(8)$; $C_1(16)$
141 $I4_1/amd$	D_{4h}^{19}	$2D_{2d}(2)$; $2C_{2h}(4)$; $C_{2v}(4)$; $2C_2(8)$; $C_s(8)$; $C_1(16)$
142 $I4_1/acd$	D_{4h}^{20}	$S_4(4)$; $D_2(4)$; $C_i(8)$; $3C_2(8)$; $C_1(16)$
143 $P3$	C_3^{1}	$3C_3(1)$; $C_1(3)$
144 $P3_1$	C_3^{2}	$C_1(3)$
145 $P3_2$	C_3^{3}	$C_1(3)$

Appendix 3 (*continued*)

Space group[1]		Site symmetries[2]
146 $R3$	C_3^4	$C_3(1)$; $C_1(3)$
147 $P\bar{3}$	C_{3i}^1	$2C_{3i}(1)$; $2C_3(2)$; $2C_i(3)$; $C_1(6)$
148 $R\bar{3}$	C_{3i}^2	$2C_{3i}(1)$; $C_3(2)$; $2C_i(3)$; $C_1(6)$
149 $P312$	D_3^1	$6D_3(1)$; $3C_3(2)$; $2C_2(3)$; $C_1(6)$
150 $P321$	D_3^2	$2D_3(1)$; $2C_3(2)$; $2C_2(3)$; $C_1(6)$
151 $P3_112$	D_3^3	$2C_2(3)$; $C_1(6)$
152 $P3_121$	D_3^4	$2C_2(3)$; $C_1(6)$
153 $P3_212$	D_3^5	$2C_2(3)$; $C_1(6)$
154 $P3_221$	D_3^6	$2C_2(3)$; $C_1(6)$
155 $R32$	D_3^7	$2D_3(1)$; $C_3(2)$; $2C_2(3)$; $C_1(6)$
156 $P3m1$	C_{3v}^1	$3C_{3v}(1)$; $C_s(3)$; $C_1(6)$
157 $P31m$	C_{3v}^2	$C_{3v}(1)$; $C_3(2)$; $C_s(3)$; $C_1(6)$
158 $P3c1$	C_{3v}^3	$3C_3(2)$; $C_1(6)$
159 $P31c$	C_{3v}^4	$2C_3(2)$; $C_1(6)$
160 $R3m$	C_{3v}^5	$C_{3v}(1)$; $C_s(3)$; $C_1(6)$
161 $R3c$	C_{3v}^6	$C_3(2)$; $C_1(6)$
162 $P\bar{3}1m$	D_{3d}^1	$2D_{3d}(1)$; $2D_3(2)$; $C_{3v}(2)$; $2C_{2h}(3)$; $C_3(4)$; $2C_2(6)$; $C_s(6)$; $C_1(12)$
163 $P\bar{3}1c$	D_{3d}^2	$D_3(2)$; $C_{3i}(2)$; $2D_3(2)$; $2C_3(4)$; $C_i(6)$; $C_2(6)$; $C_1(12)$
164 $P\bar{3}m1$	D_{3d}^3	$2D_{3d}(1)$; $2C_{3v}(2)$; $2C_{2h}(3)$; $2C_2(6)$; $C_s(6)$; $C_1(12)$
165 $P\bar{3}c1$	D_{3d}^4	$D_3(2)$; $C_{3i}(2)$; $2C_3(4)$; $C_i(6)$; $C_1(6)$; $C_2(12)$
166 $R\bar{3}m$	D_{3d}^5	$2D_{3d}(1)$; $C_{3v}(2)$; $2C_{2h}(3)$; $2C_2(6)$; $C_s(6)$; $C_1(12)$
167 $R\bar{3}c$	D_{3d}^6	$D_3(2)$; $C_{3i}(2)$; $C_3(4)$; $C_i(6)$; $C_2(6)$; $C_1(12)$
168 $P6$	C_6^1	$C_6(1)$; $C_3(2)$; $C_2(3)$; $C_1(6)$
169 $P6_1$	C_6^2	$C_1(6)$
170 $P6_5$	C_6^3	$C_1(6)$
171 $P6_2$	C_6^4	$2C_2(3)$; $C_1(6)$
172 $P6_4$	C_6^5	$2C_2(3)$; $C_1(6)$
173 $P6_3$	C_6^6	$2C_3(2)$; $C_1(6)$
174 $P\bar{6}$	C_{3h}^1	$6C_{3h}(1)$; $3C_3(2)$; $2C_s(3)$; $C_1(6)$
175 $P6/m$	C_{6h}^1	$2C_{6h}(1)$; $2C_{3h}(2)$; $C_6(2)$; $2C_{2h}(3)$; $C_3(4)$; $C_2(6)$; $2C_s(6)$; $C_1(12)$
176 $P6_3/m$	C_{6h}^2	$C_{3h}(2)$; $C_{3i}(2)$; $2C_{3h}(2)$; $2C_3(4)$; $C_i(6)$; $C_s(6)$; $C_1(12)$
173 $P622$	D_6^1	$2D_6(1)$; $2D_3(2)$; $C_6(2)$; $2D_2(3)$; $C_4(4)$; $5C_2(6)$; $C_1(12)$
178 $P6_122$	D_6^2	$2C_2(6)$; $C_1(12)$

Appendix 3 (*continued*)

Space group[1]		Site symmetries[2]
179 $P6_522$	D_6^3	$2C_2(6);\ C_1(12)$
180 $P6_222$	D_6^4	$4D_2(3);\ 6C_2(6);\ C_1(12)$
181 $P6_422$	D_6^5	$4D_2(3);\ 6C_2(6);\ C_1(12)$
182 $P6_322$	D_6^6	$4D_3(2);\ 2C_3(4);\ 2C_2(6);\ C_1(12)$
183 $P6mm$	C_{6v}^1	$C_{6v}(1);\ C_{3v}(2);\ C_{2v}(3);\ C_{2v}(3);\ 2C_s(6);\ C_1(12)$
184 $P6cc$	C_{6v}^2	$C_6(2);\ C_3(4);\ C_2(6);\ C_1(12)$
185 $P6_3cm$	C_{6v}^3	$C_{3v}(2);\ C_3(4);\ C_s(6);\ C_1(12)$
186 $P6_3mc$	C_{6v}^4	$2C_{3v}(2);\ C_s(6);\ C_1(12)$
187 $P\bar{6}m2$	D_{3h}^1	$6D_{3h}(1);\ 3C_{3v}(2);\ 2C_{2v}(3);\ 3C_s(6);\ C_1(12)$
188 $P\bar{6}c2$	D_{3h}^2	$D_3(2);\ C_{3h}(2);\ D_3(2);\ C_{3h}(2);\ D_3(2);\ C_{3h}(2);\ 3C_3(4);\ C_2(6);\ C_s(6);\ C_1(12)$
189 $P\bar{6}2m$	D_{3h}^3	$2D_{3h}(1);\ 2C_{3h}(2);\ C_{3v}(2);\ 2C_{2v}(3);\ C_3(4);\ 3C_s(6);\ C_1(12)$
190 $P\bar{6}2c$	D_{3h}^4	$D_3(2);\ 3C_{3h}(2);\ 2C_3(4);\ C_2(6);\ C_s(6);\ C_1(12)$
191 $P6/mmm$	D_{6h}^1	$2D_{6h}(1);\ 2D_{3h}(2);\ C_{6v}(2);\ 2D_{2h}(3);\ C_{3v}(4);\ 5C_{2v}(6);\ 4C_s(12);\ C_1(12)$
192 $P6/mcc$	D_{6h}^2	$D_6(2);\ C_{6h}(2);\ D_3(4);\ C_{3h}(4);\ C_6(4);\ D_2(6);\ C_{2h}(6);\ C_3(8);\ 3C_2(12);\ C_s(12);\ C_1(24)$
193 $P6_3/mcm$	D_{6h}^3	$D_{3h}(2);\ D_{3d}(2);\ C_{3h}(4);\ D_3(4);\ C_6(4);\ C_{2h}(6);\ C_{2v}(6);\ C_3(8);\ C_2(12);\ 2C_2(12);\ C_1(24)$
194 $P6_3/mmc$	D_{6h}^4	$D_{3d}(2);\ 3D_{3h}(2);\ 2C_{3v}(4);\ C_{2h}(6);\ C_{2v}(6);\ C_2(12);\ 2C_s(12);\ C_1(24)$
195 $P23$	T^1	$2T(1);\ 2D_2(3);\ C_3(4);\ 4C_2(6);\ C_1(12)$
196 $F23$	T^2	$4T(1);\ C_3(4);\ 2C_2(6);\ C_1(12)$
197 $I23$	T^3	$T(1);\ D_2(2);\ C_3(4);\ 2C_2(6);\ C_1(12)$
198 $P2_13$	T^4	$C_3(4);\ C_1(12)$
199 $I2_13$	T^5	$C_3(4);\ C_2(6);\ C_1(12)$
200 $Pm3$	T_h^1	$2T_h(1);\ 2D_{2h}(3);\ 4C_{2v}(6);\ C_3(8);\ 2C_s(12);\ C_1(24)$
201 $Pn3$	T_h^2	$T(2);\ 2C_{3i}(4);\ D_2(6);\ C_3(8);\ 2C_2(12);\ C_1(24)$
202 $Fm3$	T_h^3	$2T_h(1);\ T(2);\ C_{2h}(6);\ C_{2v}(6);\ C_3(8);\ C_2(12);\ C_s(12);\ C_1(24)$
203 $Fd3$	T_h^4	$2T(2);\ 2C_{3i}(4);\ C_3(8);C_2(12);\ C_1(24)$
204 $Im3$	T_h^5	$T_h(1);\ D_{2h}(3);\ C_{3i}(4);\ 2C_{2v}(6);\ C_3(8);\ C_s(12);\ C_1(24)$
205 $Pa3$	T_h^6	$2C_{3i}(4);\ C_3(8);\ C_1(24)$
206 $Ia3$	T_h^7	$2C_{3i}(4);\ C_3(8);\ C_2(12);\ C_1(24)$
207 $P432$	O^1	$2O(1);\ 2D_4(3);\ 2C_4(6);\ C_3(8);\ 3C_2(12);\ C_1(24)$

Appendix 3 (*continued*)

Space group[1]		Site symmetries[2]
208 $P4_232$	O^2	$T(2)$; $2D_3(4)$; $3D_2(6)$; $C_3(8)$; $5C_2(12)$; $C_1(24)$
209 $F432$	O^3	$2O(1)$; $T(2)$; $D_2(6)$; $C_4(6)$; $C_3(8)$; $3C_2(12)$; $C_1(24)$
210 $F4_132$	O^4	$2T(2)$; $2D_3(4)$; $C_3(8)$; $2C_2(12)$; $C_1(24)$
211 $I432$	O^5	$O(1)$; $D_4(3)$; $D_3(4)$; $D_2(6)$; $C_4(6)$; $C_3(8)$; $3C_2(12)$; $C_1(24)$
212 $P4_332$	O^6	$2D_3(4)$; $C_3(8)$; $C_2(12)$; $C_1(24)$
213 $P4_132$	O^7	$2D_3(4)$; $C_3(8)$; $C_2(12)$; $C_1(24)$
214 $I4_132$	O^8	$2D_3(4)$; $2D_2(6)$; $C_3(8)$; $3C_2(12)$; $C_1(24)$
215 $P\bar{4}3m$	T_d^1	$2T_d(1)$; $2D_{2d}(3)$; $C_{3v}(4)$; $2C_{2v}(6)$; $C_2(12)$; $C_s(12)$; $C_1(24)$
216 $F\bar{4}3m$	T_d^2	$4T_d(1)$; $C_{3v}(4)$; $2C_{2v}(6)$; $C_s(12)$; $C_1(24)$
217 $I\bar{4}3m$	T_d^3	$T_d(1)$; $D_{2d}(3)$; $C_{3v}(4)$; $S_4(6)$; $C_{2v}(6)$; $C_2(12)$; $C_s(12)$; $C_1(24)$
218 $P\bar{4}3n$	T_d^4	$T(2)$; $D_2(6)$; $2S_4(6)$; $C_3(8)$; $3C_2(12)$; $C_1(24)$
219 $F\bar{4}3c$	T_d^5	$2T(2)$; $2S_4(6)$; $C_3(8)$; $2C_2(12)$; $C_1(24)$
220 $I\bar{4}3d$	T_d^6	$2S_4(6)$; $C_3(8)$; $C_2(12)$; $C_1(24)$
221 $Pm3m$	O_h^1	$2O_h(1)$; $2D_{4h}(3)$; $2C_{4v}(6)$; $C_{3v}(8)$; $3C_{2v}(12)$; $3C_s(24)$; $C_1(48)$
222 $Pn3n$	O_h^2	$O(2)$; $D_4(6)$; $C_{3i}(8)$; $S_4(12)$; $C_4(12)$; $C_3(16)$; $2C_2(24)$; $C_1(48)$
223 $Pm3n$	O_h^3	$T_h(2)$; $D_{2h}(6)$; $2D_{2d}(6)$; $D_3(8)$; $3C_{2v}(12)$; $C_3(16)$; $C_2(24)$; $C_s(24)$; $C_1(48)$
224 $Pn3m$	O_h^4	$T_d(2)$; $2D_{3d}(4)$; $D_{2d}(6)$; $C_{3v}(8)$; $D_2(12)$; $C_{2v}(12)$; $3C_2(24)$; $C_1(48)$
225 $Fm3m$	O_h^5	$2O_h(1)$; $T_d(2)$; $D_{2h}(6)$; $C_{4v}(6)$; $C_{3v}(8)$; $3C_{2v}(12)$; $2C_s(24)$; $C_1(48)$
226 $Fd3c$	O_h^6	$O(2)$; $T_h(2)$; $D_{2d}(6)$; $C_{4h}(6)$; $C_{2v}(12)$; $C_4(12)$; $C_3(16)$; $C_2(24)$; $C_s(24)$; $C_1(48)$
227 $Fd3m$	O_h^7	$2T_d(2)$; $2D_{3d}(4)$; $C_{3v}(8)$; $C_{2v}(12)$; $C_s(24)$; $C_2(24)$; $C_1(48)$
228 $Fd3c$	O_h^8	$T(4)$; $D_3(8)$; $C_{3i}(8)$; $S_4(12)$; $C_3(16)$; $2C_2(24)$; $C_1(48)$
229 $Im3m$	O_h^9	$O_h(1)$; $D_{4h}(3)$; $D_{3d}(4)$; $D_{2d}(6)$; $C_{4v}(6)$; $C_{3v}(8)$; $2C_{2v}(12)$; $C_2(24)$; $2C_s(24)$; $C_1(48)$
230 $Ia3d$	O_h^{10}	$C_{3i}(8)$; $D_3(8)$; $D_2(12)$; $S_4(12)$; $C_3(16)$; $2C_2(24)$; $C_1(48)$

Note the following equivalent nomenclatures: $C_i \equiv S_2$, $C_s \equiv C_{1h}$, $D_2 \equiv V$, $D_{2h} \equiv V_h$, $D_{2d} \equiv V_d$, and $C_{3i} \equiv S_6$.

REFERENCES

1. N. F. M. Henry and K. Lonsdale (eds.), *International Tables for X-Ray Crystallography*, Vol. 1, Kynoch Press, Birmingham, England (1965).
2. R. S. Halford, *J. Chem. Phys.*, **14**:8 (1946).
3. L. Couture, *J. Chem. Phys.*, **15**:153 (1947).
4. D. E. Irish and M. H. Brooker, *Appl. Spec.*, **27**:395 (1973).

Appendix 4

CORRELATION TABLES

The correlation tables which follow are helpful in determining the site group relating to the molecular point group. The tables give the correlations between species of a group and a subgroup. We wish to express our gratitude to St. Martin's Press for permission to reproduce these tables from the book by D. M. Adams,[1] and to W. G. Fateley *et al.* for the use of their comprehensive tables.[2]

C_4	C_2
A	A
B	A
E	$2B$

C_6	C_3	C_2	C_1
A	A	A	A
B	A	B	A
E_1	E	$2B$	$2A$
E_2	E	$2A$	$2A$

D_2	C_2^z	C_2^y	C_2^x
A	A	A	A
B_1	A	B	B
B_2	B	A	B
B_3	B	B	A

D_3	C_3	C_2
A_1	A	A
A_2	A	B
E	E	$A + B$

D_4	C_2' D_2	C_2'' D_2	C_4	C_2	C_2' C_2	C_2'' C_2
A_1	A	A	A	A	A	A
A_2	B_1	B_1	A	A	B	B
B_1	A	B_1	B	A	A	B
B_2	B_1	A	B	A	B	A
E	$B_2 + B_3$	$B_2 + B_3$	E	$2B$	$A + B$	$A + B$

D_5	C_5	C_2
A_1	A	A
A_2	A	B
E_1	E_1	$A + B$
E_2	E_2	$A + B$

D_6	C_6	C_2' D_3	C_2'' D_3	D_2	C_3	C_2	C_2' C_2	C_2'' C_2
A_1	A	A_1	A_1	A	A	A	A	A
A_2	A	A_2	A_2	B_1	A	A	B	B
B_1	B	A_1	A_2	B_2	A	B	A	B
B_2	B	A_2	A_1	B_3	A	B	B	A
E_1	E_1	E	E	$B_2 + B_3$	E	$2B$	$A + B$	$A + B$
E_2	E_2	E	E	$A + B_1$	E	$2A$	$A + B$	$A + B$

C_{2v}	C_2	$\sigma(zx)$ C_s	$\sigma(yz)$ C_s
A_1	A	A'	A'
A_2	A	A''	A''
B_1	B	A'	A''
B_2	B	A''	A'

C_{3v}	C_3	C_s
A_1	A	A'
A_2	A	A''
E	E	$A' + A''$

C_{4v}	C_4	σ_v C_{2v}	σ_d C_{2v}	C_2	σ_v C_s	σ_d C_s
A_1	A	A_1	A_1	A	A'	A'
A_2	A	A_2	A_2	A	A''	A''
B_1	B	A_1	A_2	A	A'	A''
B_2	B	A_2	A_1	A	A''	A'
E	E	$B_1 + B_2$	$B_1 + B_2$	$2B$	$A' + A''$	$A' + A''$

C_{5v}	C_5	C_s
A_1	A	A'
A_2	A	A''
E_1	E_1	$A' + A''$
E_2	E_2	$A' + A''$

C_{6v}	C_6	σ_v C_{3v}	σ_d C_{3v}	$\sigma_v \to \sigma(zx)$ C_{2v}	C_3	C_2	σ_v C_s	σ_d C_s
A_1	A	A_1	A_1	A_1	A	A	A'	A'
A_2	A	A_2	A_2	A_2	A	A	A''	A''
B_1	B	A_1	A_2	B_1	A	B	A'	A''
B_2	B	A_2	A_1	B_2	A	B	A''	A'
E_1	E_1	E	E	$B_1 + B_2$	E	$2B$	$A' + A''$	$A' + A''$
E_2	E_2	E	E	$A_1 + A_2$	E	$2A$	$A' + A''$	$A' + A''$

C_{2h}	C_2	C_s	C_i
A_g	A	A'	A_g
B_g	B	A''	A_g
A_u	A	A''	A_u
B_u	B	A'	A_u

C_{3h}	C_3	C_s	C_1
A'	A	A'	A
E'	E	$2A'$	$2A$
A''	A	A''	A
E''	E	$2A''$	$2A$

C_{4h}	C_4	S_4	C_{2h}	C_2	C_s	C_i	C_1
A_g	A	A	A_g	A	A'	A_g	A
B_g	B	B	A_g	A	A'	A_g	A
E_g	E	E	$2B_g$	$2B$	$2A''$	$2A_g$	$2A$
A_u	A	B	A_u	A	A''	A_u	A
B_u	B	A	A_u	A	A''	A_u	A
E_u	E	E	$2B_u$	$2B$	$2A'$	$2A_u$	$2A$

C_{5h}	C_5	C_s	C_1
A'	A	A'	A
E_1'	E_1	$2A'$	$2A$
E_2'	E_2	$2A'$	$2A$
A''	A	A''	A
E_1''	E_1	$2A''$	$2A$
E_2''	E_2	$2A''$	$2A$

C_{6h}	C_6	C_{3h}	S_6	C_{2h}	C_3	C_2	C_s	C_i	C_1
A_g	A	A'	A_g	A_g	A	A	A'	A_g	A
B_g	B	A''	A_g	B_g	A	B	A''	A_g	A
E_{1g}	E_1	E''	E_g	$2B_g$	E	$2B$	$2A''$	$2A_g$	$2A$
E_{2g}	E_2	E'	E_g	$2A_g$	E	$2A$	$2A'$	$2A_g$	$2A$
A_u	A	A''	A_u	A_u	A	A	A''	A_u	A
B_u	B	A'	A_u	B_u	A	B	A'	A_u	A
E_{1u}	E_1	E'	E_u	$2B_u$	E	$2B$	$2A'$	$2A_u$	$2A$
E_{2u}	E_2	E''	E_u	$2A_u$	E	$2A$	$2A''$	$2A_u$	$2A$

D_{2h}	D_2	$C_2(z)$ C_{2v}	$C_2(y)$ C_{2v}	$C_2(x)$ C_{2v}	$C_2(z)$ C_{2h}	$C_2(y)$ C_{2h}	$C_2(x)$ C_{2h}
A_g	A	A_1	A_1	A_1	A_g	A_g	A_g
B_{1g}	B_1	A_2	B_2	B_1	A_g	B_g	B_g
B_{2g}	B_2	B_1	A_2	B_2	B_g	A_g	B_g
B_{3g}	B_3	B_2	B_1	A_2	B_g	B_g	A_g
A_u	A	A_2	A_2	A_2	A_u	A_u	A_u
B_{1u}	B_1	A_1	B_1	B_2	A_u	B_u	B_u
B_{2u}	B_2	B_2	A_1	B_1	B_u	A_u	B_u
B_{3u}	B_3	B_1	B_2	A_1	B_u	B_u	A_u

D_{2h} (*cont.*)	$C_2(z)$ C_2	$C_2(y)$ C_2	$C_2(x)$ C_2	$\sigma(xy)$ C_s	$\sigma(zx)$ C_s	$\sigma(yz)$ C_s	C_i
A_g	A	A	A	A'	A'	A'	A_g
B_{1g}	A	B	B	A'	A''	A''	A_g
B_{2g}	B	A	B	A''	A'	A''	A_g
B_{3g}	B	B	A	A''	A''	A'	A_g
A_u	A	A	A	A''	A''	A''	A_u
B_{1u}	A	B	B	A''	A'	A'	A_u
B_{2u}	B	A	B	A'	A''	A'	A_u
B_{3u}	B	B	A	A'	A'	A''	A_u

D_{3h}	C_{3h}	D_3	C_{3v}	$\sigma_h \rightarrow \sigma_v(zy)$ C_{2v}	C_3	C_2	σ_h C_s	σ_v C_s
A_1'	A'	A_1	A_1	A_1	A	A	A'	A'
A_2'	A'	A_2	A_2	B_2	A	B	A'	A''
E'	E'	E	E	$A_1 + B_2$	E	$A + B$	$2A'$	$A' + A''$
A_1''	A''	A_1	A_2	A_2	A	A	A''	A''
A_2''	A''	A_2	A_1	B_1	A	B	A''	A'
E''	E''	E	E	$A_2 + B_1$	E	$A + B$	$2A''$	$A' + A''$

D_{4h}	D_4	$C_2' \rightarrow C_2'$ D_{2d}	$C_2'' \rightarrow C_2'$ D_{2d}	C_{4v}	C_{4h}	C_2' D_{2h}	C_2'' D_{2h}	C_4	S_4
A_{1g}	A_1	A_1	A_1	A_1	A_g	A_g	A_g	A	A
A_{2g}	A_2	A_2	A_2	A_2	A_g	B_{1g}	B_{1g}	A	A
B_{1g}	B_1	B_1	B_2	B_1	B_g	A_g	B_{1g}	B	B
B_{2g}	B_2	B_2	B_1	B_2	B_g	B_{1g}	A_g	B	B
E_g	E	E	E	E	E_g	$B_{2g} + B_{3g}$	$B_{2g} + B_{3g}$	E	E
A_{1u}	A_1	B_1	B_1	A_2	A_u	A_u	A_u	A	B
A_{2u}	A_2	B_2	B_2	A_1	A_u	B_{1u}	B_{1u}	A	B
B_{1u}	B_1	A_1	A_2	B_2	B_u	A_u	B_{1u}	B	A
B_{2u}	B_2	A_2	A_1	B_1	B_u	B_{1u}	A_u	B	A
E_u	E	E	E	E	E_u	$B_{2u} + B_{3u}$	$B_{2u} + B_{3u}$	E	E

D_{4h} (*cont.*)	C_2' D_2	C_2'' D_2	C_2, σ_v C_{2v}	C_2, σ_d C_{2v}	C_2' C_{2v}	C_2'' C_{2v}
A_{1g}	A	A	A_1	A_1	A_1	A_1
A_{2g}	B_1	B_1	A_2	A_2	B_1	B_1
B_{1g}	A	B_1	A_1	A_2	A_1	B_1
B_{2g}	B_1	A	A_2	A_1	B_1	A_1
E_g	$B_2 + B_3$	$B_2 + B_3$	$B_1 + B_2$	$B_1 + B_2$	$A_2 + B_2$	$A_2 + B_2$
A_{1u}	A	A	A_2	A_2	A_2	A_2
A_{2u}	B_1	B_1	A_1	A_1	B_2	B_2
B_{1u}	A	B_1	A_2	A_1	A_2	B_2
B_{2u}	B_1	A	A_1	A_2	B_2	A_2
E_u	$B_2 + B_3$	$B_2 + B_3$	$B_1 + B_2$	$B_1 + B_2$	$A_1 + B_1$	$A_1 + B_1$

D_{4h} (cont.)	C_2 C_{2h}	C_2' C_{2h}	C_2'' C_{2h}	C_2 C_2	C_2' C_2	C_2'' C_2	σ_h C_s	σ_v C_s	σ_d C_s	 C_i
A_{1g}	A_g	A_g	A_g	A	A	A	A'	A'	A'	A_g
A_{2g}	A_g	B_g	B_g	A	B	B	A'	A''	A''	A_g
B_{1g}	A_g	A_g	B_g	A	A	B	A'	A'	A''	A_g
B_{2g}	A_g	B_g	A_g	A	B	A	A'	A''	A'	A_g
E_g	$2B_g$	$A_g + B_g$	$A_g + B_g$	$2B$	$A+B$	$A+B$	$2A''$	$A' + A''$	$A' + A''$	$2A_g$
A_{1u}	A_u	A_u	A_u	A	A	A	A''	A''	A''	A_u
A_{2u}	A_u	B_u	B_u	A	B	B	A''	A'	A'	A_u
B_{1u}	A_u	A_u	B_u	A	A	B	A''	A''	A'	A_u
B_{2u}	A_u	B_u	A_u	A	B	A	A''	A'	A''	A_u
E_u	$2B_u$	$A_u + B_u$	$A_u + B_u$	$2B$	$A+B$	$A+B$	$2A'$	$A' + A''$	$A' + A''$	$2A_u$

D_{5h}	D_5	C_{5v}	C_{5h}	C_5	$\sigma_h \rightarrow \sigma(zx)$ C_{2v}	C_2	σ_h C_5	σ_v C_s
A_1'	A_1	A_1	A'	A	A_1	A	A'	A'
A_2'	A_2	A_2	A'	A	B_1	B	A'	A''
E_1'	E_1	E_1	E_1'	E_1	$A_1 + B_1$	$A + B$	$2A'$	$A' + A''$
E_2'	E_2	E_2	E_2'	E_2	$A_1 + B_1$	$A + B$	$2A'$	$A' + A''$
A_1''	A_1	A_2	A''	A	A_2	A	A''	A''
A_2''	A_2	A_1	A''	A	B_2	B	A''	A'
E_1''	E_1	E_1	E_1''	E_1	$A_2 + B_2$	$A + B$	$2A''$	$A' + A''$
E_2''	E_2	E_2	E_2''	E_2	$A_2 + B_2$	$A + B$	$2A''$	$A' + A''$

D_{6h}	D_6	C_2' D_{3h}	C_2'' D_{3h}	C_{6v}	C_{6h}	C_2'' D_{3d}	C_2' D_{3d}	$\sigma_h \rightarrow \sigma(xy)$ $\sigma_v \rightarrow \sigma(yz)$ D_{2h}
A_{1g}	A_1	A_1'	A_1'	A_1	A_g	A_{1g}	A_{1g}	A_g
A_{2g}	A_2	A_2'	A_2'	A_2	A_g	A_{2g}	A_{2g}	B_{1g}
B_{1g}	B_1	A_1''	A_2''	B_2	B_g	A_{2g}	A_{1g}	B_{2g}
B_{2g}	B_2	A_2''	A_1''	B_1	B_g	A_{1g}	A_{2g}	B_{3g}
E_{1g}	E_1	E''	E''	E_1	E_{1g}	E_g	E_g	$B_{2g} + B_{3g}$
E_{2g}	E_2	E'	E'	E_2	E_{2g}	E_g	E_g	$A_g + B_{1g}$
A_{1u}	A_1	A_1''	A_1''	A_2	A_u	A_{1u}	A_{1u}	A_u
A_{2u}	A_2	A_2''	A_2''	A_1	A_u	A_{2u}	A_{2u}	B_{1u}
B_{1u}	B_1	A_1'	A_2'	B_1	B_u	A_{2u}	A_{1u}	B_{2u}
B_{2u}	B_2	A_2'	A_1'	B_2	B_u	A_{1u}	A_{2u}	B_{3u}
E_{1u}	E_1	E'	E'	E_1	E_{1u}	E_u	E_u	$B_{2u} + B_{3u}$
E_{2u}	E_2	E''	E''	E_2	E_{2u}	E_u	E_u	$A_u + B_{1u}$

D_{6h} (cont.)	C_6	C_{3h}	C_2' D_3	C_2'' D_3	σ_v C_{3v}	σ_d C_{3v}	S_6	D_2
A_{1g}	A	A'	A_1	A_1	A_1	A_1	A_g	A
A_{2g}	A	A'	A_2	A_2	A_2	A_2	A_g	B_1
B_{1g}	B	A''	A_1	A_2	A_2	A_1	A_g	B_2
B_{2g}	B	A''	A_2	A_1	A_1	A_2	A_g	B_3
E_{1g}	E_1	E''	E	E	E	E	E_g	$B_2 + B_3$
E_{2g}	E_2	E'	E	E	E	E	E_g	$A + B_1$
A_{1u}	A	A''	A_1	A_1	A_2	A_2	A_u	A
A_{2u}	A	A''	A_2	A_2	A_1	A_1	A_u	B_1
B_{1u}	B	A'	A_1	A_2	A_1	A_2	A_u	B_2
B_{2u}	B	A'	A_2	A_1	A_2	A_1	A_u	B_3
E_{1u}	E_1	E'	E	E	E	E	E_u	$B_2 + B_3$
E_{2u}	E_2	E''	E	E	E	E	E_u	$A + B_1$

D_{6h} (cont.)	C_2 C_{2v}	C_2' C_{2v}	C_2'' C_{2v}	C_2 C_{2h}	C_2' C_{2h}	C_2'' C_{2h}	C_3	C_2 C_2
A_{1g}	A_1	A_1	A_1	A_g	A_g	A_g	A	A
A_{2g}	A_2	B_1	B_1	A_g	B_g	B_g	A	A
B_{1g}	B_1	A_2	B_2	B_g	A_g	B_g	A	B
B_{2g}	B_2	B_2	A_2	B_g	B_g	A_g	A	B
E_{1g}	$B_1 + B_2$	$A_2 + B_2$	$A_2 + B_2$	$2B_g$	$A_g + B_g$	$A_g + B_g$	E	$2B$
E_{2g}	$A_1 + A_2$	$A_1 + B_1$	$A_1 + B_1$	$2A_g$	$A_g + B_g$	$A_g + B_g$	E	$2A$
A_{1u}	A_2	A_2	A_2	A_u	A_u	A_u	A	A
A_{2u}	A_1	B_1	B_2	A_u	B_u	B_u	A	A
B_{1u}	B_2	A_1	B_1	B_u	A_u	B_u	A	B
B_{2u}	B_1	B_2	A_1	B_u	B_u	A_u	A	B
E_{1u}	$B_2 + B_1$	$A_1 + B_2$	$A_1 + B_1$	$2B_u$	$A_u + B_u$	$A_u + B_u$	E	$2B$
E_{2u}	$A_2 + A_1$	$A_2 + B_1$	$A_2 + B_2$	$2A_u$	$A_u + B_u$	$A_u + B_u$	E	$2A$

D_{6h}	C'_2	C''_2	σ_h	σ_d	σ_v	
(*cont.*)	C_2	C_2	C_s	C_s	C_s	C_i
A_{1g}	A	A	A'	A'	A'	A_g
A_{2g}	B	B	A'	A''	A''	A_g
B_{1g}	A	B	A''	A'	A''	A_g
B_{2g}	B	A	A''	A''	A'	A_g
E_{1g}	$A+B$	$A+B$	$2A''$	$A'+A''$	$A'+A''$	$2A_g$
E_{2g}	$A+B$	$A+B$	$2A'$	$A'+A''$	$A'+A''$	$2A_g$
A_{1u}	A	A	A''	A''	A''	A_u
A_{2u}	B	B	A''	A'	A'	A_u
B_{1u}	A	B	A'	A''	A'	A_u
B_{2u}	B	A	A'	A'	A''	A_u
E_{1u}	$A+B$	$A+B$	$2A'$	$A'+A''$	$A'+A''$	$2A_u$
E_{2u}	$A+B$	$A+B$	$2A''$	$A'+A''$	$A'+A''$	$2A_u$

D_{7h}	A'_1	A'_2	E'_1	E'_2	E'_3	A''_1	A''_2	E''_1	E''_2	E''_3
C_{7v}	A_1	A_2	E_1	E_2	E_3	A_2	A_1	E_1	E_2	E_3

		$C_2 \rightarrow C_2(z)$		C_2	C'_2	
D_{2d}	S_4	D_2	C_{2v}	C_2	C_2	C_s
A_1	A	A	A_1	A	A	A'
A_2	A	B_1	A_2	A	B	A''
B_1	B	A	A_2	A	A	A''
B_2	B	B_1	A_1	A	B	A'
E	E	B_2+B_3	B_1+B_2	$2B$	$A+B$	$A'+A''$

D_{3d}	D_3	C_{3v}	S_6	C_3	C_{2h}	C_2	C_s	C_i
A_{1g}	A_1	A_1	A_g	A	A_g	A	A'	A_g
A_{2g}	A_2	A_2	A_g	A	B_g	B	A''	A_g
E_g	E	E	E_g	E	A_g+B_g	$A+B$	$A'+A''$	$2A_g$
A_{1u}	A_1	A_2	A_u	A	A_u	A	A''	A_u
A_{2u}	A_2	A_1	A_u	A	B_u	B	A'	A_u
E_u	E	E	E_u	E	A_u+B_u	$A+B$	$A'+A''$	$2A_u$

D_{4d}	D_4	C_{4v}	S_8	C_4	C_{2v}	C_2 C_2	C_2' C_2	C_s
A_1	A_1	A_1	A	A	A_1	A	A	A'
A_2	A_2	A_2	A	A	A_2	A	B	A''
B_1	A_1	A_2	B	A	A_2	A	A	A''
B_2	A_2	A_1	B	A	A_1	A	B	A'
E_1	E	E	E_1	E	$B_1 + B_2$	$2B$	$A + B$	$A' + A''$
E_2	$B_1 + B_2$	$B_1 + B_2$	E_2	$2B$	$A_1 + A_2$	$2A$	$A + B$	$A' + A''$
E_3	E	E	E_3	E	$B_1 + B_2$	$2B$	$A + B$	$A' + A''$

D_{5d}	D_5	C_{5v}	C_5	C_2	C_s	C_i
A_{1g}	A_1	A_1	A	A	A'	A_g
A_{2g}	A_2	A_2	A	B	A''	A_g
E_{1g}	E_1	E_1	E_1	$A + B$	$A' + A''$	$2A_g$
E_{2g}	E_2	E_2	E_2	$A + B$	$A' + A''$	$2A_g$
A_{1u}	A_1	A_2	A	A	A''	A_u
A_{2u}	A_2	A_1	A	B	A'	A_u
E_{1u}	E_1	E_1	E_1	$A + B$	$A' + A''$	$2A_u$
E_{2u}	E_2	E_2	E_2	$A + B$	$A' + A''$	$2A_u$

D_{6d}	D_6	C_{6v}	C_6	D_{2d}	D_3	C_{3v}
A_1	A_1	A_1	A	A_1	A_1	A_1
A_2	A_2	A_2	A	A_2	A_2	A_2
B_1	A_1	A_2	A	B_1	A_1	A_2
B_2	A_2	A_1	A	B_2	A_2	A_1
E_1	E_1	E_1	E_1	E	E	E
E_2	E_2	E_2	E_2	$B_1 + B_2$	E	E
E_3	$B_1 + B_2$	$B_1 + B_2$	$2B$	E	$A_1 + A_2$	$A_1 + A_2$
E_4	E_2	E_2	E_2	$A_1 + A_2$	E	E
E_5	E_1	E_1	E_1	E	E	E

D_{6d} (cont.)	D_2	C_{2v}	S_4	C_3	C_2 C_2	C_2' C_2	C_s
A_1	A	A_1	A	A	A	A	A'
A_2	B_1	A_2	A	A		B	A''
B_1	A	A_2	B	A	A	A	A''
B_2	B_1	A_1	B	A	A	B	A'
E_1	$B_2 + B_3$	$B_1 + B_2$	E	E	$2B$	$A + B$	$A' + A''$
E_2	$A + B_1$	$A_1 + A_2$	$2B$	E	$2A$	$A + B$	$A' + A''$
E_3	$B_2 + B_3$	$B_1 + B_2$	E	$2A$	$2B$	$A + B$	$A' + A''$
E_4	$A + B_1$	$A_1 + A_2$	$2A$	E	$2A$	$A + B$	$A' + A''$
E_5	$B_2 + B_1$	$B_1 + B_2$	E	E	$2B$	$A + B$	$A' + A''$

S_4	C_2	C_1
A	A	A
B	A	A
E	$2B$	$2A$

S_6	C_3	C_i	C_1
A_g	A	A_g	A
E_g	E	$2A_g$	$2A$
A_u	A	A_u	A
E_u	E	$2A_u$	$2A$

S_8	C_4	C_2	C_1
A	A	A	A
B	A	A	A
E_1	E	$2B$	$2A$
E_2	$2B$	$2A$	$2A$
E_3	E	$2B$	$2A$

T	D_2	C_3	C_2	C_1
A	A	A	A	A
E	$2A$	E	$2A$	$2A$
F	$B_1 + B_2 + B_3$	$A + E$	$A + 2B$	$3A$

T_h	T	D_{2h}	S_6	D_2
A_g	A	A_g	A_g	A
E_g	E	$2A_g$	E_g	$2A$
F_g	F	$B_{1g} + B_{2g} + B_{3g}$	$A_g + E_g$	$B_1 + B_2 + B_3$
A_u	A	A_u	A_u	A
E_u	E	$2A_u$	E_u	$2A$
F_u	F	$B_{1u} + B_{2u} + B_{3u}$	$A_u + E_u$	$B_1 + B_2 + B_3$

T_h (cont.)	C_{2v}	C_{2h}	C_3	C_2	C_s	C_i	C_1
A_g	A_1	A_g	A	A	A'	A_g	A
E_g	$2A_1$	$2A_2$	E	$2A$	$2A'$	$2A_g$	$2A$
F_g	$A_2 + B_1 + B_2$	$A_g + 2B_g$	$A + E$	$A + 2B$	$A' + 2A''$	$3A_g$	$3A$
A_u	A_2	A_u	A	A	A''	A_u	A
E_u	$2A_2$	$2A_u$	E	$2A$	$2A''$	$2A_u$	$2A$
F_u	$A_1 + B_1 + B_2$	$A_u + 2B_u$	$A + E$	$A + 2B$	$2A' + A''$	$3A_u$	$3A$

T_d	T	D_{2d}	C_{3v}	S_4	D_2	C_{2v}
A_1	A	A_1	A_1	A	A	A_1
A_2	A	B_1	A_2	B	A	A_2
E	E	$A_1 + B_1$	E	$A + B$	$2A$	$A_1 + A_2$
F_1	F	$A_2 + E$	$A_2 + E$	$A + E$	$B_1 + B_2 + B_3$	$A_2 + B_1 + B_2$
F_2	F	$B_2 + E$	$A_1 + E$	$B + E$	$B_1 + B_2 + B_3$	$B_1 + B_2 + B_3$

T_d (cont.)	C_3	C_2	C_s
A_1	A	A	A'
A_2	A	A	A''
E	E	$2A$	$A' + A''$
F_1	$A + E$	$A + 2B$	$A' + 2A''$
F_2	$A + E$	$A + 2B$	$2A' + A''$

O	T	D_4	D_3	C_4	$3C_2$ D_2	$C_2, 2C_2'$ D_2
A_1	A	A_1	A_1	A	A	A
A_2	A	B_1	A_2	B	A	B_1
E	E	$A_1 + B_1$	E	$A + B$	$2A$	$A + B_1$
F_1	F	$A_2 + E$	$A_2 + E$	$A + E$	$B_1 + B_2 + B_3$	$B_1 + B_2 + B_3$
F_2	F	$B_2 + E$	$A_1 + E$	$B + E$	$B_1 + B_2 + B_3$	$A + B_2 + B_3$

O (cont.)	C_3	C_2	C_2
A_1	A	A	A
A_2	A	A	B
E	E	$2A$	$A + B$
F_1	$A + E$	$A + 2B$	$A + 2B$
F_2	$A + E$	$A + 2B$	$2A + B$

O_h *	O	T_d	T_h	T	D_{4h}	D_{3d}	D_{4d}	C_{3v}	D_3	$D_{3i} \equiv S_6$
A_{1g}	A_1	A_1	A_g	A	A_{1g}	A_{1g}	A_{1g}	A_1	A_1	A_g
A_{2g}	A_2	A_2	A_g	A	B_{1g}	A_{2g}	B_{1g}	A_2	A_2	A_g
E_g	E	E	E_g	E	$A_{1g} + B_{1g}$	E_g	$A_{1g} + B_{1g}$	E	E	E_g
F_{1g}	F_1	F_1	F_g	F	$A_{2g} + E_g$	$A_{2g} + E_g$	$A_{2g} + E_g$	$A_2 + E$	$A_2 + E$	$A_g + E_g$
F_{2g}	F_2	F_2	F_g	F	$B_{2g} + E_g$	$A_{1g} + E_g$	$B_{2g} + E_g$	$A_1 + E$	$A_1 + E$	$A_g + E_g$
A_{1u}	A_1	A_2	A_u	A	A_{1u}	A_{1u}	A_{1u}	A_2	A_1	A_u
A_{2u}	A_2	A_1	A_u	A	B_{1u}	A_{2u}	B_{1u}	A_1	A_2	A_u
E_u	E	E	E_u	E	$A_{1u} + B_{1u}$	E_u	$A_{1u} + B_{1u}$	E	E	E_u
F_{1u}	F_1	F_2	F_u	F	$A_{2u} + E_u$	$A_{2u} + E_u$	$A_{2u} + E_u$	$A_1 + E$	$A_2 + E$	$A_u + E_u$
F_{2u}	F_2	F_1	F_u	F	$B_{2u} + E_u$	$A_{1u} + E_u$	$B_{2u} + E_u$	$A_2 + E$	$A_1 + E$	$A_u + E_u$

O_h (cont.)	C_3	C_2, σ_d D_{2d}	C'_2, σ_h D_{2d}	C_{4v}	D_4	C_{4h}	S_4	C_4
A_{1g}	A	A_1	A_1	A_1	A_1	A_g	A	A
A_{2g}	A	B_1	B_2	B_1	B_1	B_g	B	B
E_g	E	$A_1 + B_1$	$A_1 + B_2$	$A_1 + B_1$	$A_1 + B_1$	$A_g + B_g$	$A + B$	$A + B$
F_{1g}	$A + E$	$A_2 + E$	$A_2 + E$	$A_2 + E$	$A_2 + E$	$A_g + E_g$	$A + E$	$A + E$
F_{2g}	$A + E$	$B_2 + E$	$B_1 + E$	$B_2 + E$	$B_2 + E$	$B_g + E_g$	$B + E$	$B + E$
A_{1u}	A	B_1	B_1	A_2	A_1	A_u	B	A
A_{2u}	A	A_1	A_2	B_2	B_1	B_u	A	B
E_u	E	$A_1 + B_1$	$A_2 + B_1$	$A_2 + B_2$	$A_1 + B_1$	$A_u + B_u$	$A + B$	$A + B$
F_{1u}	$A + E$	$B_2 + E$	$B_2 + E$	$A_1 + E$	$A_2 + E$	$A_u + E_u$	$B + E$	$B + E$
F_{2u}	$A + E$	$A_2 + E$	$A_1 + E$	$B_1 + E$	$B_2 + E$	$B_u + E_u$	$A + E$	$B + E$

* To find correlations with smaller subgroups, carry out the correlation in two steps; for example, if the correlation of O_h with C_{2v} is desired, use the table to pass from O_h to T_d and then employ the table for T_d to go on to C_{2v}.

O_h (cont.)	$3C_2$ D_{2h}	C_2, $2C_2'$ D_{2h}	C_2, σ_h C_{2v}	C_2, σ_d C_{2v}
A_{1g}	A_g	A_g	A_1	A_1
A_{2g}	A_g	B_{1g}	A_1	A_2
E_g	$2A_g$	$A_g + B_{1g}$	$2A_1$	$A_1 + A_2$
F_{1g}	$B_{1g} + B_{2g} + B_{3g}$	$B_{1g} + B_{2g} + B_{3g}$	$A_2 + B_1 + B_2$	$A_2 + B_1 + B_2$
F_{2g}	$B_{1g} + B_{2g} + B_{3g}$	$A_{1g} + B_{2g} + B_{3g}$	$A_2 + B_1 + B_2$	$A_1 + B_1 + B_2$
A_{1u}	A_u	A_u	A_2	A_2
A_{2u}	A_u	B_{2u}	A_2	A_1
E_u	$2A_u$	$A_u + B_{1u}$	$2A_2$	$A_1 + A_2$
F_{1u}	$B_{1u} + B_{2u} + B_{3u}$	$B_{1u} + B_{2u} + B_{3u}$	$A_1 + B_1 + B_2$	$A_1 + B_1 + B_2$
F_{2u}	$B_{1u} + B_{2u} + B_{3u}$	$A_u + B_{2u} + B_{3u}$	$A_1 + B_1 + B_2$	$A_2 + B_1 + B_2$

O_h (cont.)	C_2', σ_h C_{2v}	$3C_2$ D_2	C_2, $2C_2'$ D_2	C_2, σ_h C_{2h}	C_2', σ_h C_{2h}
A_{1g}	A_1	A	A	A_g	A_g
A_{2g}	B_1	A	B_1	A_g	B_g
E_g	$A_1 + B_1$	$2A$	$A + B_1$	$2A_g$	$A_g + B_g$
F_{1g}	$A_2 + B_1 + B_2$	$B_1 + B_2 + B_3$	$B_1 + B_2 + B_3$	$A_g + 2B_g$	$A_g + 2B_g$
F_{2g}	$A_1 + A_2 + B_2$	$B_1 + B_2 + B_3$	$A + B_2 + B_3$	$A_g + 2B_g$	$2A_g + B_g$
A_{1u}	A_2	A	A	A_u	A_u
A_{2u}	B_2	A	B_1	A_u	B_u
E_u	$A_2 + B_2$	$2A$	$A + B_1$	$2A_u$	$A_u + B_u$
F_{1u}	$A_1 + B_1 + B_2$	$B_1 + B_2 + B_3$	$B_1 + B_2 + B_3$	$A_u + 2B_u$	$A_u + 2B_u$
F_{2u}	$A_1 + A_2 + B_1$	$B_1 + B_2 + B_3$	$A + B_2 + B_3$	$A_u + 2B_u$	$2A_u + B_u$

O_h (cont.)	σ_h C_s	σ_d C_s	C_2 C_2	C_2' C_2	C_i	C_1
A_{1g}	A'	A'	A	A	A_g	A
A_{2g}	A'	A''	A	B	A_g	A
E_g	$2A'$	$A' + A''$	$2A$	$A + B$	$2A_g$	$2A$
F_{1g}	$A' + A''$	$A' + 2A''$	$A + 2B$	$A + 2B$	$3A_g$	$3A$
F_{2g}	$A' + 2A''$	$2A' + A''$	$A + 2B$	$2A + B$	$3A_g$	$3A$
A_{1u}	A''	A''	A	A	A_u	A
A_{2u}	A''	A'	A	B	A_u	A
E_u	$2A''$	$A' + A''$	$2A$	$A + B$	$2A_u$	$2A$
F_{1u}	$2A' + A''$	$2A' + A''$	$A + 2B$	$A + 2B$	$3A_u$	$3A$
F_{2u}	$2A' + A''$	$A' + 2A''$	$A + 2B$	$2A + B$	$3A_u$	$3A$

I_h	I	C_5	C_3	C_2	C_1
A_g	A	A	A	A	A
A_u	A	A	A	A	A
F_{1g}	F_1	$A + E_1$	$A + E$	$A + 2B$	$3A$
F_{1u}	F_1	$A + E_1$	$A + E$	$A + 2B$	$3A$
F_{2g}	F_2	$A + E_2$	$A + E$	$A + 2B$	$3A$
F_{2u}	F_2	$A + E_2$	$A + E$	$A + 2B$	$3A$
G_{1g}	G_1	$E_1 + E_2$	$2A + E$	$2A + 2B$	$4A$
G_{1u}	G_1	$E_1 + E_2$	$2A + E$	$2A + 2B$	$4A$
H_g	H	$A + E_1 + E_2$	$A + 2E$	$3A + 2B$	$5A$
H_u	H	$A + E_1 + E_2$	$A + 2E$	$3A + 2B$	$5A$

REFERENCES

1. D. M. Adams, *Metal–Ligand and Related Vibrations,* St. Martin's Press, New York (1968).
2. W. G. Fateley, F. R. Dollish, N. T. McDevitt, and F. F. Bentley, *Infrared and Raman Selection Rules for Molecular and Lattice Vibrations: The Correlation Method,* Wiley–Interscience, New York (1972).

Appendix 5

ELEMENTARY MATHEMATICS

This appendix will attempt to present the fundamental definitions and theorems necessary for an understanding of group theory. The presentation will not be detailed, for the approach to group theory followed in this book will be empirical rather than mathematical. For a more detailed discussion of the subject see the texts by Margenau and Murphy[1] and others.[2–4]

DEFINITION OF A GROUP

The group has been defined in Section 1-4B. In summary, a set of elements $A, B, C, \ldots$ is said to be a group if for every pair of elements (e.g., A and B) a binary operation exists that yields the product AB which belongs to the set; if the associative law holds for the combination of elements; if the set contains the identity element; and if there is an inverse for each element.

FINITE AND INFINITE GROUPS

Groups containing a limited number of elements are called finite, while groups containing an unlimited number of elements are called infinite. The number of elements, g, in a finite group determines the order of the group. All of the groups that we will encounter will be finite groups of order g, with the exception of those for linear molecules, of which there are two ($C_{\infty v}$, $D_{\infty h}$).

SUBGROUPS

Inspection of a group will show that within the group there are smaller groups with the same operation. In the group C_{3v}, which is of order 6, the following smaller groups will be found; E by itself; σ_v, of order 2; and C_3, of order 3. If the order of the group is g, then the order of the subgroup, h, must be an integral divisor of g.

CLASSES

If A and B are elements of a group, then $B^{-1}AB$ will be equal to some element Y of the group. Thus

$$Y = B^{-1}AB \tag{A5-1}$$

Y is called the transform of A by B, or we say that A is conjugate to Y. The following are properties of conjugate elements:

1) every element is a conjugate to itself;

2) if A is conjugate to Y, then Y is conjugate to A; e.g., $A = B^{-1}YB$;

3) if A is conjugate to Y and Y is conjugate to X, then A is conjugate to X and A, Y, and X belong to the same class.

A complete set of elements conjugate to each other is called a class of the group.

The method of arranging the elements of a group into classes exhibits the structure under the relation of conjugation. The result is that the symmetry of the molecule can be presented as a set of disjoint sets of geometric (symmetry) elements. For C_{3v} the complete set of elements conjugate to each other is E, C_3, C_3^2; σ_{v1}, σ_{v2}, σ_{v3}. For C_{4v}: E, C_4, C_4^3; $C_2 \equiv C_4^2$; σ_{v1}, σ_{v2}; σ_{d1}, σ_{d2}.

DEFINITION OF A MATRIX

A collection of real or complex quantities displayed in a table of rows and columns is called an array. The most familiar type of array is the determinant, which always has the same number of rows and columns, and is always a number. It can be written as

$$A = |A| = \begin{vmatrix} A_{11} & A_{12} & A_{13} & \cdots & A_{1n} \\ A_{21} & A_{22} & A_{23} & \cdots & A_{2n} \\ A_{31} & A_{32} & A_{33} & \cdots & A_{3n} \\ \cdots & \cdots & \cdots & \cdots & \cdots \\ A_{n1} & A_{n2} & A_{n3} & \cdots & A_{nn} \end{vmatrix} \tag{A5-2}$$

A matrix, on the other hand, is an array in which the number of rows and columns can differ. It is an element from a set of matrices with a specific (row-by-column) multiplication (unlike determinants, which have different multiplications in the sense that the determinant is a number and this number is invariant under the interchange of rows and columns). The matrix product is not a number. However, a matrix product can have a set of determinants of various orders. We may represent a matrix as

$$A = [A_{i,j}] = \begin{bmatrix} A_{11} & A_{12} & A_{13} & \ldots & A_{1m} \\ A_{21} & A_{22} & A_{23} & \ldots & A_{2m} \\ A_{31} & A_{32} & A_{33} & \ldots & A_{3m} \\ \ldots & \ldots & \ldots & \ldots & \ldots \\ A_{n1} & A_{n2} & A_{n3} & \ldots & A_{nm} \end{bmatrix} \tag{A5-3}$$

Here n and m determine the order of the matrix, n giving the number of the rows and m the number of columns. When $n = m$, the matrix is called square.

MULTIPLICATION OF MATRICES

A matrix A having three rows and three columns is to be multiplied by a matrix B having three rows and two columns. The row elements of matrix A are multiplied by the corresponding column elements of B. The following example will illustrate this operation:

$$A = \begin{bmatrix} A_{11} & A_{12} & A_{13} \\ A_{21} & A_{22} & A_{23} \\ A_{31} & A_{32} & A_{33} \end{bmatrix} \qquad B = \begin{bmatrix} B_{11} & B_{12} \\ B_{21} & B_{22} \\ B_{31} & B_{32} \end{bmatrix} \tag{A5-4}$$

(3×3 matrix) (3×2 matrix)

$$\begin{bmatrix} A_{11} & A_{12} & A_{13} \\ A_{21} & A_{22} & A_{23} \\ A_{31} & A_{32} & A_{33} \end{bmatrix} \begin{bmatrix} B_{11} & B_{12} \\ B_{21} & B_{22} \\ B_{31} & B_{32} \end{bmatrix}$$

$$= \begin{bmatrix} A_{11}B_{11} + A_{12}B_{21} + A_{13}B_{31} & A_{11}B_{12} + A_{12}B_{22} + A_{13}B_{32} \\ A_{21}B_{11} + A_{22}B_{21} + A_{23}B_{31} & A_{21}B_{12} + A_{22}B_{22} + A_{23}B_{32} \\ A_{31}B_{11} + A_{32}B_{21} + A_{33}B_{31} & A_{31}B_{12} + A_{32}B_{22} + A_{33}B_{32} \end{bmatrix} = C \tag{A5-5}$$

Thus, $AB = C$. Here the product is a matrix of three rows and two columns.

TRANSPOSE OF A MATRIX

Consider the matrix

$$A = \begin{bmatrix} A_{11} & A_{12} & A_{13} \\ A_{21} & A_{22} & A_{23} \\ A_{31} & A_{32} & A_{33} \end{bmatrix} \tag{A5-6}$$

Its transpose is

$$A' = \begin{bmatrix} A_{11} & A_{21} & A_{31} \\ A_{12} & A_{22} & A_{32} \\ A_{13} & A_{23} & A_{33} \end{bmatrix} \quad \text{(A5-7)}$$

REPRESENTATION OF GROUPS

The elements of a group, such as the symmetry operations of a molecule, can be represented by matrices. For true representations, the multiplication of the numbers representing A and B of the group must, if $AB = C$, lead to the number which represents the element C. A set of numbers or matrices which can be assigned to the elements of a group and which can properly represent the multiplications of the elements of this group is said to constitute a representation of the group. This can be illustrated by considering the molecule NH_3 in the C_{3v} point group (Fig. A5-1). The following treatment is taken from an article by Ziomek.[2]

The following internal coordinates can be written:

ΔD_1 change in bond distance XY_1

ΔD_2 change in bond distance XY_2

ΔD_3 change in bond distance XY_3

$\Delta\alpha_{12}$ change in angle Y_1XY_2

$\Delta\alpha_{13}$ change in angle Y_1XY_3

$\Delta\alpha_{23}$ change in angle Y_2XY_3

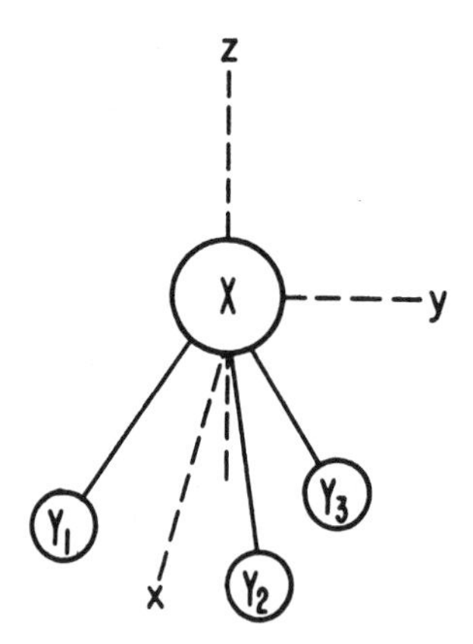

Fig. A5-1. The NH_3 (XY_3) molecule showing the x, y, z coordinates.

If the C_3^+ (120° clockwise) symmetry operation is carried out for the XY_3 molecule, the following shifts occur:

	C_3^+
ΔD_1	ΔD_3
ΔD_2	ΔD_1
ΔD_3	ΔD_2
$\Delta \alpha_{12}$	$\Delta \alpha_{13}$
$\Delta \alpha_{13}$	$\Delta \alpha_{23}$
$\Delta \alpha_{23}$	$\Delta \alpha_{12}$

(A5-8)

If the resulting shifts are written for all the symmetry operations in the C_{3v} point group, the following table is obtained:

C_{3v}	E	C_3^+	C_3^-	σ_{v1}	σ_{v2}	σ_{v3}
ΔD_1	ΔD_1	ΔD_3	ΔD_2	ΔD_1	ΔD_3	ΔD_2
ΔD_2	ΔD_2	ΔD_1	ΔD_3	ΔD_3	ΔD_2	ΔD_1
ΔD_3	ΔD_3	ΔD_2	ΔD_1	ΔD_2	ΔD_1	ΔD_3
$\Delta \alpha_{12}$	$\Delta \alpha_{12}$	$\Delta \alpha_{13}$	$\Delta \alpha_{23}$	$\Delta \alpha_{13}$	$\Delta \alpha_{23}$	$\Delta \alpha_{12}$
$\Delta \alpha_{13}$	$\Delta \alpha_{13}$	$\Delta \alpha_{23}$	$\Delta \alpha_{12}$	$\Delta \alpha_{12}$	$\Delta \alpha_{13}$	$\Delta \alpha_{23}$
$\Delta \alpha_{23}$	$\Delta \alpha_{23}$	$\Delta \alpha_{12}$	$\Delta \alpha_{13}$	$\Delta \alpha_{23}$	$\Delta \alpha_{12}$	$\Delta \alpha_{13}$

(A5-9)

Each column can be considered a vector. If we take column C_3^+, the vector $r_{C_3^+}$ whose components are given under C_3^+ is a transform of the vector whose components are under C_{3v}. Symbolically $r_{C_3^+} = D(C_3^+)r$, where $D(C_3^+)$ is a matrix used to transform r into $r_{C_3^+}$, and in more detail this becomes

$$\begin{bmatrix} \Delta D_3 \\ \Delta D_1 \\ \Delta D_2 \\ \Delta \alpha_{13} \\ \Delta \alpha_{23} \\ \Delta \alpha_{12} \end{bmatrix} = \begin{bmatrix} 0 & 0 & 1 & 0 & 0 & 0 \\ 1 & 0 & 0 & 0 & 0 & 0 \\ 0 & 1 & 0 & 0 & 0 & 0 \\ 0 & 0 & 0 & 0 & 1 & 0 \\ 0 & 0 & 0 & 0 & 0 & 1 \\ 0 & 0 & 0 & 1 & 0 & 0 \end{bmatrix} \begin{bmatrix} \Delta D_1 \\ \Delta D_2 \\ \Delta D_3 \\ \Delta \alpha_{12} \\ \Delta \alpha_{13} \\ \Delta \alpha_{23} \end{bmatrix} \qquad \text{(A5-10)}$$

This procedure can be repeated for each of the symmetry operations. A set of 6×6 matrices that is a group is obtained, and it is a six-dimensional representation. This set is displayed below in such a way that the element R of the group is given first, its corresponding $D(R)$ second, the sum $\chi(R)$ of the terms along the diagonal (called trace) third, and the value of $\chi(R)$ last:

$$E \rightarrow \begin{bmatrix} 1 & 0 & 0 & 0 & 0 & 0 \\ 0 & 1 & 0 & 0 & 0 & 0 \\ 0 & 0 & 1 & 0 & 0 & 0 \\ 0 & 0 & 0 & 1 & 0 & 0 \\ 0 & 0 & 0 & 0 & 1 & 0 \\ 0 & 0 & 0 & 0 & 0 & 1 \end{bmatrix} \equiv D(E) \qquad \chi(E) = 6 \qquad \text{(A5-11)}$$

$$C_3^+ \rightarrow \begin{bmatrix} 0 & 0 & 1 & 0 & 0 & 0 \\ 1 & 0 & 0 & 0 & 0 & 0 \\ 0 & 1 & 0 & 0 & 0 & 0 \\ 0 & 0 & 0 & 0 & 1 & 0 \\ 0 & 0 & 0 & 0 & 0 & 1 \\ 0 & 0 & 0 & 1 & 0 & 0 \end{bmatrix} \equiv D(C_3^+) \qquad \chi(C_3^+) = 0 \qquad \text{(A5-12)}$$

$$C_3^- \rightarrow \begin{bmatrix} 0 & 1 & 0 & 0 & 0 & 0 \\ 0 & 0 & 1 & 0 & 0 & 0 \\ 1 & 0 & 0 & 0 & 0 & 0 \\ 0 & 0 & 0 & 0 & 0 & 1 \\ 0 & 0 & 0 & 1 & 0 & 0 \\ 0 & 0 & 0 & 0 & 1 & 0 \end{bmatrix} \equiv D(C_3^-) \qquad \chi(C_3^-) = 0 \qquad \text{(A5-13)}$$

$$\sigma_{v1} \rightarrow \begin{bmatrix} 1 & 0 & 0 & 0 & 0 & 0 \\ 0 & 0 & 1 & 0 & 0 & 0 \\ 0 & 1 & 0 & 0 & 0 & 0 \\ 0 & 0 & 0 & 0 & 1 & 0 \\ 0 & 0 & 0 & 1 & 0 & 0 \\ 0 & 0 & 0 & 0 & 0 & 1 \end{bmatrix} \equiv D(\sigma_{v1}) \qquad \chi(\sigma_{v1}) = 2 \qquad \text{(A5-14)}$$

$$\sigma_{v2} \rightarrow \begin{bmatrix} 0 & 0 & 1 & 0 & 0 & 0 \\ 0 & 1 & 0 & 0 & 0 & 0 \\ 1 & 0 & 0 & 0 & 0 & 0 \\ 0 & 0 & 0 & 0 & 0 & 1 \\ 0 & 0 & 0 & 0 & 1 & 0 \\ 0 & 0 & 0 & 1 & 0 & 0 \end{bmatrix} \equiv D(\sigma_{v2}) \qquad \chi(\sigma_{v2}) = 2 \qquad \text{(A5-15)}$$

$$\sigma_{v3} \rightarrow \begin{bmatrix} 0 & 1 & 0 & 0 & 0 & 0 \\ 1 & 0 & 0 & 0 & 0 & 0 \\ 0 & 0 & 1 & 0 & 0 & 0 \\ 0 & 0 & 0 & 1 & 0 & 0 \\ 0 & 0 & 0 & 0 & 0 & 1 \\ 0 & 0 & 0 & 0 & 1 & 0 \end{bmatrix} \equiv D(\sigma_{v3}) \qquad \chi(\sigma_{v3}) = 2 \qquad \text{(A5-16)}$$

The matrices given in (A5-11) to (A5-16) are of a special form when they are partitioned as 2×2 matrices. For example,

$$C_3^+ \rightarrow \left[\begin{array}{ccc|ccc} 0 & 0 & 1 & 0 & 0 & 0 \\ 1 & 0 & 0 & 0 & 0 & 0 \\ 0 & 1 & 0 & 0 & 0 & 0 \\ \hline 0 & 0 & 0 & 0 & 1 & 0 \\ 0 & 0 & 0 & 0 & 0 & 1 \\ 0 & 0 & 0 & 1 & 0 & 0 \end{array}\right] = \left[\begin{array}{c|c} A & 0 \\ \hline 0 & C \end{array}\right] \tag{A5-17}$$

where

$$A = \begin{bmatrix} 0 & 0 & 1 \\ 1 & 0 & 0 \\ 0 & 1 & 0 \end{bmatrix} \quad 0 = \begin{bmatrix} 0 & 0 & 0 \\ 0 & 0 & 0 \\ 0 & 0 & 0 \end{bmatrix} \quad C = \begin{bmatrix} 0 & 1 & 0 \\ 0 & 0 & 1 \\ 1 & 0 & 0 \end{bmatrix} \tag{A5-18}$$

Here the matrix representing C_3^+ is said to be in the reducible form. If a set of matrices can be presented in this form, it, too, is said to be in the reducible form. Since the set of matrices in (A5-11) to (A5-16) is called a representation and since its matrices can be presented in reduced form, the representation is reducible. This statement implies that a transformation (called similarity) can be employed on the original set to display the matrices in the reduced form. The representation (set of matrices) so treated is called a reducible representation. If no similarity transformation exists, the representation is said to be irreducible

Another criterion is the following. If

$$\sum_R |\chi(R)|^2 > g \tag{A5-19}$$

the representation is reducible, and if

$$\sum_R |\chi(R)|^2 = g \tag{A5-20}$$

the representation is irreducible.

The trace of the 6×6 matrices of Eqs. (A5-11) to (A5-16) is the sum of the diagonal terms. Thus, the traces for the transformations of the displacement coordinates are

$$\chi(E) = 6, \quad \chi(C_3^+) = 0, \quad \chi(C_3^-) = 0, \quad \chi(\sigma_{v1}) = 2, \quad \chi(\sigma_{v2}) = 2, \quad \chi(\sigma_{v3}) = 2$$

The set of traces is called the character of the representation. It may be summarized as follows:

$$\begin{array}{lcccccc} & E & C_3^+ & C_3^- & \sigma_{v1} & \sigma_{v2} & \sigma_{v3} \\ \text{vib}\ (NH_3) & 6 & 0 & 0 & 2 & 2 & 2 \end{array} \qquad \text{(A5-21)}$$

$$\sum_R |\ \chi(R)^2 = 6^2 + 2^2 + 2^2 + 2^2 = 48$$

Since $48 > 6$, where $g = 6$ (the number of elements in a C_{3v} group), the representation is reducible. This reducible representation can be decomposed into a sum of irreducible representations. For the purpose of decomposition the characters of the irreducible representations are required. These characters are conveniently given in tabular form, the character table for C_{3v}, for instance, being written as follows:

C_{3v}	E	$2C_3$	$3\sigma_v$
A_1	$\chi_{A_1}(E)$	$\chi_{A_1}(C_3)$	$\chi_{A_1}(\sigma_v)$
A_2	$\chi_{A_2}(E)$	$\chi_{A_2}(C_3)$	$\chi_{A_2}(\sigma_v)$
E	$\chi_E(E)$	$\chi_E(C_3)$	$\chi_E(\sigma_v)$

Since, from the character table for C_{3v},

$$\begin{array}{llllccc} 2\chi_{A_1}(E) & 2\chi_{A_1}(C_3) & 2\chi_{A_1}(\sigma_v) & 2 & 2 & 2 & \\ 2\chi_E(E) & 2\chi_E(C_3) & 2\chi_E(\sigma_v) & 4 & -2 & 0 & \\ \hline 2\chi_{A_1}(E)+2\chi_E(E) & 2\chi_{A_1}(C_3)+2\chi_E(C_3) & 2\chi_{A_1}(\sigma_v)+2\chi_E(\sigma_v) & 6 & 0 & 2 & \text{add} \end{array}$$

we may write

$$\text{vib}\ (NH_3) = 2A_1 + 2E$$

PROBLEMS

1. Multiply the following matrices:

a) $$\begin{bmatrix} A & B & C \\ D & E & F \\ G & H & I \end{bmatrix} \times \begin{bmatrix} J \\ K \\ L \end{bmatrix}$$

b) $$\begin{bmatrix} A \\ B \\ C \end{bmatrix} \times [D\ E\ K]$$

c) $$[4\ 5\ 6] \times \begin{bmatrix} 2 \\ 3 \\ 1 \end{bmatrix}$$

d) $$\begin{bmatrix} 2 \\ 3 \\ -1 \end{bmatrix} \times [4\ 5\ 6]$$

e) $$[1\ 2\ 3] \times \begin{bmatrix} 4 & -6 & 9 & 6 \\ 0 & -7 & 6 & 7 \\ 5 & 8 & -11 & -8 \end{bmatrix}$$

f) $$\begin{bmatrix} 2 & 3 & 4 \\ 1 & 5 & 6 \end{bmatrix} \times \begin{bmatrix} 1 \\ 2 \\ 3 \end{bmatrix}$$

g) $$\begin{bmatrix} 1 & 2 & 1 \\ 4 & 0 & 2 \end{bmatrix} \times \begin{bmatrix} 3 & -4 \\ 1 & 5 \\ -2 & 2 \end{bmatrix}$$

h) $$\begin{bmatrix} A & B & C \\ D & E & F \\ G & H & I \end{bmatrix} \times \begin{bmatrix} J & K \\ L & M \\ N & O \end{bmatrix}$$

i) $$\begin{bmatrix} A & B & C \\ D & E & F \end{bmatrix} \times \begin{bmatrix} J & K \\ L & M \\ N & O \end{bmatrix}$$

Answers

a) $$\begin{bmatrix} AJ + BK + CL \\ DJ + EK + FL \\ GJ + HK + IL \end{bmatrix}$$

b) $$\begin{bmatrix} AD + AE + AK \\ BD + BE + BK \\ CD + CE + CK \end{bmatrix}$$

c) [29]

d) $$\begin{bmatrix} 8 & 10 & 12 \\ 12 & 15 & 18 \\ -4 & -5 & -6 \end{bmatrix}$$

e) $[19\quad 4\quad -12\quad -4]$

f) $$\begin{bmatrix} 20 \\ 29 \end{bmatrix}$$

g) $\begin{bmatrix} 3 & 8 \\ 8 & -12 \end{bmatrix}$

h) $\begin{bmatrix} AJ + BL + CN & AK + BM + CO \\ DJ + EL + FN & DK + EM + FO \\ GJ + HL + IN & GK + HM + IO \end{bmatrix}$

i) $\begin{bmatrix} AJ + BL + CN & AK + BM + CO \\ DJ + EL + FN & DK + EM + FO \end{bmatrix}$

REFERENCES

1. H. Margenau and G. M. Murphy, *The Mathematics of Physics and Chemistry*, D. Van Nostrand Co., Inc., New York (1956).
2. J. S. Ziomek, "Group Theory" in: *Progress in Infrared Spectroscopy*, Vol. 1 (H. A. Szymanski, ed.) Plenum Press, New York (1962).
3. F. A. Cotton, *Chemical Applications of Group Theory*, Interscience Publishers, New York (1963).
4. G. Stephenson, *Mathematical Methods for Science Students*, Longmans, London (1962).

Appendix 6

THE g ELEMENTS

It has been mentioned in Chapter 4 that the elements of the g matrix are given in terms of the internal coordinates, the bond stretching, r, and the angle deformation, φ. Wilson, Decius, and Cross[1] have suggested a method by which the g elements can be described. Figure A6-1 shows the schematic representations of the g elements. The superscript always indicates the number of common atoms. Atoms common to both coordinates are indicated by double circles, and are always put in a horizontal line. The noncommon atoms are indicated along a 45° diagonal. When a single com-

Table A6-1. g Elements—General Case

g^2_{rr}	$\mu_1 + \mu_2$
g^1_{rr}	$\mu_1 \cos\varphi$
$g^2_{r\varphi}$	$-\varrho_{23}\mu_2 \sin\varphi$
$g^1_{r\varphi}\binom{1}{2}$	$\varrho_{13}\mu_1 \sin\varphi \cos\tau$
$\binom{1}{1}$	$-(\varrho_{13} \sin\varphi_{213} \cos\psi_{234} + \varrho_{14} \sin\varphi_{214} \cos\psi_{243})\mu_1$
$g^3_{\varphi\varphi}$	$\varrho^2_{12}\mu_1 + \varrho^2_{23}\mu_3 + (\varrho^2_{12} + \varrho^2_{23} - 2\varrho_{12}\varrho_{23} \cos\varphi)\mu_2$
$g^2_{\varphi\varphi}\binom{1}{1}$	$(\varrho^2_{12} \cos\psi_{314})\mu_1 + [(\varrho_{12} - \varrho_{23} \cos\varphi_{123} - \varrho_{24} \cos\varphi_{124})\varrho_{12} \cos\psi_{314}$
	$+ \sin\varphi_{123} \sin\varphi_{124} \sin^2\psi_{314} + \cos\varphi_{324} \cos\psi_{314})\varrho_{23}\varrho_{24}]\mu_2$
$\binom{1}{0}$	$-\varrho_{12} \cos\tau[(\varrho_{12} - \varrho_{14} \cos\varphi_1)\mu_1 + (\varrho_{12} - \varrho_{23} \cos\varphi_2)\mu_2]$
$g^1_{\varphi\varphi}\binom{2}{2}$	$- \sin\tau_{25} \sin\tau_{34} + \cos\tau_{25} \cos\tau_{34} \cos\varphi_1)\varrho_{12}\varrho_{14}\mu_1$
$\binom{2}{1}$	$(\sin\varphi_{214} \cos\varphi_{415} \cos\tau_{34} - \sin\varphi_{215} \cos\tau_{35})\varrho_{14}$
	$+ \sin\varphi_{215} \cos\varphi_{415} \cos\tau_{35} - \sin\varphi_{214} \cos\tau_{34})\varrho_{15} \dfrac{\varrho_{12}\mu_1}{\sin\varphi_{415}}$
$\binom{1}{1}$	$[(\cos\varphi_{415} - \cos\varphi_{314} \cos {}_{315} - \cos\varphi_{214} \cos\varphi_{215} + \cos\varphi_{213} \cos\varphi_{214} \cos\varphi_{315})$
	$\times \varrho_{12}\varrho_{13} + (\cos\varphi_{413} - \cos\varphi_{514} \cos\varphi_{513} - \cos\varphi_{214} \cos\varphi_{213}$
	$+ \cos\varphi_{215} \cos\varphi_{214} \cos\varphi_{513})\varrho_{12}\varrho_{15} + \cos\varphi_{215} - \cos\varphi_{312} \cos\varphi_{315}$
	$- \cos\varphi_{412} \cos\varphi_{415} + \cos\varphi_{413} \cos\varphi_{412} \cos\varphi_{315})\varrho_{14}\varrho_{13}$
	$+ (\cos\varphi_{213} - \cos\varphi_{512} \cos\varphi_{513} - \cos\varphi_{412} \cos\varphi_{413} + \cos\varphi_{415} \cos\varphi_{412} \cos\varphi_{513})$
	$\times \varrho_{14}\varrho_{15}] \dfrac{\mu_1}{\sin\varphi_{214} \sin\varphi_{315}}$

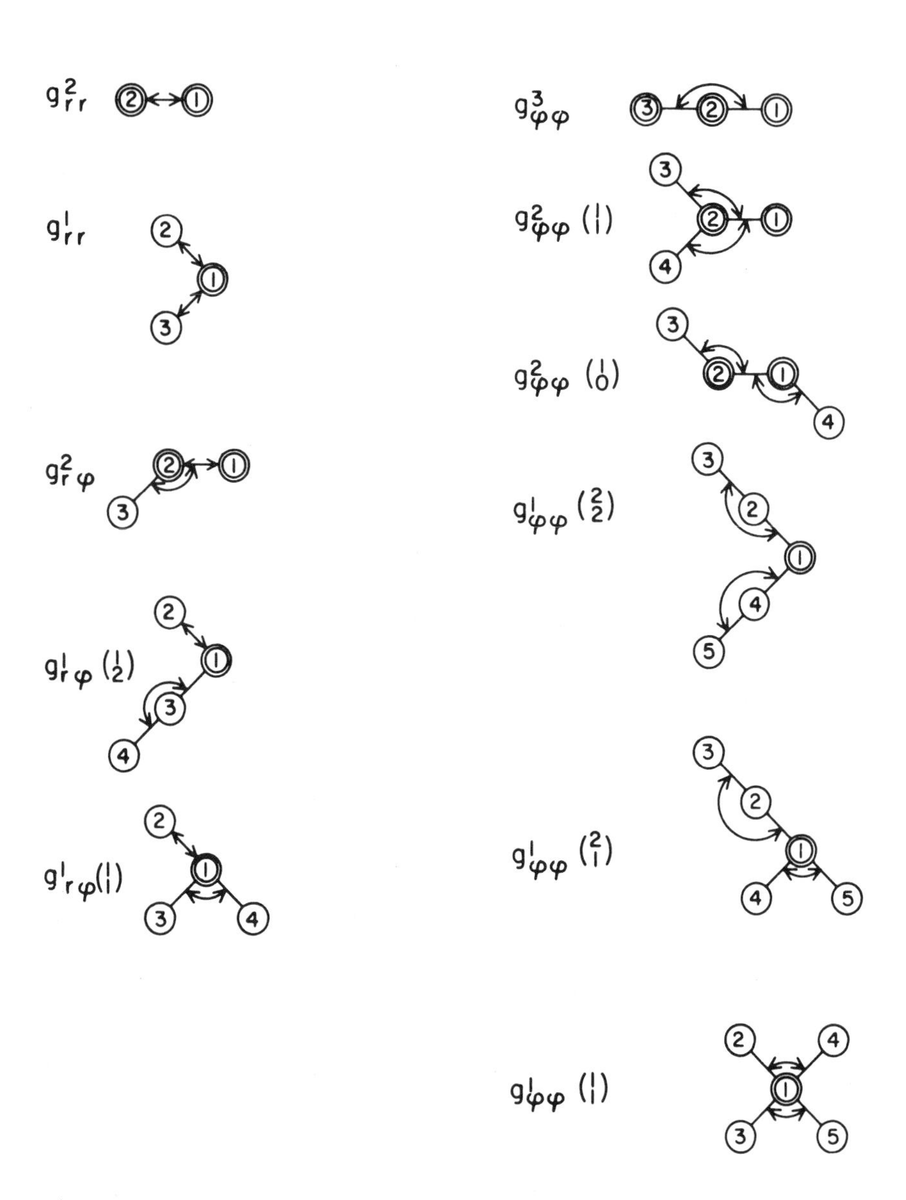

Fig. A6-1. Schematic representations of *g* elements for nonlinear molecules. [From *Molecular Vibrations* by E. B. Wilson, J. C. Decius, and P. C. Cross, McGraw–Hill, New York (1955), as modified in *Introduction to Infrared and Raman Spectroscopy* by N. B. Colthup, L. H. Daly, and S. E. Wiberley, Academic Press, New York (1975). Used by permission of McGraw–Hill Book Company and Academic Press.]

Table A6-2. g Elements for $\varphi = 109°28'$

g_{rr}^{2}	$\mu_1 + \mu_2$
g_{rr}^{1}	$-\frac{1}{3}\mu_1$
$g_{r\varphi}^{2}$	$-\frac{2}{3}\sqrt{2}\,\varrho_{23}\mu_2$
$g_{r\varphi}^{1}\binom{1}{2}$	$\frac{2}{3}\sqrt{2}\,\varrho_{13}\mu_1 \cos\tau$
$g_{r\varphi}^{1}\binom{1}{1}$	$\frac{1}{3}\sqrt{2}\,(\varrho_{13} + \varrho_{14})\mu_1$
$g_{\varphi\varphi}^{3}$	$\varrho_{12}^2\mu_1 + \varrho_{23}^2\mu_3 + \frac{1}{3}(3\varrho_{12}^2 + 3\varrho_{23}^2 + 2\varrho_{12}\varrho_{23})\mu_2$
$g_{\varphi\varphi}^{2}\binom{1}{1}$	$-\frac{1}{6}\{3\varrho_{21}^2\mu_1 + [3\varrho_{21}^2 + (\varrho_{23} + \varrho_{24})\varrho_{21} - 5\varrho_{23}\varrho_{24}]\mu_2\}$
$g_{\varphi\varphi}^{2}\binom{1}{0}$	$-\frac{1}{3}\varrho_{12}\cos\tau[(3\varrho_{12} + \varrho_{14})\mu_1 + (3\varrho_{12} + \varrho_{23})\mu_2]$
$g_{\varphi\varphi}^{1}\binom{2}{2}$	$-\frac{1}{3}(3\sin\tau_{25}\sin\tau_{34} - \cos\tau_{25}\cos\tau_{34})\varrho_{12}\varrho_{14}\mu_1$
$g_{\varphi\varphi}^{1}\binom{2}{1}$	$-\frac{1}{3}[(3\cos\tau_{35} + \cos\tau_{34})\varrho_{14} + (3\cos\tau_{34} + \cos\tau_{35})\varrho_{15}]\varrho_{12}\mu_1$
$g_{\varphi\varphi}^{1}\binom{1}{1}$	$-\frac{2}{3}(\varrho_{12} + \varrho_{14})(\varrho_{13} + \varrho_{15})\mu_1$

mon atom is the terminal atom in a bending vibration, the notation $g_{r\varphi}^{1}\binom{1}{2}$ is used; when it is the central atom in a bending vibration, the notation $g_{r\varphi}^{1}\binom{1}{1}$ is used. Here, the pair of numbers in parentheses has the following significance: the top number gives the number of atoms in the top left line, and the bottom number gives the number of atoms in the bottom left line of the schematic representation. Tables A6-1 to A6-3 record the g elements for the general case, the case where $\varphi = 109°28'$, and the case

Table A6-3. g Elements for $\varphi = 120°$

g_{rr}^{2}	$\mu_1 + \mu_2$
g_{rr}^{1}	$-\frac{1}{2}\mu_1$
$g_{r\varphi}^{2}$	$-\frac{1}{2}\sqrt{3}\,\varrho_{23}\mu_2$
$g_{r\varphi}^{1}\binom{1}{2}$	$\frac{1}{2}\sqrt{3}\,\varrho_{13}\mu_1 \cos\tau$
$g_{r\varphi}^{1}\binom{1}{1}$	$\frac{1}{2}\sqrt{3}\,(\varrho_{13} + \varrho_{14})\mu_1$
$g_{\varphi\varphi}^{3}$	$\varrho_{12}^2\mu_1 + \varrho_{23}^2\mu_3 + (\varrho_{12}^2 + \varrho_{23}^2 + \varrho_{12}\varrho_{23})\mu_2$
$g_{\varphi\varphi}^{2}\binom{1}{1}$	$-\frac{1}{2}\{2\varrho_{12}^2\mu_1 + [\varrho_{12}(2\varrho_{12} + \varrho_{23} + \varrho_{24}) - \varrho_{23}\varrho_{24}]\mu_2\}$
$g_{\varphi\varphi}^{2}\binom{1}{0}$	$-\frac{1}{2}\varrho_{12}\cos\tau[(2\varrho_{12} + \varrho_{14})\mu_1 + (2\varrho_{12} + \varrho_{23})\mu_2]$
$g_{\varphi\varphi}^{1}\binom{2}{2}$	$\frac{1}{2}(\cos\tau_{25}\cos\tau_{34} - 2\sin\tau_{25}\sin\tau_{34})\varrho_{12}\varrho_{14}\mu_1$
$g_{\varphi\varphi}^{1}\binom{2}{1}$	$\frac{1}{2}\cos\tau_{34}(\varrho_{14} - \varrho_{15})\varrho_{12}\mu_1$
$g_{\varphi\varphi}^{1}\binom{1}{1}$	This type of g element impossible when $\varphi = 120°$

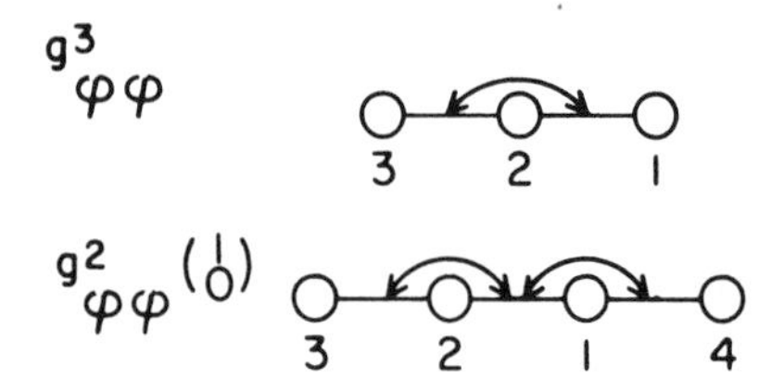

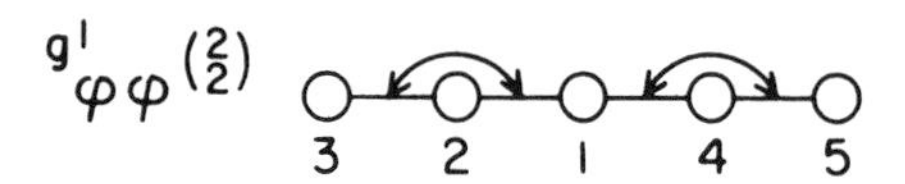

Fig. A6-2. Schematic representations of g elements for linear molecules.

where $\varphi = 120°$. For results including torsions, out-of-plane bending, and cyclic structures, see Decius.[2] Ferigle and Meister[3] have determined formulas for g elements for linear cases, and these are illustrated in Fig. A6-2.

For the use of the formulas in Tables A6-1 and A6-3, it is necessary to define certain terms; μ and ϱ are the reciprocals of mass and bond distance, respectively. The formulas also involve an angle ψ, which is defined as

$$\cos \psi_{\alpha\beta\gamma} = \frac{\cos \varphi_{\alpha\zeta\gamma} - \cos \varphi_{\alpha\zeta\beta} \cos \varphi_{\beta\zeta\gamma}}{\sin \varphi_{\alpha\zeta\beta} \sin \varphi_{\beta\zeta\gamma}}$$

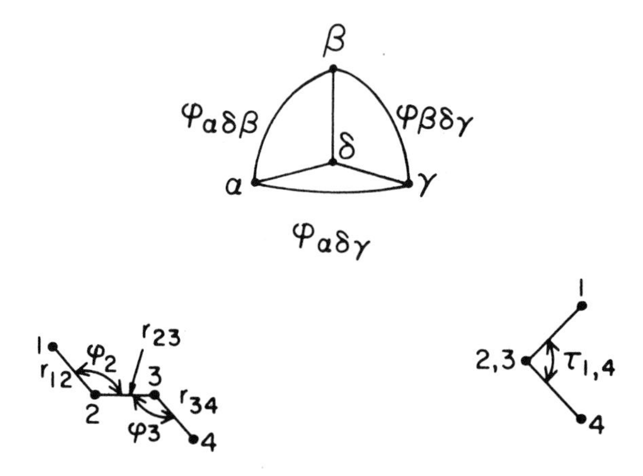

Fig. A6-3. Definitions of notation used in Appendix 6.

the notation being defined by the top diagram in Fig. A6-3. The torsion angle τ is defined by the two bottom diagrams in Fig. A6-3.

REFERENCES

1. E. B. Wilson, J. C. Decius, and P. C. Cross, *Molecular Vibrations*, McGraw–Hill, New York (1955).
2. J. C. Decius, *J. Chem. Phys.*, **16**: 1025 (1948).
3. S. M. Ferigle and A. G. Meister, *J. Chem. Phys.*, **19**: 982 (1951).

Appendix 7

GENERAL METHOD OF OBTAINING MOLECULAR SYMMETRY COORDINATES

A general method for obtaining molecular symmetry coordinates has been developed by Nielsen and Berryman.[1] According to them, the prescription for determining symmetry coordinates is given as follows.

1. Choose a linear combination L_k of the internal coordinates from a set of equivalent coordinates.
2. Then the sum

$$S^j_{k\alpha} = (l_j/h) \sum_R D^j(R)_{\alpha\alpha} RL_k \tag{A7-1}$$

will be a symmetry coordinate belonging to the αth row of the irreducible representation D^j and its $l_j - 1$ partners will be given by

$$S_k{}^j = (l_j/h) \sum_R D^j(R)_{\beta\alpha} RS^j_{k\alpha} \tag{A7-2}$$

Here the subscript k labels the number of symmetry coordinates in the jth irreducible representations, $D^j(R)_{\alpha\alpha}$ and $D^j(R)_{\beta\alpha}$ are matrix elements from the αth row and αth column and from the βth row and αth column of the matrix $D^j(R)$ for the jth representation, respectively, R represents an element of the group, and RL_k represents the transform of L_k by R. The order of the molecular symmetry group is h.

Consider the molecule XY_3 (NH_3), whose symmetry is given by C_{3v}. Figure 7-1 illustrates the geometric configuration and the set of internal coordinates chosen. According to their prescription, one requires the matrix elements for the irreducible representations of C_{3v}, the chosen set of internal coordinates, the transforms of the internal coordinates to form the transforms of the linear combinations of internal coordinates, the dimensions of the irreducible representations, and the order of the group. First, the chosen set of internal coordinates is Δd_1, Δd_2, Δd_3, $\Delta\alpha_{12}$, $\Delta\alpha_{13}$, and $\Delta\alpha_{23}$.

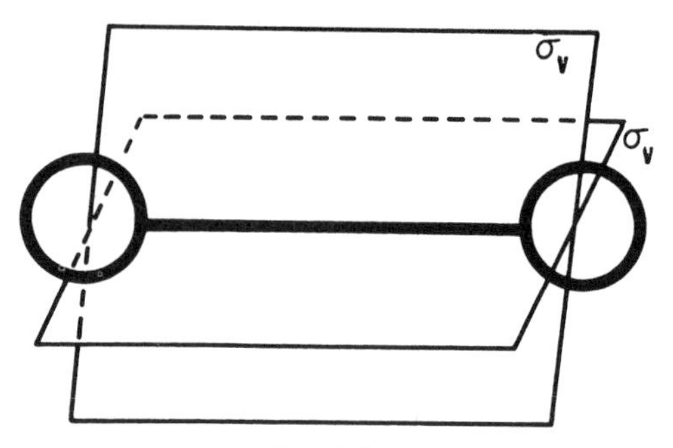

Fig. A7-1.

Second, the following table gives the transforms of these internal coordinates:

r_k	$R_1 r_k$	$R_2 r_k$	$R_3 r_k$	$R_4 r_k$	$R_5 r_k$	$R_6 r_k$		R
C_{3v}	E	C_3^+	C_3^-	σ_{v_1}	σ_{v_2}	σ_{v_3}		
Δd_1	Δd_1	Δd_3	Δd_2	Δd_1	Δd_3	Δd_2	r_k	Rr_k
Δd_2	Δd_2	Δd_1	Δd_3	Δd_3	Δd_2	Δd_1		
Δd_3	Δd_3	Δd_2	Δd_1	Δd_2	Δd_1	Δd_3		
$\Delta\alpha_{12}$	$\Delta\alpha_{12}$	$\Delta\alpha_{13}$	$\Delta\alpha_{23}$	$\Delta\alpha_{13}$	$\Delta\alpha_{23}$	$\Delta\alpha_{12}$		
$\Delta\alpha_{23}$	$\Delta\alpha_{23}$	$\Delta\alpha_{12}$	$\Delta\alpha_{13}$	$\Delta\alpha_{23}$	$\Delta\alpha_{12}$	$\Delta\alpha_{13}$		
$\Delta\alpha_{13}$	$\Delta\alpha_{13}$	$\Delta\alpha_{23}$	$\Delta\alpha_{12}$	$\Delta\alpha_{12}$	$\Delta\alpha_{13}$	$\Delta\alpha_{23}$		

where the legend is

	R
r_k	Rr_k

where r_k is the kth internal coordinate and Rr_k is the transform of r_k by R. Third, the selection rules state that there are two fundamental modes of motion in the A_1 and two in the E irreducible representations. This implies that k will take on values 1 and 2 in A_1 species and 1 and 2 in E species. Therefore for $S_{k\alpha}^j$ for A_1 one has $S_{1\alpha}^{A_1}$ and $S_{2\alpha}^{A_1}$; and for $S_{k\alpha}^j$ for E one has $(S_{1\alpha}^E, S_{1\beta}^E)$ and $(S_{2\alpha}^E, S_{2\beta}^E)$. Fourth, the matrices for the irreducible representation of C_{3v} are given by the table

	R_1	R_2	R_3	R_4	R_5	R_6
j	$D^j(R_1)$ E	$D^j(R_2)$ C_3^+	$D^j(R_3)$ C_3^-	$D^j(R_4)$ σ_{v_1}	$D^j(R_5)$ σ_{v_2}	$D^j(R_6)$ σ_{v_3}
A_1	1	1	1	1	1	1
A_2	1	1	1	-1	-1	-1
E	$\begin{pmatrix} 1 & 0 \\ 0 & 1 \end{pmatrix}$	$\begin{pmatrix} -\frac{1}{2} & \frac{1}{2}\sqrt{3} \\ -\frac{1}{2}\sqrt{3} & -\frac{1}{2} \end{pmatrix}$	$\begin{pmatrix} -\frac{1}{2} & \frac{1}{2}\sqrt{3} \\ \frac{1}{2}\sqrt{3} & -\frac{1}{2} \end{pmatrix}$	$\begin{pmatrix} 1 & 0 \\ 0 & -1 \end{pmatrix}$	$\begin{pmatrix} -\frac{1}{2} & -\frac{1}{2}\sqrt{3} \\ -\frac{1}{2}\sqrt{3} & \frac{1}{2} \end{pmatrix}$	$\begin{pmatrix} -\frac{1}{2} & -\frac{1}{2}\sqrt{3} \\ \frac{1}{2}\sqrt{3} & \frac{1}{2} \end{pmatrix}$

For the A_1 representation first choose

$$L_1 = \Delta d_1 \qquad \text{and} \qquad L_2 = \Delta\alpha_{13}$$

Then

$$S_{11}^{A_1} = \tfrac{1}{6} \sum_R D^{A_1}(R)_{11} R L_1$$

$$\begin{aligned} S_{11}^{A_1} &= \tfrac{1}{6}[D^{A_1}(E)_{11} E L_1 + D^{A_1}(C_3^+) C_3^+ L_1 + D^{A_1}(C_3^-) C_3^- L_1 \\ &\quad + D^{A_1}(\sigma_{v_1})\sigma_{v_1} L_1 + D^{A_1}(\sigma_{v_2})\sigma_{v_2} L_1 + D^{A_1}(\sigma_{v_3})\sigma_{v_3} L_1] \\ &= \tfrac{1}{6}(1\,\Delta d_1 + 1\,\Delta d_3 + 1\,\Delta d_2 + 1\,\Delta d_1 + 1\,\Delta d_3 + 1\,\Delta d_2) \\ &= \tfrac{2}{6}(\Delta d_1 + \Delta d_2 + \Delta d_3) \end{aligned}$$

Thus $S_{11}^{A_1} = \frac{1}{3}(\Delta d_1 + \Delta d_2 + \Delta d_3)$ and one then chooses

$$S_1^{A_1} = (1/\sqrt{3})\,(\Delta d_1 + \Delta d_2 + \Delta d_3)$$

as one of the normalized symmetry coordinates of A_1 type. There is no partner to this coordinate since the A_1 representation is nondegenerate and hence one dimensional.

In a similar manner one obtains

$$S_2^{A_1} = \tfrac{1}{6} \sum_R D^{A_1}(R) R\alpha_{12} = \tfrac{2}{6}(\Delta\alpha_{12} + \Delta\alpha_{13} + \Delta\alpha_{23})$$

and then the choice is

$$S_2^{A_1} = (1/\sqrt{3})(\Delta\alpha_{12} + \Delta\alpha_{13} + \Delta\alpha_{23})$$

For the E species one has (S_{11}^E, S_{12}^E) and (S_{21}^E, S_{22}^E). First consider $L_1 = \Delta d$, $l_1 = 2$, and $h = 6$. Then

$$\begin{aligned} S_{11}^E &= \tfrac{2}{6}[D^E(E)_{11}L_1 + D^E(C_3^+)_{11}L_1 + D^E(C_3^-)L_1 \\ &\quad + D^E(\sigma_{v_1})L_1 + D^E(\sigma_{v_2})L_1 + D^E(\sigma_{v_3})L_1] \\ &= \tfrac{1}{3}(1\ \Delta d_1 - \tfrac{1}{2}\ \Delta d_3 - \tfrac{1}{2}\ \Delta d_2 + 1\ \Delta d_1 - \tfrac{1}{2}\ \Delta d_3 - \tfrac{1}{2}\ \Delta d_2) \\ S_{11}^E &= \tfrac{1}{3}(2\ \Delta d_1 - \Delta d_2 - \Delta d_3) \end{aligned}$$

Normalized, we have

$$S_{11}^E = (1/\sqrt{6})(2\ \Delta d_1 - \Delta d_2 - \Delta d_3)$$

The first and only partner of S_{11}^E is S_{12}^E and is obtained from the following expression:

$$S_{12}^E = \tfrac{2}{6} \sum_R D^E(R)_{21} R S_{11}^E$$

which becomes

$$\begin{aligned} S_{12}^E &= \tfrac{2}{6}[D^E(E)_{21} E S_{11}^E + D^E(C_3^+)_{21} C_3^+ S_{11}^E + D^E(C_3^-)_{21} C_3^- S_{11}^E \\ &\quad + D^E(\sigma_{v_1})\sigma_{v_1} S_{11}^E + D^E(\sigma_{v_2})\sigma_{v_2} S_{11}^E + D^E(\sigma_{v_3})\sigma_{v_3} S_{11}^E] \\ &= 3\sqrt{3}\,(\Delta d_2 - \Delta d_3) \end{aligned}$$

Then S_{12}^E is taken to be

$$S_{12}^E = (1/\sqrt{2})(\Delta d_2 - \Delta d_3)$$

Hence the normalized and orthogonal pair selected for the first set of degenerate symmetry coordinates is

$$S_{11}^E = (1/\sqrt{6})(2\ \Delta d_1 - \Delta d_2 - \Delta d_3)$$

$$S_{12}^E = (1/\sqrt{2})(\Delta d_2 - \Delta d_3)$$

For (S_{21}^E, S_{22}^E) one takes $L_2 = \Delta\alpha_{12}$ and $k = 2$, and

$$S_{21}^E = \tfrac{1}{6} \sum_R D^E(R)_n R L_2$$

becomes

$$S_{12}^E = \tfrac{1}{3}(\Delta\alpha_{12} + \Delta\alpha_{13} - 2\ \Delta\alpha_{23})$$

while

$$S_{22}^{E} = \tfrac{2}{6} \sum_{R} D^{E}(R)_{21} R S_{21}^{E}$$

becomes

$$S_{22}^{E} = \sqrt{3}\,(\Delta\alpha_{12} - \Delta\alpha_{13})$$

Then for (S_{21}^{E}, S_{22}^{E}) to be symmetry coordinates one chooses the following orthogonal and normalized pair:

$$S_{21}^{E} = (1/\sqrt{6})(\Delta\alpha_{12} + \Delta\alpha_{13} - 2\,\Delta\alpha_{23})$$

$$S_{22}^{E} = (1/\sqrt{2})(\Delta\alpha_{12} - \Delta\alpha_{13})$$

REFERENCE

1. J. R. Nielsen and L. H. Berryman, *J. Chem. Phys.*, **17**:659 (1949).

Appendix 8

CALCULATION OF THERMODYNAMIC FUNCTIONS FROM VIBRATIONAL–ROTATIONAL SPECTRA

From the vibrational–rotational spectra of a molecule, it is possible to determine the thermodynamic functions of the molecule. A brief discussion follows. For a more detailed discussion, see the books by Herzberg[1] or Glasstone.[2]

It can be shown that the total energy in a molecule is the sum of the translational energy and the internal energy, where the internal energy is the sum of the vibrational, rotational, and electronic energies.

$$E_{\text{total}} = E_{\text{tr}} + E_{\text{int}} \tag{A8-1}$$

$$E_{\text{int}} = E_{\text{vib}} + E_{\text{rot}} + E_{\text{elec}} \tag{A8-2}$$

therefore,

$$E_{\text{total}} = E_{\text{tr}} + E_{\text{vib}} + E_{\text{rot}} + E_{\text{elec}} \tag{A8-3}$$

The total energy of a molecule is found to be related to the partition function Q as follows:

$$Q = \Sigma\, g_i e^{-E_{\text{tot}}/kT} \tag{A8-4}$$

where g_i is the total statistical weight (degeneracy), k is the Boltzmann constant, and T the absolute temperature. Therefore

$$Q_{\text{total}} = Q_{\text{tr}} \cdot Q_{\text{vib}} \cdot Q_{\text{rot}} \cdot Q_{\text{elec}} \tag{A8-5}$$

and Q_{elec} is usually neglected since it is very small. Q_{total} is related to all the thermodynamic functions as follows.

Enthalpy or heat content, (H°)

$$\frac{H^0 - E_0}{T} = RT\frac{d \ln Q_{\text{tot}}}{dT} + R \tag{A8-6}$$

where R the molar gas constant, equal to 1.987 cal/deg · mole, and E_0 is the total energy at absolute zero for one mole of an ideal gas.

Heat capacity at constant volume (C_v^0)

$$C_v^0 = \frac{R}{T^2}\left[\frac{d^2Q_{\text{tot}}/d(1/T)^2}{Q_{\text{tot}}} - \left(\frac{dQ_{\text{tot}}/d(1/T)}{Q_{\text{tot}}}\right)^2\right] \tag{A8-7}$$

Entropy (S^0)

$$S^0 = \frac{RT\,d\ln Q_{\text{tot}}}{dT} + R\ln Q_{\text{tot}} - R\ln N + R \tag{A8-8}$$

where N is the number of distinguishable particles in a gas, and can be assumed to be the number of molecules in a mole, or Avogadro's number.

Free energy (G^0)

$$\frac{G^0 - E_0}{T} = -R\ln\frac{Q_{\text{tot}}}{N} \tag{A8-9}$$

Q_{tot} is related to the other Q's as in Eq. (A8-5). Since Q_{rot} and Q_{vib} can be related to spectroscopic data, it follows that thermodynamic functions can be obtained from spectroscopic data. For example, for a linear molecule,

$$Q_{\text{rot}} = \frac{8\pi^2 IkT}{\sigma h^2} \tag{A8-10}$$

where σ is the symmetry number and is equal to the number of equivalent orientations of the molecule which can be obtained by rotation, h is Planck's constant, and I is the moment of inertia. The value of I can be obtained from the equation

$$I = \frac{h}{4\pi^2 c\Delta\nu} \tag{A8-11}$$

where $\Delta\nu$ is the spacing between absorptions in the rotational fine structure of the molecule, and c is the velocity of light (2.9978×10^{10} cm/sec). Thus, the Q_{rot} contribution to all the thermodynamic functions can be determined from spectroscopic data. For the Q_{vib} contributions, assuming a harmonic oscillator and no coupling between rotation and vibration,

$$Q_{\text{vib}} = e^{-h\nu c/2kT}\,\frac{1}{1 - e^{-h\nu c/kT}} \tag{A8-12}$$

where ν is the frequency in cm^{-1} and Q_{vib} can be determined for each fundamental frequency (ν) and summed. The Q_{vib} contribution to each thermodynamic function can then also be determined. The Q_{tr} contribution can be calculated from known formulas and added to the rotational and vibrational contributions. The thermodynamic functions are thus determined as functions of the temperature.

REFERENCES

1. G. Herzberg, *Molecular Spectra and Molecular Structure, II, Infrared and Raman Spectra of Polyatomic Molecules*, Van Nostrand, New York (1945).
2. S. Glasstone, *Theoretical Chemistry*, Van Nostrand, New York (1944).

Appendix 9

DIAGRAMS OF NORMAL VIBRATIONS FOR COMMON POINT GROUPS

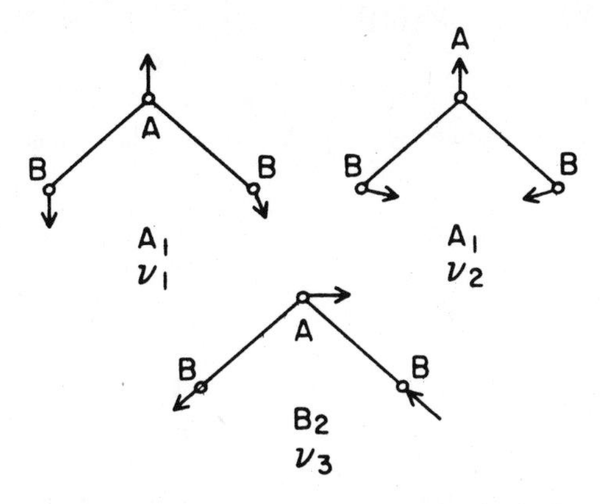

Normal modes of vibration of bent AB_2 molecule—point group C_{2v}.

$\nu_1(\Sigma^+)$ A B C

$\nu_2(\pi)$

$\nu_3(\Sigma^+)$

Normal modes of vibration of linear ABC molecule—point group $C_{\infty v}$.

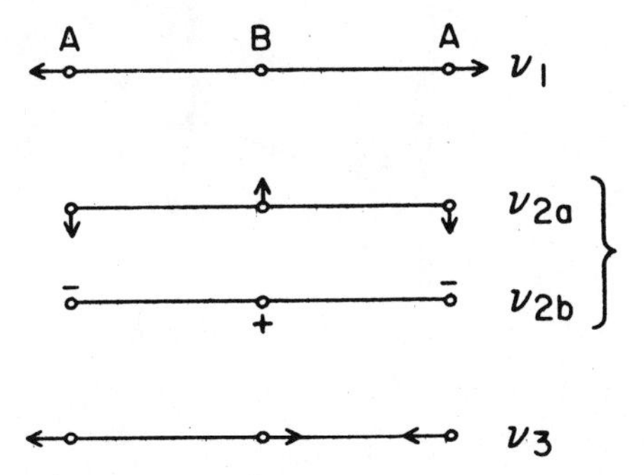

Normal modes of vibration of linear A_3 or ABA molecule—point group $D_{\infty h}$.

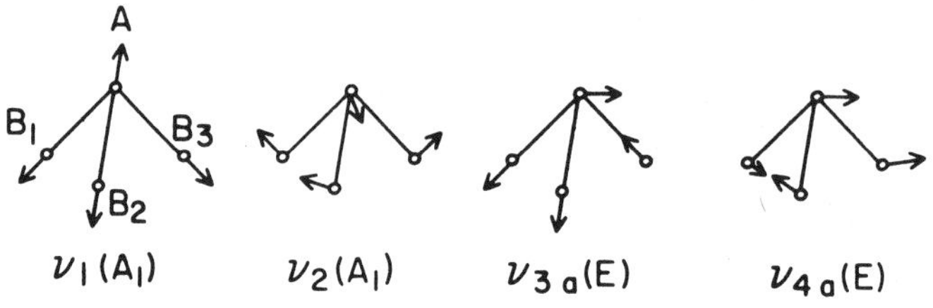

A B1 B2 B3

$\nu_1(A_1)$ $\nu_2(A_1)$ $\nu_{3a}(E)$ $\nu_{4a}(E)$

Normal modes of vibration of pyramidal AB_3 molecule—point group C_{3v}.

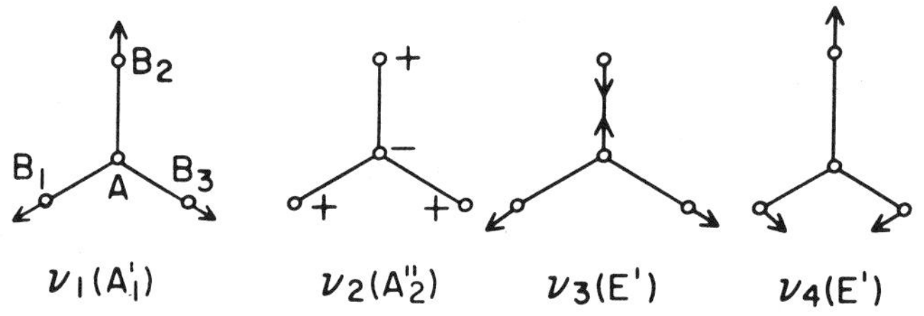

B2 B1 A B3

+ − + +

$\nu_1(A_1')$ $\nu_2(A_2'')$ $\nu_3(E')$ $\nu_4(E')$

Normal modes of vibration of planar AB_3 molecule—point group D_{3h}.

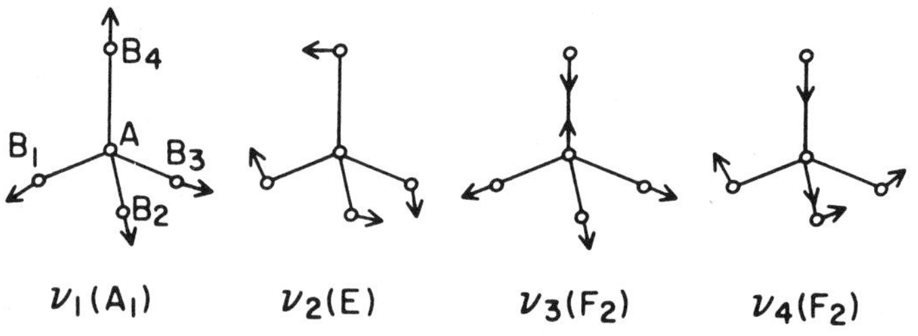

B4 B1 A B3 B2

$\nu_1(A_1)$ $\nu_2(E)$ $\nu_3(F_2)$ $\nu_4(F_2)$

Normal modes of vibration of tetrahedral AB_4 molecule—point group T_d.

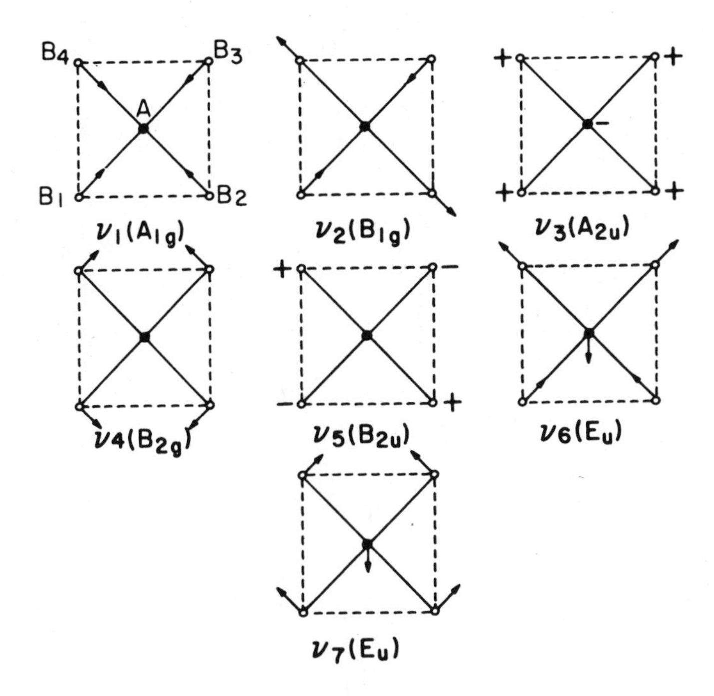

Normal modes of vibrations of square planar AB_4 molecule—point group D_{4h}.

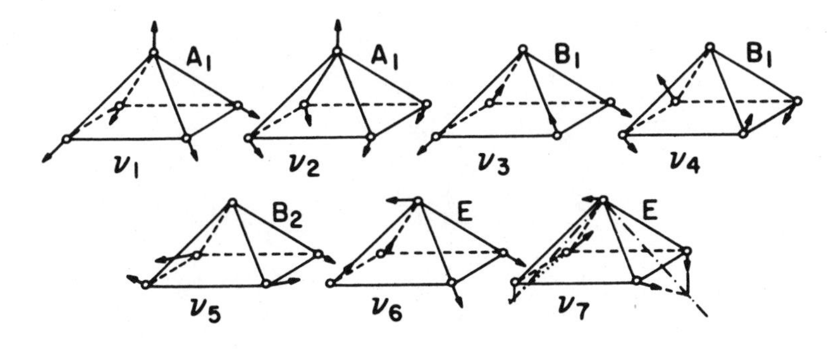

Normal modes of vibrations of pyramidal AB_4 molecule—point group C_{4v}.

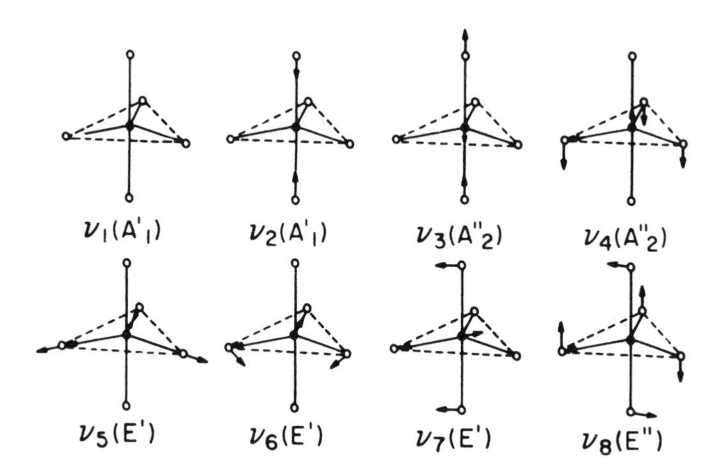

Normal modes of vibrations of trigonal bipyramidal AB_5 molecule—point group D_{3h}.

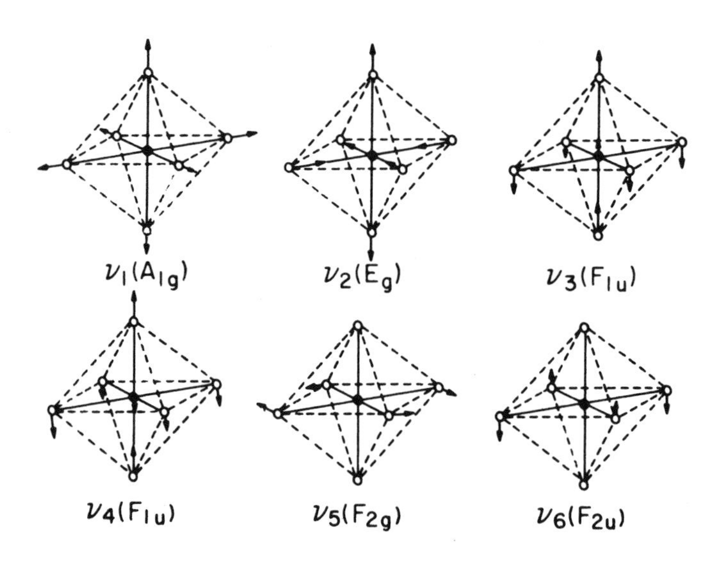

Normal modes of vibrations of octahedral AB_6 molecule—point group O_h.

Appendix 10

DERIVATION OF THE CHARACTERS NECESSARY FOR SELECTION RULES

In the development of selection rules for isolated molecules (see Chapter 2), it is necessary to derive the character of the dipole moment to be used in determining the infrared activity, the character to represent the polarizability for the determination of Raman activity, and the character to determine the number of fundamentals belonging to each vibration type. The derivation of these characters follows.

DERIVATION OF THE CHARACTER OF THE DIPOLE MOMENT $\chi_M(R)$

Consider a coordinate system (x, y, z) and assume a molecule whose center of gravity is located at p (Fig. A10-1). If one performs a proper operation along the z axis such as a clockwise rotation of angle $\Theta(C_\Theta^+)$, x is displaced from the equilibrium position to x', y is displaced to y', and z remains invariant. The relationships between x and x', y and y', and z and z' are

$$\begin{aligned} x' &= x \cos \Theta + y \sin \Theta \\ y' &= -x \sin \Theta + y \cos \Theta \\ z' &= z \end{aligned} \tag{A10-1}$$

In matrix notation this becomes

$$\begin{bmatrix} x' \\ y' \\ z' \end{bmatrix} = C_\Theta^+ \begin{bmatrix} x \\ y \\ z \end{bmatrix} = \begin{bmatrix} \cos \Theta & \sin \Theta & 0 \\ -\sin \Theta & \cos \Theta & 0 \\ 0 & 0 & 1 \end{bmatrix} \begin{bmatrix} x \\ y \\ z \end{bmatrix} \tag{A10-2}$$

The character of the matrix is the sum of the elements along the diagonals (also called the trace or spur) and is given by

$$\chi_M(R) = \chi_M(C_\Theta^+) = (1 + 2 \cos \Theta) \tag{A10-3}$$

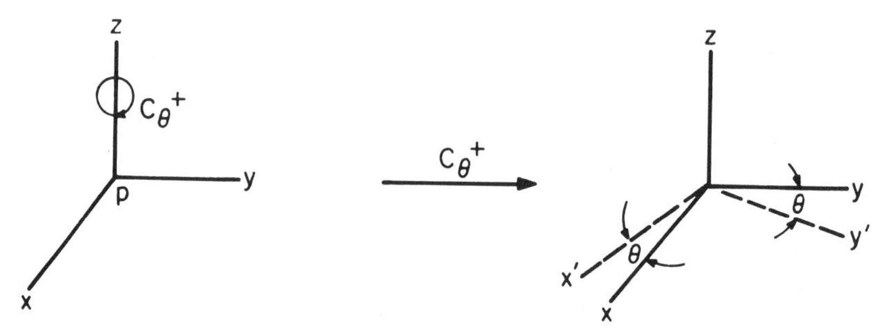

Fig. A10-1.

The same results are obtained if the rotation operation is performed counterclockwise (C_{Θ}^{-}). Equation (A10-3) can be written for the general case as follows:

$$\chi_M(R) = \mu_R(1 + 2\cos\Theta) \tag{A10-4}$$

where μ_R is the number of atoms left unchanged by the operation.

If one considers an improper operation such as a rotation–reflection (S), the x, y coordinates are again displaced to x', y' but in this operation the z coordinate also changes (Fig. A10-2). The relationships between the equilibrium positions and the displacements can now be written

$$\begin{aligned} x' &= x\cos\Theta + y\sin\Theta \\ y' &= -x\sin\Theta + y\cos\Theta \\ z' &= -z \end{aligned} \tag{A10-5}$$

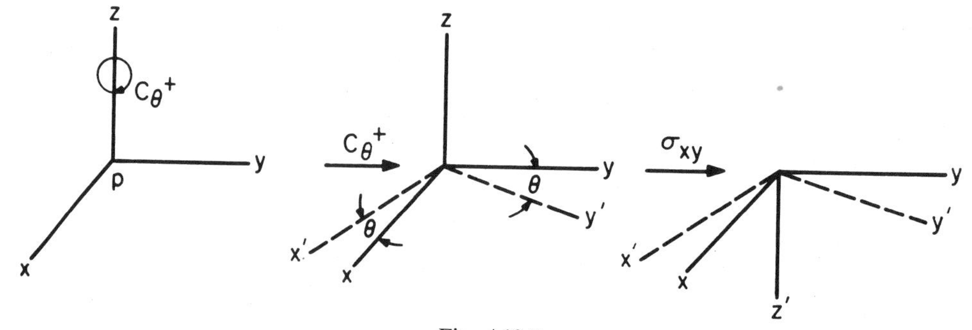

Fig. A10-2.

Written in matrix notation, the expressions become

$$\begin{bmatrix} x' \\ y' \\ z' \end{bmatrix} = S \begin{bmatrix} x \\ y \\ z \end{bmatrix} \begin{bmatrix} \cos\Theta & \sin\Theta & 0 \\ -\sin\Theta & \cos\Theta & 0 \\ 0 & 0 & -1 \end{bmatrix} \begin{bmatrix} x \\ y \\ z \end{bmatrix} \qquad \text{(A10-6)}$$

The characters become

$$\chi_M(R) = \chi_M(S) = (-1 + 2\cos\Theta) \qquad \text{(A10-7)}$$

For the general case the character is written

$$\chi_M(R) = \mu_R(-1 + 2\cos\Theta) \qquad \text{(A10-8)}$$

Thus the character for the dipole moment necessary to calculate the infrared activity is

$$\chi_M(R) = \mu_R(\pm 1 + 2\cos\Theta) \qquad \text{(A10-9)}$$

where the plus sign is used for proper operations and the minus sign for improper operations.

DERIVATION OF THE CHARACTER OF THE POLARIZABILITY $\chi_\alpha(R)$

For most molecules the polarizability α may be different in the x, y, and z directions. For the general case the following equations apply:

$$\begin{aligned} \mu_x &= \alpha_{xx}E_x + \alpha_{yx}E_y + \alpha_{xz}E_z \\ \mu_y &= \alpha_{yx}E_x + \alpha_{yy}E_y + \alpha_{yz}E_z \\ \mu_z &= \alpha_{zx}E_x + \alpha_{zy}E_y + \alpha_{zz}E_z \end{aligned} \qquad \text{(A10-10)}$$

where E represents a vector of the electronic field and μ is the induced dipole moment. Since the polarizability tensor is symmetric, $\alpha_{xy} = \alpha_{yx}$, $\alpha_{yz} = \alpha_{zy}$, and $\alpha_{xz} = \alpha_{zx}$. Thus, there are six components to the symmetric tensor of polarizability. If a particular operation such as C_Θ is performed, the components of polarizability will undergo the following changes:

$$\alpha_{xx} \to \alpha_{x'x'}, \qquad \alpha_{yy} \to \alpha_{y'y'}, \qquad \alpha_{zz} \to \alpha_{z'z'}$$

$$\alpha_{xy} \to \alpha_{x'y'}, \qquad \alpha_{xz} \to \alpha_{x'z'}, \qquad \alpha_{yz} \to \alpha_{y'z'}$$

The relationships between the equilibrium polarizabilities and the changed polarizabilities are given in matrix notation as follows:

$$\begin{bmatrix} \alpha_{x'x'}\alpha_{x'y'}\alpha_{x'z'} \\ \alpha_{y'x'}\alpha_{y'y'}\alpha_{y'z'} \\ \alpha_{z'x'}\alpha_{z'y'}\alpha_{z'z'} \end{bmatrix} = \underset{A}{\begin{bmatrix} C_{x'x}C_{x'y}C_{x'z} \\ C_{y'x}C_{y'y}C_{y'z} \\ C_{z'z}C_{z'y}C_{z'z} \end{bmatrix}} \begin{bmatrix} \alpha_{xx}\alpha_{xy}\alpha_{xz} \\ \alpha_{yx}\alpha_{yy}\alpha_{yz} \\ \alpha_{zx}\alpha_{zy}\alpha_{zz} \end{bmatrix} \underset{B}{\begin{bmatrix} C_{x'x}C_{y'x}C_{z'x} \\ C_{x'y}C_{y'y}C_{z'y} \\ C_{x'z}C_{y'z}C_{z'z} \end{bmatrix}} \quad \text{(A10-11)}$$

where $C_{x'x}$ denote the direction cosine between the axes x' and x, etc. (Note B is the transpose of A.) For a rotation (proper operation) through an angle Θ, the relationship becomes

$$C_\Theta \begin{bmatrix} \alpha_{xx}\alpha_{xy}\alpha_{xz} \\ \alpha_{yx}\alpha_{yy}\alpha_{yz} \\ \alpha_{zx}\alpha_{zy}\alpha_{zz} \end{bmatrix} = \begin{bmatrix} \cos\Theta & \sin\Theta & 0 \\ -\sin\Theta & \cos\Theta & 0 \\ 0 & 0 & 1 \end{bmatrix} \begin{bmatrix} \alpha_{xx}\alpha_{xy}\alpha_{xz} \\ \alpha_{yx}\alpha_{yy}\alpha_{yz} \\ \alpha_{zx}\alpha_{zy}\alpha_{zz} \end{bmatrix} \begin{bmatrix} \cos\Theta & -\sin\Theta & 0 \\ \sin\Theta & \cos\Theta & 0 \\ 0 & 0 & 1 \end{bmatrix} \quad \text{(A10-12)}$$

which can be written

$$C_\Theta \begin{bmatrix} \alpha_{xx} \\ \alpha_{yy} \\ \alpha_{zz} \\ \alpha_{xy} \\ \alpha_{xz} \\ \alpha_{yz} \end{bmatrix} = \begin{bmatrix} \cos^2\Theta & \sin^2\Theta & 0 & 2\sin\Theta\cos\Theta & 0 & 0 \\ \sin^2\Theta & \cos^2\Theta & 0 & -2\sin\Theta\cos\Theta & 0 & 0 \\ 0 & 0 & 1 & 0 & 0 & 0 \\ -\sin\Theta\cos\Theta & \sin\Theta\cos\Theta & 0 & 2\cos^2\Theta - 1 & 0 & 0 \\ 0 & 0 & 0 & 0 & \cos\Theta & \sin\Theta \\ 0 & 0 & 0 & 0 & -\sin\Theta & \cos\Theta \end{bmatrix} \times \begin{bmatrix} \alpha_{xx} \\ \alpha_{yy} \\ \alpha_{zz} \\ \alpha_{xy} \\ \alpha_{yz} \\ \alpha_{yz} \end{bmatrix} \quad \text{(A10-13}$$

The character of this representation is

$$\chi_\alpha(C_\Theta) = 4\cos^2\Theta + 2\cos\Theta \quad \text{(A10-14)}$$

Since

$$\cos 2\Theta = 2\cos^2\Theta - 1$$

and

$$\tfrac{1}{2}(\cos 2\Theta + 1) = \cos^2\Theta$$

substituting in Eq. (A10-14), one obtains

$$\chi_\alpha(C_\Theta) = 2 + 2\cos\Theta + 2\cos 2\Theta \qquad \text{(A10-15)}$$

For an improper operation

$$\chi_\alpha(S) = 2 - 2\cos\Theta + 2\cos 2\Theta \qquad \text{(A10-16)}$$

Thus, for proper and improper operations one obtains the character

$$\chi_\alpha(R) = 2 \pm 2\cos\Theta + 2\cos 2\Theta \qquad \text{(A10-17)}$$

This is the character to be used for the determination of the Raman activity.

DERIVATION OF THE CHARACTER FOR THE DETERMINATION OF NUMBER OF FUNDAMENTALS $\chi_\Xi(R)$

The character previously developed for the dipole moment refers to $3n$ variables, where n is the number of atoms in the molecule. For characters applying to representations for $3n - 6$ coordinates, the characters for the motions of translation and rotation must be subtracted.

The n vectors giving the displacements of the nuclei are, by the laws of mechanics, equivalent to the resultant vector acting at the center of gravity of a molecule. The three components of this vector transform for the translation like that of the dipole moment. Therefore

$$\chi_T(R) = \chi_M(R) = \mu_R(\pm 1 + 2\cos\Theta) \qquad \text{(A10-18)}$$

For displacements involving rotation one must consider angular momentum changes for each coordinate. For a proper rotation such as a rotation about the z axis (C_Θ), the changes are represented by

$$\begin{aligned} lx' &= lx\cos\Theta + ly\sin\Theta \\ ly' &= -lx\sin\Theta + ly\cos\Theta \\ lz' &= lz \end{aligned} \qquad \text{(A10-19)}$$

For an improper operation S

$$
\begin{aligned}
ls'' &= -lx \cos \Theta - ly \sin \Theta \\
ly'' &= lx \sin \Theta - ly \cos \Theta \\
lz'' &= lz
\end{aligned}
\tag{A10-20}
$$

Therefore $\chi_R(R) = 1 + 2 \cos \Theta$ for the C_Θ operation, and $\chi_R(R) = 1 - 2 \cos \Theta$ for the S operation, and for $3n - 6$ coordinates the character for the number of fundamentals $\chi\Xi(R)$ becomes

$$\chi\Xi(R) = \chi_M(R) - \chi_T(R) - \chi_R(R) \tag{A10-21}$$

For a proper operation (C_Θ)

$$\chi\Xi(C_\Theta) = \mu_R(1 + 2 \cos \Theta) - (1 + 2 \cos \Theta) - (1 + 2 \cos \Theta) \tag{A10-22}$$

$$\chi\Xi(C_\Theta) = \mu_R(1 + 2 \cos \Theta) - 2(1 + 2 \cos \Theta)$$

$$\chi\Xi(C_\Theta) = (\mu_R - 2)(1 + 2 \cos \Theta) \tag{A10-23}$$

For an improper operation (S)

$$\chi\Xi(S) = \mu_R[(-1 + 2 \cos \Theta) - (-1 + 2 \cos \Theta) - (1 - 2 \cos \Theta)] \tag{A10-24}$$

$$\chi\Xi(S) = \mu_R(-1 + 2 \cos \Theta) \tag{A10-25}$$

REFERENCE

J. E. Rosenthal and G. M. Murphy, *Rev. Mod. Phys.*, **8**:317 (1936).

Appendix 11

UPDATED BIBLIOGRAPHY

In the time elapsed since the submission of the manuscript for this edition, many publications relevant to the subject matter have appeared in the open literature. The most important of these are listed below.

1. J. D. H. Donnay and G. Turrell, *Chem. Phys.* **6**:1 (1974).
2. G. Turrell, *Infrared and Raman Spectra of Crystals*, Academic Press, London (1972).
3. L. L. Boyle, *Spec. Acta*, **28A**:1347 (1972).
4. L. L. Boyle, *Acta Cryst.* **28A**:172 (1972).
5. L. L. Boyle, *Spec. Acta*, **28A**:1355 (1972).
6. C. Hsu and M. Orchin, *J. Chem. Ed.* **51**:725 (1974).
7. G. Davidson, *Introductory Group Theory for Chemists*, Elsevier, London (1971).
8. S. D. Ross, *Inorganic Infrared and Raman Spectra*, McGraw-Hill, New York (1972).
9. R. L. Carter, *J. Chem. Ed.* **48**:297 (1971).
10. W. L. Jolly, *The Synthesis and Characterization of Inorganic Compounds*, Prentice-Hall, Englewood Cliffs, N.J. (1970).
11. D. Steele, *The Theory of Vibrational Spectroscopy*, W. B. Saunders, Philadelphia (1971).
12. M. St. C. Flett, *The Theoretical Basis of Infrared Spectroscopy*, in *An Introduction of Organic Compounds*, F. Scheinmann (ed.), Vol. I, Pergamon Press, Oxford (1970).
13. K. N. Rao and C. W. Mathews (eds.), *Molecular Spectroscopy—Modern Research*, Academic Press, New York (1972).
14. J. A. Salthouse and M. J. Ware, *Point Group Character Tables and Related Data*, Cambridge University Press, London (1972).
15. G. W. Chantry, *Submillimetre Spectroscopy*, Academic Press, London (1971).
16. P. M. A. Sherwood, *Vibrational Spectroscopy of Solids*, Cambridge University Press, London (1972).
17. V. C. Farmer, "Site Group to Factor Group Correlation Tables" and "Symmetry and Crystal Vibrations," in *The Infrared Spectra of Minerals*, The Mineralogical Society, London (1974), pp. 515–525, 51–67.
18. D. F. Irish and M. H. Brooker, *Appl. Spec.*, 27:395 (1973).
19. W. G. Fateley, *Appl. Spec.*, **27**:305 (1973).
20. J. R. Ferraro, *Appl. Spec.*, **29**:354, 418 (1975).
21. D. M. Adams, *Coord. Chem. Rev.*, **10**:183 (1973).
22. C. F. Shaw and L. G. Newbury, *Can. J. Spec.*, **20**:65 (1975).
22. M. H. Brooker, *Appl. Spec.*, **29**: 528 (1975).

INDEX